Analytical Chemistry II

Ulf Ritgen

Analytical Chemistry II

With Contributions by Christina Oligschleger

Ulf Ritgen
FB 05 – Angewandte Naturwissenschaft
Hochschule Bonn-Rhein-Sieg
Rheinbach, Nordrhein-Westfalen, Germany

ISBN 978-3-662-68709-3   ISBN 978-3-662-68710-9 (eBook)
https://doi.org/10.1007/978-3-662-68710-9

This book is a translation of the original German edition "Analytische Chemie II" by Ulf Ritgen and Christina Oligschleger, published by Springer-Verlag GmbH, DE in 2020. The translation was done with the help of an artificial intelligence machine translation tool. A subsequent human revision was done primarily in terms of content, so that the book will read stylistically differently from a conventional translation. Springer Nature works continuously to further the development of tools for the production of books and on the related technologies to support the authors.

Translation from the German language edition: "Analytische Chemie II" by Ulf Ritgen and Christina Oligschleger, © 2020 2020. Published by Springer Berlin Heidelberg. All Rights Reserved.

© Springer-Verlag GmbH Germany, part of Springer Nature 2025
This work is subject to copyright. All rights are solely and exclusively licensed by the Publisher, whether the whole or part of the material is concerned, specifically the rights of reprinting, reuse of illustrations, recitation, broadcasting, reproduction on microfilms or in any other physical way, and transmission or information storage and retrieval, electronic adaptation, computer software, or by similar or dissimilar methodology now known or hereafter developed.
The use of general descriptive names, registered names, trademarks, service marks, etc. in this publication does not imply, even in the absence of a specific statement, that such names are exempt from the relevant protective laws and regulations and therefore free for general use.
The publisher, the authors and the editors are safe to assume that the advice and information in this book are believed to be true and accurate at the date of publication. Neither the publisher nor the authors or the editors give a warranty, expressed or implied, with respect to the material contained herein or for any errors or omissions that may have been made. The publisher remains neutral with regard to jurisdictional claims in published maps and institutional affiliations.

Editorial Contact: Sinem Toksabay

This Springer imprint is published by the registered company Springer-Verlag GmbH, DE, part of Springer Nature.
The registered company address is: Heidelberger Platz 3, 14197 Berlin, Germany

If disposing of this product, please recycle the paper.

Preface

Just like the textbook *Analytical Chemistry I*, this second volume of the series is also intended as a "lecture to read up on," and just as for Volume I, there is also an authoritative textbook here, which is referred to again and again (and again…) in the following pages. It is the work *Instrumental Analysis* by D. A. Skoog, F. J. Holler, and S. R. Crouch, or "Skoog" for short. You should have this handy when working through the present book, because, for example, frequent reference is made to illustrations there. In addition, the topics dealt with in this book are understandably presented much more comprehensively there, so that this book should also serve accordingly to deepen your understanding and to look things up.

Skoog

This does not mean, however, that the chapters do not occasionally refer to the ones by D. C. Harris, which you, if you already dealt with *Analytical Chemistry I* (should this be the case: Welcome back!), already know, because now and then you will also find particularly concise examples and the like there. "Skoog" and "Harris" are complementary to one another in many ways. So please do not be too surprised if there are even "double references" to some topics: These again serve to deepen your understanding.

Harris

And because the title *Analytical Chemistry II* already suggests that you are holding a "sequel" in your hands, there will also be repeated references to *Analytical Chemistry I*. (If you have *not* dealt with it, this should not be a problem; this hint is only there so that you are not surprised about any occasional corresponding reference).

For the more advanced areas of analytics, the fundamentals of both general chemistry and inorganic chemistry are needed, as well as of organic chemistry and selected areas of physical chemistry, and the line at which chemistry merges inseparably with other scientific disciplines is also crossed here—where, for example, would you want to draw the line between organic chemistry, biochemistry, and molecular biology?

Accordingly, the field of analytics has also expanded considerably—and yet this book will also repeatedly refer back to things that you are certainly already familiar with from other teaching texts (maybe including *Analytical Chemistry I*).

In Part I we return to *molecular spectroscopy*, the basics of which you may already have become familiar with in Part IV of *Analytical Chemistry I*—we will allow our analytes to interact with electromagnetic radiation so that the data obtained in this way allow conclusions to be drawn about the properties of the substance(s) under investigation, in particular their (molecular) structure. Of particular importance here are the radio waves—by no means very energetic, of course—which, in combination with a powerful magnet, lead to *nuclear magnetic resonance spectroscopy* (NMR), for which there are not only numerous applications, but also different variants. In the same part of the book we also deal with *mass spectrometry* (MS), the basics of which have already touched upon in Part V of *Analytical Chemistry I*. Depending on the technical approach, both methods are equally suitable for comparatively small (organic) molecules as for the investigation of significantly larger atomic assemblies, be they higher-molecular natural substances (e.g., starch, etc.) or analytes that are commonly assigned to biochemistry, such as proteins or DNA/RNA strands.

In Part II, we will then deal with *electroanalytical* methods, the basic principles of which readers of *Analytical Chemistry I* will already be familiar with from Part II there. This time, however, we will delve a little deeper into the subject matter and also deal with more specific methods of analysis. The main focus is on how the analytes (which can also be monoatomic cations or anions!) change by accepting or releasing one or more electrons and to what extent this can be detected macroscopically. In other words: What macroscopic properties change when the analyte is reduced or oxidised? Thus, since we are primarily concerned with the transfer of electrons, in the methods presented here we also consider properties that are directly correlated with this: We deal with electrochemical potentials or their change, and the resulting voltage, but also with techniques that depend on precisely these factors but are actually intended to allow conclusions to be drawn about other properties.

The same applies to Part III of this book. Here, various principles are combined—more so than has been the case so far anyway—so that, in addition to more advanced *routine analytical techniques*, you also gain a somewhat deeper insight into *bioanalytics*. Some methods have been compiled here, where aspects already covered are deepened and, above all, combined with each other. (So, should you feel compelled to look up one or the other principle again briefly while reading this part or to recall it in some other way: by all means go ahead!) With these introductory words, I do not intend to worry you in the least bit: It does not get *really complicated in* this part of the book. This chapter is just more than ever about the interplay of different techniques, principles, procedures, and the like.

As you may already have noticed in the volumetric analysis (Part II of *Analytical Chemistry I*), analytical chemistry increasingly makes use of quite complex instruments, where students all too easily run the risk of regarding them as "black boxes" and only consider (or *want* to consider) the measured values supplied by them. Even in the case of the simple pH meter, the thought of what is actually behind it "chemically speaking" may quickly recede into the background, when you "only just" want to determine a pH value, after all there *is* such a nice pH-sensitive sensor available in the first place. In Part IV of this book you will get to know selected other *sensors* commonly used in analysis—but also the respective (physico-)chemical background. It also explains how selected methods can be combined with each other (so that you may also experience the desired "aha! effect" here and there when reading this part of the book). In addition, the advantages of the increasing "instrumentalisation" of analytics are discussed, especially in combination with the computer-controlled automation of corresponding processes. At the same time, the topic of "sensors" will be used to show you how close the fields of "chemistry" and "biology" have come to one another in the meantime.

Finally, the last part of this book, which was written together with Prof. Oligschleger, is dedicated to the topic of *statistics*: You may have learned that *statistics* plays a considerable role in analytics not only in the first part of *Analytical Chemistry I*, but also in part III (in connection with the quality of a chromatographic separation—keywords: theoretical ground height, peak width, resolution, etc.). Just like all other parts of the textbooks, Part V will not deal exclusively with the associated mathematical considerations, but will be specifically oriented towards the necessities given in the context of analytics. So you will learn about the use of standards in the various methods of analytics in addition to appropriate basic principles. Why do you need this?—Almost nothing is easier to construct (or infer) false correlations with than misunderstood or misinterpreted statistical data. Not for nothing is there the beautiful quote (usually attributed to Winston Churchill), "Don't trust a statistic you haven't fudged yourself!" And just as the aim of the previous parts of the book was not to impart pure factual knowledge, but to develop a certain understanding of the methods of analytics described in each case, in this part of the book we want to show you that the mathematical tools for the statistical evaluation of measured values are not an end in themselves either (nor are they *black boxes* that magically transform a multitude of individual values into a reasonable whole), but that there are well-thought-out principles and ideas behind each individual method. You may be wondering why this section is so extensive. The reason is not that this topic were to be more important than all other areas of this book, but rather that in many (but not all) cases the *mathematical derivations* have also been included—just so that the various formulas you will get to know there do not "fall from the heavens". Some of these formulas may seem a little daunting at first glance (and are not compulsory reading for those of you who "just want to *apply* statistics"), but they do show quite clearly what considerations actually lie behind the various methods of statistical evaluation. In the end, it is important to note: Statistics is not magic, and if you know what you are actually doing, it is not "just dry mathematics", but a real tool. And the meaning of any tool only unfolds in its application, which is why we have included numerous examples. We hope to have packed a useful toolbox for you!

The whole preface can be summarised briefly: A large part of what you have learned so far, you will meet again here, deepened and interconnected.

In fact, "deepening" and "linking" are the key words for this entire book: Similar to *Analytical Chemistry I*, which built on the fundamentals of general, inorganic and organic as well as physical chemistry and thus ultimately represented a deepening and linking of the previous teaching material, *Analytical Chemistry II* is also a continuation of already familiar principles and approaches. The more often you have the feeling "I already know that!", the better, and the more often the thought "Oh, *that's* how it's connected!" crosses your mind, the greater the gain in knowledge.

And now I wish you to have both success and also fun on your further journey into the world of analytical chemistry!

Ulf Ritgen
Rheinbach, North Rhine-Westphalia, Germany

Contents

I Molecular Spectroscopy

1	**General**	5
2	**Mass Spectrometry (MS)**	7
2.1	Masses	8
2.2	Mass Spectrometer	11
2.3	Ionisation Methods	12
2.4	Fragmentations	16
2.4.1	Cleaving of Bonds	17
2.4.2	Rearrangements	20
2.4.3	And Now: Step by Step	22
3	**Nuclear Magnetic Resonance Spectroscopy (NMR)**	27
3.1	**Physical Principles**	28
3.1.1	Spin States in the Magnetic Field	29
3.1.2	Energy Considerations	30
3.2	**First NMR Spectra**	32
3.2.1	Influence of the Electron Density	33
3.2.2	Multiplet	36
3.2.3	The Chemical Shift δ, Considered in More Detail	39
3.2.4	Anisotropy Effects	44
3.3	**^{1}H-NMR**	46
3.3.1	Chemical Equivalence	46
3.3.2	Couplings	48
3.4	**^{13}C-NMR**	57
3.4.1	Shifts, Couplings, Spectra	57
3.4.2	Two-dimensional NMR	61
3.5	**Other Usable Cores**	63
	Further Reading	77

II Electroanalytical Methods

4	**General Aspects**	81
5	**Potentiometry**	93
5.1	**Electroactive Analytes**	94
5.1.1	Direct Potentiometry	97
5.1.2	Potentiometric Titrations	101
5.2	**Ion Selective Electrodes (ISE)**	102
5.2.1	pH Measurement with the Glass Electrode	102
5.2.2	Other Ion Sensitive Electrodes (ISE)	105
6	**Coulometry**	113
6.1	**The Karl Fischer Titration**	115
6.2	**Karl Fischer Titration—Classic**	115
6.3	**Karl Fischer Titration—Coulometric**	116

7	**Amperometry**	117
8	**Voltammetry**	121
	Further Reading	132

III Other Analytical Methods

9	**Gravimetric Analyses**	137
9.1	Electrogravimetry	138
9.2	Thermal Methods: Thermogravimetry (TG)	143
10	**Thermal Processes**	147
10.1	Differential Thermal Analysis	148
10.2	Calorimetry	150
11	**Use of Radioactive Nuclides**	153
11.1	Radiochemical Analysis: Neutron Activation Methods	155
11.2	Radioactive *Tracers*	157
11.3	Radioactive Age Determination	159
11.4	Radioimmunoassay (RIA)	161
12	**Fluorescence Methods**	169
12.1	Basics of Fluorescence: A Brief Review and Some Additional Information	170
12.2	Fluorescence Spectrometry	176
12.3	Fluorescence Microscopy	181
12.4	Summary	183
	Further Reading	188

IV Sensors and Automation Technologies

13	**General Information About Sensors**	191
14	**Electrochemical Sensors**	193
14.1	Classical Inorganic Sensors	194
14.2	Amperometric and Voltammetric Biosensors	197
15	**Optical Sensors (Optodes)**	203
15.1	An Inorganic Example	204
15.2	A Bio-Organic Example	206
15.3	An Inorganic Example in Living Cells	206
16	**Flow Injection Analysis (FIA)**	209
16.1	Summary	214
	Further Reading	216

V Statistics

17	**Experimental Errors**	219
18	**Statistical Analysis**	221
18.1	Mean Value ($\bar{x}$)	222
18.2	Standard Deviation (s)	226
18.3	Confidence Interval	232

Contents

19	**Gaussian Error Propagation**	235
19.1	Linear Regression/Equilibrium Line	237
19.2	Adjustment of Fit Parameters for Equalisation Curves/Parabolas	241
20	**Measured Value Distribution**	245
20.1	Discrete Uniform Distributions	246
20.2	Two-Point Distribution	247
20.3	Binomial Distribution	249
20.4	Hypergeometric Distribution	254
20.5	Poisson Distribution	259
20.6	Continuous Uniform Distribution	264
20.7	Exponential Distribution	267
20.8	Gaussian Normal Distribution	269
20.9	Logarithmic Normal Distribution	274
21	**Parameter Estimates**	277
21.1	Chi-Square Distribution (χ^2-Distribution)	278
21.2	Student t-Distribution	281
21.3	Estimation Methods	284
21.4	Maximum Likelihood Method	288
21.5	Confidence Intervals for the Unknown Parameters ϑ of a Distribution	291
21.6	Parameter Tests	299
22	**Validation of Methods**	309
22.1	Standard Addition	310
22.2	Internal Standard and External Standard	312
23	**Outlier Tests**	319
23.1	Dixon Q-Test	320
23.2	4σ-Environment	322
23.3	Grubbs Test	322
23.4	Summary: This Time in Keywords	323
23.4.1	Experimental Errors	323
23.4.2	Error Propagation	324
23.4.3	Measured Value Distribution	324
23.4.4	Parameter Estimates	326
23.4.5	Validation of Methods	326
23.5	Outlier Tests	326
	Further Reading	332
Glossary		334
Index		353

Molecular Spectroscopy

Requirements

As already explained in Part IV of "Analytical Chemistry I", (molecular) spectroscopy is about the interaction of our analytes with electromagnetic radiation. Even though nuclear magnetic resonance (NMR) spectroscopy "only" uses radio waves, you should be familiar with the associated basics (already assumed in the aforementioned Part IV):
- the quantisation of energy
- the concept of the photon
- the electromagnetic spectrum
- the relationship between wavelength and energy content

In addition, the concept of electron density distribution within analyte molecules is required, which is inextricably linked to electronegativity as well as the HSAB concept.

In addition, a strong magnetic field is required for nuclear resonance, so the relevant basics (certainly known from courses dealing with physical phenomena) are also required (here, however, really only the basics).

In mass spectrometry (MS), on the other hand, we need above all:
- atomic masses
- isotopes

Since in mass spectrometry the analytes are ionised and then fragmented, i.e. bonds are broken, it is also helpful—especially in the case of analytes regarded as organic—to have a good sense of which bonds can be broken particularly easily (i.e. are rather unstable), which do not break so easily, which of the resulting ions will be particularly stable and which (more or less stable) uncharged fragments can be produced in the process. In addition, depending on the structure of the analyte in question, there are various possibilities for rearrangement. Here, a solid knowledge of organic chemistry is very useful.

Learning Objectives

In this part you will get to know different molecular spectroscopic methods which allow conclusions to be drawn about the (molecular) structure of the analytes under consideration, i.e. which can be used for structure elucidation.

In mass spectrometry, you will become familiar with characteristic isotope patterns as well as the fragmentation and rearrangement processes common in this method of analysis. This is the basis for the evaluation of mass spectra.

For nuclear magnetic resonance spectroscopy, we first look at the atomic nuclei that can be used for this purpose and the theoretical background of the associated resonance experiments. You will learn what kind of excitation can take place and what information can be obtained by the interaction of the analytes with electromagnetic radiation from the radio wave range in a magnetic field. We will deal with the structural features of individual analytes so that you can independently estimate what kind of signals to expect in nuclear magnetic resonance spectra and, conversely, learn to draw conclusions from these to structural elements of the respective analytes on the basis of selected ^{1}H- and ^{13}C-NMR (partial) spectra. In addition, you will see that—and why—interacting atomic nuclei lead to multi-line systems in the nuclear magnetic resonance spectra. Thus, after reading this part, you will in turn be able to both predict how corresponding signals are split and use this information to evaluate spectra.

In more in-depth sections, you will also become familiar with multidimensional NMR spectroscopy, which has become an indispensable part of modern analytics.

As in the book "Analytical Chemistry I", these parts also aim to combine the respective theoretical principles of the various analytical methods ("What is behind them in each case?") with their application in practice ("What can the respective methods be used for?")—will be linked.

— For mass spectrometry, basic knowledge of organic chemistry in particular is essential: Which bonds are particularly easy to cleave in each case—and why? Which intramolecular/interfragmentary stabilisation effects come into play? Which rearrangements are to be expected? Which fragments are formed preferentially, and how do any isotopes affect the mass spectra obtained?
Even if the basics of this technique are explained very well comprehensibly in Skoog, which—as already mentioned—is the authoritative textbook for the book "Analytical Chemistry II", the evaluation of these spectra is predominantly a matter of practice. For this reason, a classic is recommended to you, which, although in the meantime other authors have also contributed to it, is generally referred to as Hesse–Meier–Zeeh.

— In nuclear magnetic resonance spectroscopy, although the excitability of atomic nuclei is doubtlessly the main issue, electronic conditions also have to be taken into account: Where do hetero-atoms (i.e., everything except carbon and hydrogen) provide increased or decreased electron density, and how does this affect the particular atomic nucleus under consideration? What influence do multiple bonds have? Which interactions of the respective atomic nuclei lead to further information that can be taken from the resulting nuclear magnetic resonance spectrum and allow additional conclusions to be drawn about the (molecular) structure of the respective analyte?
On this subject, in addition to Skoog, we refer in particular to the work of Breitmaier, and in the active application of precisely these principles—i.e. the independent evaluation of NMR spectra—the above-mentioned Hesse–Meier–Zeeh will again serve you well.

Bienz S, Bigler L, Fox T, Meier H (2016) Hesse - Meier - Zeeh: Spektroskopische Methoden in der organischen Chemie

Breitmaier E, Vom NMR-Spektrum zur Strukturformel organischer Verbindungen.

Contents

Chapter 1   General – 5

Chapter 2   Mass Spectrometry (MS) – 7

Chapter 3   Nuclear Magnetic Resonance Spectroscopy (NMR) – 27

General

As we have already discussed in Part I, "Analytical Chemistry I", in general on the subject of 'Analytical Methods', a distinction between non-destructive methods and spectroscopic methods in which the analyte is altered or even completely destroyed is also made in molecular spectroscopy. This part of the book will deal on the one hand with different methods of mass spectrometry, in which the analytes are fragmented (i.e. "torn to pieces"), and on the other hand with nuclear magnetic resonance spectroscopy, in which, depending on the methodology, it is quite possible to recover the analyte intact after the measurements have been completed (even if this can occasionally be a little tricky).

> **Confusion of Terms**
> Depending on which textbooks you work with in addition to Skoog (and Harris), there is a good chance that some will consistently use the term "spectroscopy", while in others you will only ever find the term "spectrometry". Commonly, these two terms are used interchangeably—but this is not entirely correct, as the term spectroscopy is derived from the Greek word σκοπειν (skopein) = to see. Thus, one should only speak of "spectroscopy" when the human eye is used as a measuring instrument (in this respect, even the term "UV/VIS spectroscopy" is at least dubious, since the human eye does not respond to UV radiation). This already shows that said definition would be a bit too narrow, which is why—at the instigation of IUPAC—another subdivision has become established:
> - **Spectroscopy** refers to methods of analysis in which the interaction of analytes with electromagnetic radiation or the electromagnetic radiation emitted by the analytes after excitation is observed.
> - In **spectrometry,** on the other hand, quantitative measurements are made, because the ending—metrie comes from the Greek μετρον (metron) = the measure. Here, other quantities also come into play, which are not resolved according to their energy content; an example of this is the number of particles in mass spectrometry.

Cave: Some examiners put a lot of emphasis on the correct choice of words, in that regard.

Mass Spectrometry (MS)

Contents

2.1 **Masses – 8**

2.2 **Mass Spectrometer – 11**

2.3 **Ionisation Methods – 12**

2.4 **Fragmentations – 16**
2.4.1 Cleaving of Bonds – 17
2.4.2 Rearrangements – 20
2.4.3 And Now: Step by Step – 22

© Springer-Verlag GmbH Germany, part of Springer Nature 2025
U. Ritgen, *Analytical Chemistry II*, https://doi.org/10.1007/978-3-662-68710-9_2

Skoog, Chapter 20: Molecular mass spectrometry

Harris, Chapter 21: Mass Spectrometry

The basic principle of mass spectrometry has already been explained in Part V of "Analytical Chemistry I", so it will only be briefly summarised again here before we deal with the individual steps of this procedure one after the other and look at how such a spectrum is generated and evaluated. An introduction is provided by both ▶ Skoog and Harris.

First, the analytes are ionised: There are various methods for this, which are presented in turn in ▶ Sect. 2.3; they all have in common that electrically neutral analytes are converted to (ideally singly positively charged) radical cations. These cations then (usually) decay into various fragments (supported by a magnetic field; we will briefly discuss the general set-up of a mass spectrometer in ▶ Sect. 2.2). These fragments are then separated according to their respective mass. The fragments obtained (more on this in ▶ Sect. 2.4) then allow conclusions to be drawn about the original structure of the analyte.

If a mixture of different analytes is present at the beginning of the analysis, these must be separated from each other before their respective fragmentation. (You have become familiar with various methods for this in Part III, "Analytical Chemistry I".)

There are two basic things you should keep in mind with respect to mass spectrometry (again, this has already been addressed in Part V of "Analytical Chemistry I", so here it will be covered only briefly):

— In mass spectrometry, the analyte molecules are fragmented *individually*; since there are different possibilities for fragmentation, different fragments are produced accordingly. (It should be borne in mind that each analyte molecule can only be fragmented *once*.) On the other hand, a really large number of analyte molecules are introduced into the analysis at the same time, so a mixture of different fragments is produced, the composition of which obeys statistics on the one hand, but on the other depends on the kinds of bonds present in the analyte in question: Fragments resulting from the breaking of more easily cleavable bonds will occur correspondingly more frequently, others significantly less frequently.
— The resulting fragments are separated according to their **mass-to-charge ratio m/z** (in the course of ionisation, polyvalent cations, such as X^{2+} or even X^{3+} may also be formed). If different **isotopes** are involved, the individual fragments can be distinguished accordingly.

If one keeps these two basic facts in mind, the handling of mass spectrometry and especially the evaluation of the spectra obtained is actually not that difficult.

2.1 Masses

You have been familiar with the concept of molar mass since general and inorganic chemistry, but there is still a danger of confusing some things that—especially in mass spectrometry—are better kept neatly separate. In MS, this is especially true with regard to the term "mass".

Of course you know: If a substance (e.g. ethanol, CH_3–CH_2–OH) is said to have a molar mass of 46.07 g/mol, then the **relative molar mass** of a single molecule is 46.07 **u** (for *unified atomic mass unit*).

— In older English-language textbooks, the unit **amu** *(atomic mass unit)* is used, although the numerical values actually differ slightly (1 u = 1.000 037 amu).

- In biochemistry, when describing the relative mass of a protein or similar, the unit **Dalton (Da)** is usually used instead of u; here, too, however, the numerical value remains identical: 1 u = 1 Da.

But this information is an *average value of all **isotopes*** relevant for the respective types of atoms involved. In order to determine the *actual mass of a single molecule*, one need not only know the sum formula, but also which isotopes are present in each case.

Let us take a closer look at ethanol as an example: The sum formula C_2H_6O tells us that the elements carbon, hydrogen and oxygen are involved here (which definitely should not surprise you …):

- In the case of hydrogen, it must be taken into account that in addition to the isotope ^1H (which of course massively predominates with a natural abundance of 99.988 %), the isotope ^2H (deuterium—one of the few isotopes with its own element symbol: D) also contributes to the relative total mass (albeit only minimally with an abundance of 0.012 %). The radioactive tritium (^3H, with the element symbol T) does not plays a role here.
- In the case of carbon, the isotopic influence already becomes clearer: In addition to the "standard carbon" ^{12}C, which occurs with an abundance of 98.93 %, there is also ^{13}C with an abundance of 1.07 %.
- In the case of oxygen, the isotope ^{16}O dominates with an abundance of 99.632 %, but ^{17}O (0.038 %) and ^{18}O (0.205 %) must not be completely ignored either.

(The natural abundance of the most important isotopes, especially in organic chemistry, can be found in Table 20.3 of ▶ Skoog; more detailed is Table 21.1 of ▶ Harris .)

Skoog, Section. 20.2.1: The electron impact ion source

Harris, Section 21.2: Oh, mass spectrum, speak to me!

> This means that in a sample containing 50 ethanol molecules, i.e. a total of 100 C atoms, there should be, purely statistically, *one* atom of the isotope ^{13}C. Accordingly, one of our 50 ethanol molecules has a higher molecular mass than the other 49.

There is another point to consider: Even if the **mass number m** of each isotope is always an integer (because it is simply the sum of the number of protons and neutrons, i.e. it corresponds to the **atomic number** of the respective atom), this is by no means true for the *actual mass* of an atom, because neither protons nor neutrons have a mass of exactly 1 u. Only in the case of the carbon isotope ^{12}C, one can rely on the fact that its mass is exactly 12.000000 u—but this is precisely because this value was *defined this way*. The hydrogen isotope deuterium (^2H), for example, with mass number m = 2 (nucleons present: 1 proton, 1 neutron) has an actual mass of 2.014 u (or 2.014 Da, if you prefer).

In mass spectrometry, where the type of isotope present in each case is of indispensable importance, the **nominal mass** is also used: This results from the *sum of the mass numbers of all atoms involved in the specific molecule* under consideration.

> **For the Example of Our Fifty Ethanol Molecules:**
> (a) 49 of the molecules consist of: $6 \times {}^1H$, $2 \times {}^{12}C$, $1 \times {}^{16}O$; therefore, the total nominal mass is 46.
> (b) 1 of the molecules consists of $6 \times {}^1H$, $1 \times {}^{12}C$, $1 \times {}^{13}C$, $1 \times {}^{16}O$; the nominal mass of this ethanol molecule is 47.

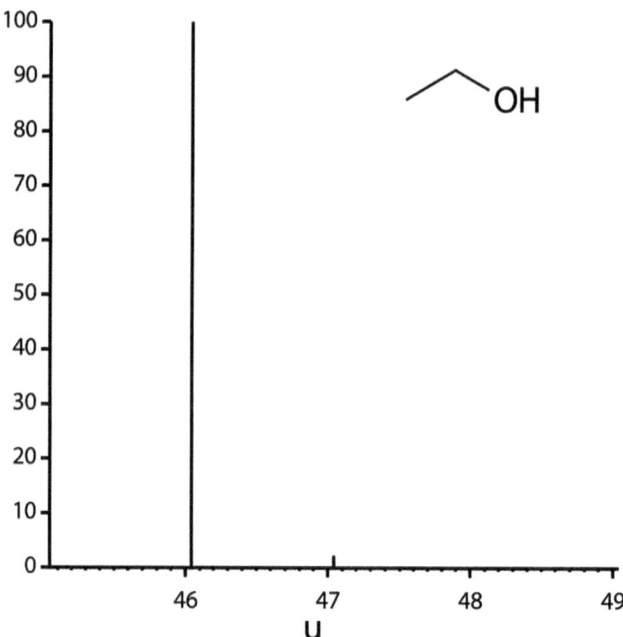

◘ Fig. 2.1 Mass spectrum of ethanol (without fragmentation)

If we now plot the nominal masses of the various molecules present against their relative abundance (in %), we already obtain a first mass spectrum (◘ Fig. 2.1, again related to our ethanol example):

You can see: The peak for the molecule with nominal mass 46 is distinctly larger than that for the one with nominal mass 47. If you look closely, you can also see that the peaks do not lie *exactly* at 46 and 47, but are shifted a little to larger masses, because after all the relative atomic mass of hydrogen is not exactly 1.0, but is a little larger (1.008).

And if you look *really closely*, you will (maybe) see another peak, albeit tiny, a little ways off from 48. (If it does not seem to be there, it is been lost in print – it is *really* tiny). But where does it come from?—So far we have assumed that there are only a few molecules (in our example there were just 50). Under this assumption, we have been able to neglect the influence of the other isotopes (^{2}H, ^{17}O, ^{18}O). But what if that does not work? Let us return to our ethanol example:

> ► **Example**
> The isotope ^{18}O, for example, occurs with a frequency of 0.205%, so—again purely statistically!—it is to be expected that in a larger ethanol sample (which does not consist of just fifty molecules, but several thousands, at least) about every five hundredth O atom is an ^{18}O. Correspondingly, there is now—purely theoretically—another possible nominal mass (without statement about the relative abundance). Besides the "normal case" (6 × ^{1}H, 2 × ^{12}C, 1 × ^{16}O; nominal mass 46; this corresponds to (a) from above) there are also.
> (c) 6 × ^{1}H, 2 × ^{12}C, 1 × ^{18}O; nominal mass: 48.
>   But if we already have five hundred ethanol molecules in the sample, this already contains—purely statistically (!)—10 carbon atoms of the isotope ^{13}C, after all 500 ethanol molecules also mean 1000 carbon atoms, and every hundredth of them has—purely statistically (!)—an additional neutron, because there is nothing to prevent the heavy O atom from being part of a molecule that also contains a heavier C isotope:

2.2 · Mass Spectrometer

(d) So in addition to the nominal mass 47 of (b) with $6 \times {}^1H$, $1 \times {}^{12}C$, $1 \times {}^{13}C$, $1 \times {}^{16}O$ there is also the possibility to consider $6 \times {}^1H$, $1 \times {}^{12}C$, $1 \times {}^{13}C$, $1 \times {}^{18}O$ with the nominal mass 49.

Now, it should be emphasised once again that we are dealing with such a large number of molecules that statistically even rarer cases are possible: If every one hundredth carbon atom is a representative of the category ${}^{13}C$, then the probability that a molecule composed of two C atoms actually has *two* carbon-13 atoms is only (0.01×0.01) to 1, but this also means that in a sample that contains a practically uncountable number of these molecules, every ten thousandth molecule actually contains *two* ${}^{13}C$ atoms:

(e) $6 \times {}^1H$, $2 \times {}^{13}C$, $1 \times {}^{16}O$ (nominal mass: 48)

Occasionally, two or even more peaks coincide in an "ordinary" mass spectrometry spectrum, although they belong to different analytes—like the two peaks of mass number 48 (c and e) in our ethanol example. (If you at this point spontaneously thought of **isobaric interference**—known from Part V, "Analytical Chemistry I"—I am deeply impressed, but this is exactly what we are talking about here!) In a high-resolution spectrum (such a thing also exists, but it is much more complicated and therefore much less common than "routine analysis"), even peaks with the same nominal mass can be distinguished from each other, because only the *sums of the mass numbers* match, but not the *actual atomic masses*. (Nevertheless, high-resolution MS goes much too far here.) ◄

► Example

One last time back to the ethanol example: Even more improbable, but still possible (and therefore, according to all laws of statistics, also present in a sufficiently large sample) is the occurrence of the following molecule:

(f) $6 \times {}^1H$, $2 \times {}^{13}C$, $1 \times {}^{18}O$; $m = 50$

The corresponding peak at 50 on the x-axis is expected to be so small that it will disappear in the background noise of the spectrum (you will certainly not see it in ◘ Fig. 2.1), but statistics says: It *must* be there.

In routine MS analysis, such rarely occurring particles do not play any role at all, but it is important for me to show you the immense importance of isotopes in mass spectrometry.

And if you feel like it, you can now think about the fact that with a sufficiently large sample, deuterium should no longer be neglected, either. After all, statistically, every ten thousandth hydrogen atom is a representative of the isotope ${}^2H$, and even the small molecule ethanol contains six times as many H atoms as O atoms. ◄

❓ Questions
1. Given the abundance of the various isotopes, say something about the nominal mass of the following simple molecules:
   (a) Bromomethane (CH_3Br)
   (b) Chloromethane (CH_3Cl)
   (c) Bromochloromethane (CH_2BrCl)

   Refer to Table 21.1 of ► Harris for information on the relative abundance of the various isotopes; feel free to neglect the hydrogen isotopes.

2.2 Mass Spectrometer

Mass spectrometry is based on two fundamental facts:
- Charged particles experience an acceleration in an electric field; this is, of course, especially true for charged particles in the gaseous phase.

– If these particles are then brought into a magnetic field, they can be separated in proportion to their mass or, more precisely, their mass-to-charge ratio (m/z).

Skoog, Section 20.3.3: Mass analysers

Mass inertia plays an important role here: The more massive a charged particle is, the less strongly it is deflected from its original trajectory by the magnetic field. Schematically, the setup of a mass spectrometer is shown as Fig. 20.13 in Skoog. (Various methods of converting the normally electrically neutral analytes to corresponding charge carriers are discussed in ▶ Sect. 2.3.)

Various analysers are used, but the *quadrupole mass spectrometer* is by far the most common today. Here, the analyte ions pass through four parallel rod-electrodes arranged at the corners of a square, with opposite electrodes having the same charge, resulting in an alternating field. Depending on their m/z ratio, the analyte ions are deflected from their trajectory to different degrees; the location of their impact at the detector (which we will not discuss further here) then allows corresponding conclusions to be drawn about their mass-to-charge ratio. (Nowadays, this separation is done practically without exception by computers, which is why we do not want to go into this point any further.)

2.3 Ionisation Methods

The fragmentation of analytes in mass spectrometry is based on the initial ionisation of the analytes, which often causes secondary processes that ultimately lead to (further) fragmentation of the individual analyte molecules. (The different possibilities of fragmentation are discussed further in ▶ Sect. 2.4.) A distinction is made between different ionisation methods, with one or the other being more advantageous depending on the type of analyte.

In all cases, however, it is essential that molecules of the compound to be investigated are introduced as a fine particle beam (in the gas phase) into a high vacuum in order to avoid undesired interactions/collisions of the molecules (or of the ions then produced by the ionisation) as far as possible. (We will not go into the technical necessities here; if they are of importance to you, you will have to consult further literature anyway.) This procedure requires, of course, that the analytes to be examined can be *evaporated without decomposition*. (If this is not possible, other ionisation techniques must be chosen; these will be briefly discussed at the end of this section.)

Of particular importance in ionisation are mainly:
– **electron impact ionisation** (short: **EI**) *and*
– **chemical ionisation (CI)**

However, it should be noted that other ionisation methods are known (and occasionally used); some of these will be mentioned at least briefly in this section, and the principle behind each will be explained.

■ **Electron Impact Ionisation (EI)**

In this process, a fine particle beam of the analyte particles is directed perpendicularly onto an electron beam. The interaction of the (currently still uncharged) analytes with the high-energy electrons leads to ionisation by knocking an electron out of the analyte molecules:

$$M + e^- \rightarrow M^{+\bullet} + 2e^-$$

Here M generally stands for the uncharged analyte, the "parent substance" (if you want to, consider the M to stand for "mother", even though it actually

2.3 · Ionisation Methods

stands for "original **M**olecule); during ionisation a (radical) molecular cation $M^{+\cdot}$ is formed, which is then might be further fragmented (more on this in ▶ Sect. 2.4). However, since the mass of the now missing electron has no influence on the total mass of the corresponding ion in the context of measurement accuracy (remember that the mass of an electron is only about one two-thousandth of the mass of a proton or neutron), it may be assumed that the mass of this **molecular ion** corresponds to that of the neutral analyte.

Since charged particles are now present, they can and will be accelerated in an electric field—and the more massive the particles in question are, the less strongly they can be accelerated. This is the basis for the separation of the resulting ions (or fragmentation products), to which we will also return in ▶ Sect. 2.4.

- **Slightly More Precise**

As already mentioned in the introduction to ▶ Chap. 2, *double ionisation* can also occur, in which case a doubly positively charged molecular ion M^{2+} is formed. Since the separation of the molecular ions obtained in the course of mass spectrometry is not carried out according to their absolute mass, but according to the mass-to-charge ratio (m/z), it is recommended to always work with exactly that. (For a *monovalent* molecular ion [with $z = 1$], m/z naturally corresponds to m.) This also applies to any molecular *fragment ions* subsequently formed by fragmentation, which may occasionally also have a higher charge than just 1.

- **Chemical Ionisation (CI)**

An alternative to electron impact ionisation is chemical ionisation, in which the analytes do not interact directly with an electron beam. Instead a reactant is first ionised by electron impact—light hydrocarbons, such as methane and ethane, or other simple compounds like hydrogen, water or ammonia are usually used for this purpose.

Let us take methane as an example, which is ionised in a first step (quite analogous to the EI from the previous section) to the methane radical cation $CH_4^{+\cdot}$:

$$CH_4 + e^- \rightarrow CH_4^{+\bullet} + 2e^-$$

(Yes, the bonding conditions in this radical cation are ... interesting—and they go beyond the scope of this book. Should you come across *multi-center bonds* in advanced courses of organic or inorganic chemistry, I would be pleased if you then think back to the radical cations from the MS, because exactly such bonds are present here.)

But there is more: In most cases, the resulting methane radical cation ($CH_4^{+\bullet}$) does not react immediately with the analyte (M), but first with another methane molecule. In addition to a methyl radical ($^{\bullet}CH_3$), a perhaps even more strange-looking intermediate is formed: a protonated methane cation (CH_5^+—no, that is really not a typographical error, here the bonding conditions are even *more* interesting!).

$$CH_4^{+\cdot} + CH_4 \rightarrow CH_5^+ + {}^{\cdot}CH_3$$

The radical character of the methane radical cation is, thus, transferred to a "conventional" methane carbon by the radical cation taking over an uncharged (!) hydrogen atom from this methane, so that there is *no* unpaired electron left at this now (formally) pentavalent carbon atom. The by-product is a methyl radical—we only have an *odd* number of electrons available, and the radical character has to end up somewhere, after all.

It is doubtlessly understandable that the protonated methane cation (CH_5^+) resulting from this reaction is extremely reactive. In the subsequent step, it then interacts with the actual analyte (M), whereby the latter is in turn ionised (and methane is released). This ionisation, thus, "actually" takes place only by simple protonation, but the proton donor in this acid/base reaction (and nothing else is happening here) is remarkable after all; the pK_A-value of CH_5^+ (which can only be determined by calculation) is $\ll -11!$)):

$$M + CH_5^+ \rightarrow M-H^+ + CH_4$$

It should not be concealed that there are other intermediate reactions, but these will not be discussed further here. The most important point of chemical ionisation is:

> Unlike EI, CI does not yield a molecular ion whose mass would be equal to that of the analyte itself—i.e. $[M]^+$. Instead a *protonated molecular ion* is formed, whose nominal mass is increased by 1: $[M–H]^+$.

Because ionisation of the analytes by appropriately energetic molecular ions is considerably less aggressive than ionisation by direct interaction with the electron beam, chemical ionisation generally yields considerably fewer fragments; correspondingly, fewer peaks are obtained in the associated spectra—and it can be advantageous, particularly with more complicated analytes, if the number of fragments is kept within limits. Nevertheless, electron impact ionisation represents the "standard" nowadays; CI is usually resorted to only in case of concrete need. For this reason, in the following we will mainly deal with EI spectra.

However, there are other ionisation methods that are also suitable for analytes that can be transferred into the gaseous phase without decomposition:

— In **electron spray ionisation (ESI),** not the pure analyte but an *analyte solution* is introduced into an electric field and comes—in counter current—into contact with a drying gas. This way, charged drops are formed, precisely because of the electric field, whereby the solvent gradually evaporates so that these drops become steadily smaller. (One could claim that the analyte is not *evaporated* here, but rather "tricked" into the gas phase.) These droplets then enter the actual mass spectrometer. The decisive factor here is that this procedure, which is suitable for particularly large analytes, leads to the formation of multiply-charged ions (up to several dozen positive or also negative charges!); here too, of course, the m/z ratio is of immense importance.

— **Field ionisation (FI)** requires the application of a considerably strong electric field; this leads primarily to the corresponding molecular ions, while fragmentation hardly occurs. Especially for the determination of the absolute mass (or the m/z ratio) of high-mass analytes this method is used quite frequently.

- **Analytes That Cannot Be Vaporised Without Decomposition**

For analytes that cannot be evaporated without decomposing—which is often the case with high-molecular analytes such as proteins and the like—other methods must be used. Various variants are available:

— In **field desorption (FD),** the analyte, which is present as a solid, is applied to a surface (usually a wire); the application of a strong electric field (the similarity to field ionisation is remarkable) causes ionisation, whereby the resulting ions are then desorbed from the surface.

2.3 · Ionisation Methods

- In **laser desorption (LD)** variant, desorption is promoted by exciting the analyte with a (high-energy) UV laser: Cations and anions originating from the analyte material are obtained, which are then fed into the mass spectrometer.
- Currently, the most common application of FD is **MALDI-TOF** *(Matrix-Assisted Laser Desorption/Ionisation—Time Of Flight)*, in which the analyte is first crystallised together with a suitable matrix material, with the matrix material used in huge excess. The decisive factor for the matrix material used (usually: smaller organic molecules) is that it can be easily excited by the selected laser. This leads to the explosive ablation of particle mixtures at the surface, which thus enter the gas phase. Here, too, the transfer of hydrogen atoms yields analyte cations and anions, which can then be investigated by mass spectrometry. Again, fragmentation of the analytes is almost completely eliminated; information about their mass is obtained from the time required for these ions to reach the corresponding detector: One determines their "time of flight" in the mass spectrometer, from which the mass of the analyte molecules can then be determined—using calibration substances as standards. MALDI-TOF is of particular importance for high-molecular substances, such as proteins or polymers.
- The **fast atom bombardment (FAB)** technique is particularly suitable for non-evaporable analytes. Here, a thin layer of analyte molecules is bombarded with accelerated neutral molecules so that ionisation occurs. Even though FAB sounds like a variant of chemical ionisation, this is not quite true, because here *two consecutive steps* are required:
  1. First, argon or xenon atoms are converted to the corresponding cations by charge separation. The resulting radical cations ($Ar^{+\bullet}$ or $Xe^{+\bullet}$) are then accelerated in an electric field and collide with neutral atoms *of the same species* in a collision chamber. The collision leads to an exchange of charges, so that the accelerated particles become neutral atoms again without significantly losing speed.
  2. An atomic beam of these—still remarkably fast, but now uncharged—noble gas atoms is then directed onto the sample itself, where it leads to ionisation. (The details of this ionisation process are once again beyond the scope of such an introductory text.)

  FAB is particularly suitable for organic acids of any mass and also for analytes with high molecular mass.
- Finally, in the **thermospray method**, mass spectrometry is combined with liquid chromatography (LC, known from Part III, "Analytical Chemistry I"), using aqueous solutions buffered with sodium or ammonium acetate. Accordingly, the ionisation is based on the interaction of the analytes with these cations, so that in the actual MS one usually obtains $[M+NH_4]^+$ and $[M+Na]^+$ as the "molecular ion". This method is also suitable for very polar and thermally unstable analytes.

❓ Questions

2. What is the m/z ratio obtained in the mass spectrometric study of acetic acid ionised by EI for (a) the analyte radical cation $CH_3COOH^{+\bullet}$ (b) for the double ionised dication CH_3COOH^{2+} (c) for the radical cation $CH_3COOD^{+\bullet}$ (d) for the radical cation $CD_3COOD^{+\bullet}$ (e) for the double ionised dication CD_3COOH^{2+}?
3. What would be the m/z-values for the five analytes from task 2 if the ionisation had been carried out by CI?

4. What would be the m/z ratio for a relatively high molecular weight analyte X with molar mass M(X) = 1442 g/mol, which would be (a) singly and (b) doubly ionised by thermospray ionisation from a solution buffered with sodium acetate? What would be the m/z values for (c) single and (d) double ionisation if ammonium acetate were used as buffer?

However, since the most common method in mass spectrometry is EI, we will limit ourselves to this method in the following.

One of the advantages of this ionisation variant is that not only the **molecular peak (parent peak)** itself, i.e. M$^+$, can be detected, but often a large number of **fragments**. And since the nature of the detectable fragments allows us to draw conclusions about the structure of the analyte before it was torn apart by ionisation and the effects of electric acceleration and magnetic field, we will take a closer look at the possible (and especially particularly probable!) fragmentations.

2.4 Fragmentations

It should be emphasised again which factors have to be taken into account in mass spectrometry:
- Not only is a single analyte molecule ionised and possibly fragmented, but even the smallest sample quantities still contains a considerable number of them. (Perhaps you would like to recall the basic principle of MS, which you already learned about in Part V, "Analytical Chemistry I"? At that time, it was illustrated using a simile based on puppets.)
- *Individual* molecular ions or molecule-fragment ions are encountered at the detector, i.e. with regard to the mass-to-charge ratio (m/z) it has a clear effect which isotopes are involved in each case: A methyl fragment (CH_3^+), for example, has different nominal masses depending on whether the carbon atom is the isotope ^{12}C or ^{13}C. With $z = 1$ the result is
    - m/z ($^{12}CH_3^+$) = 15
    - m/z ($^{13}CH_3^+$) = 16

  The same then applies, of course, to the isotopes of other elements: A methyl fragment containing a deuterium atom (^2H = D, i.e. $^{12}CDH_2$) also has a nominal mass of 16, etc.

> **If One Wants to Be Quite Precise**
> Actually, one should always write $CH_3^{+\bullet}$ for these methyl fragments, because they are not "simply cations", but *radical cations* (with an unpaired electron, which is exactly what the superscript dot stands for). In many textbooks on mass spectrometry, this is also consistently done, but it is not particularly conducive to general readability. You learned about the principle of ionisation in ▶ Sect. 2.3, so you should be aware that *radical cations* are preferentially formed and that the notation CH_3^+ is a bit ... imprecise. We will use it anyway.

Furthermore—and this makes the handling of mass spectra resulting from mass spectrometry immensely easier—some bonds are clearly more "labile" than others, i.e. there are characteristic fragmentations.

2.4 · Fragmentations

2.4.1 Cleaving of Bonds

Quite often the molecular peak (also known as the *parent peak*) is found in a mass spectrum, i.e. the peak which belongs to the mass of the molecular ion $M^{+\bullet}$ that has only been formed by electron loss. (Actually, one should now consistently mention not the *mass* but the mass-to-charge ratio m/z every time, but since in electron impact ionisation mainly only single ionisation takes place, so that $z = 1$, we will save ourselves that trouble here.) If no further fragmentations took place, i.e. if there were no bond cleavages in the MS, the corresponding spectrum of ethanol would look like ◘ Fig. 2.1.

However, in mass spectrometry, some—often very simple—bond cleavages usually have to be taken into account. This results in corresponding cleavage products with reduced mass—precisely because part of the molecule "was broken off". The product ions obtained this way thus lead to further peaks. However, since the positive charge can remain on the "mother molecule" during the cleavage of a molecule fragment just as it can end up on the corresponding cleaved fragment, a peak of the corresponding broken-off molecule fragment is often *also* found in the spectrum. Which of the two respective fragments takes over the charge (and thus also: Which of the fragments appears as a peak in a mass spectrum) depends primarily on which of the two fragments can better stabilise the charge. This becomes very clear in the first fission variant, which we will now deal with:

- **α-Cleavage**

In molecules with a hetero atom (O, N, etc.), α-cleavage refers to the cleavage of the bond that exists, or existed in the case of cleavage, in the unchanged analyte between the atoms α- and β-standing to the hetero atom (i.e. the C directly bonded to the hetero atom and its immediate neighbour). As a reminder, although you are certainly already familiar with this term from organic chemistry: ◘ Fig. 2.2 shows, relative to a heteroatom labeled X, the α-, β-, and γ-positions and the associated bonds.

In particular, if the hetero-atom (in ◘ Fig. 2.2: X) has one or more free electron pairs, this negative charge density can stabilise the positive charge occurring during fragmentation; accordingly, the associated peaks are found quite often (thus: with quite a large abundance) in mass spectra: These peaks turn out to be quite high.

> We have already seen in ◘ Fig. 2.1 that the relative abundances of the various ions are plotted on the y-axis of the mass spectra; the most abundant ion is then assigned the value 100 %: Commonly, mass spectra are thus *normalised* to the fragment ion with the highest abundance. The ion with the highest abundance is by no means always the *parent peak* belonging to the molecular ion!

◘ **Fig. 2.2** Relative position and bond designations within a molecule

☐ **Fig. 2.3** α-cleavages on butanone

> ▶ **Example**
>
> ☐ Figure 2.3 shows an example of the possible α-cleavages of the simple molecule butanone (CH_3–C(=O)–CH_2–CH_3) with mass number 72. (Because of their relative abundance, only the most abundant isotopes in each case are considered in these examples, i.e. ^1H and ^{12}C.)
> - There may be a loss of the methyl radical due to α-cleavage; correspondingly, cleavage of $^•CH_3$ results in a mass loss of 15, leaving a fragment ion of the nominal mass 57.
> - Alternatively, the α-cleavage may lead to the loss of the ethyl residue ($^•CH_2CH_3$, mass: 29), so that an acylium ion with m/z = 43 appears in the mass spectrum.
>
> However, the latter is more likely because the (radical) cation CH_3–$CO^{+•}$ is remarkably stable: The peak with m/z = 43 will occur with by far the greatest frequency in the mass spectrum of butanone. Accordingly, the spectrum will be normalised to this peak.
>
> There are several freely accessible databases on the Internet that contain spectroscopic data on a large number of compounds, including mass spectra (mostly obtained by electron impact ionisation). As an example, we refer to the database of the US *National Institute of Standards and Technology* (NIST), which you already know from Part IV, "Analytical Chemistry I". (Butanone can also be found there.)

▶ http://webbook.nist.gov/chemistry/

Since—as already mentioned several times—not only one butanone molecule is fragmented, but always a considerable number of them, *both* fragment ions will always be found in the spectrum of this compound. In addition, there is still the possibility that the positive charge does not remain with the molecule fragment with the hetero atom—even if this is much more likely due to its stabilising effect—but with the fragments that have been split off (here: methyl and ethyl radical). Accordingly, the occurrence of peaks at m/z = 15 (methyl) and m/z = 29 (ethyl) is to be expected in the corresponding spectra. ◀

- **σ-Cleavage**

Fragmentation also occurs in the mass spectrometer for compounds without any heteroatoms at all, such as hydrocarbons: In this case, simple C–C bonds are cleaved. The resulting ions are less stable, precisely because they are not stabilised by heteroatoms, and therefore do not occur as often. The corresponding peaks will therefore not be very high, but they will occur nevertheless.

2.4 · Fragmentations

> **▶ Example**
>
> In the fragmentation of butane, for example ($M(C_4H_{10}) = 58$), in addition to the molecular ion peak $M^{+\cdot}$ (with m/z = 58), the "residual molecular ion" CH_3–CH_2–$CH_2^{+\cdot}$ (with m/z = 43) will appear, which was formed by splitting off one of the two methyl groups. Likewise, cleavage of the central C–C bond is possible, yielding as fragments an (uncharged and therefore non-detectable) ethyl radical and an ethyl radical cation (CH_3–$CH_2^{+\cdot}$), and for the latter, m/z = 29.
>
> Again, only the most abundant isotopes were used for the calculation: Of course, besides the molecule-ion peak with m/z = 58, there will also be a peak at c 59. However, because statistically only every twenty-fifth butane molecule contains a ^{13}C atom, this peak will only have an intensity corresponding to one twenty-fifth of the abundance of the "normal" molecule-ion peak. The peak for the few molecular ions that contain even *two* carbon-13 atoms (i.e. $^{12}C_2\,^{13}C_2H_{10}$, m/z = 60) will probably disappear in the background noise. ◀

Especially in the case of cyclic compounds, σ-cleavage often leads to quite informative peaks, because the substituents bonded to a ring are quite easily separated in the course of σ-cleavage and are then either detected as radical cations—if they themselves take over the charge in the process—or—if the charge remains on the ring—the ring system remaining after the cleavage of the substituent appears in the spectrum. Especially if the number of substituents is kept within limits, characteristic peaks are obtained, for example, for a cyclohexane radical cation (m/z = 83) or a methylcyclohexane radical cation (m/z = 97).

■ **Benzyl and Allyl Cleavage**

Special cases of σ-cleavage are **benzyl** and **allyl cleavage**: Here, the C–C single bond "next but one" to an aromatic ring or a C=C double bond is cleaved.

> **▶ Example**
>
> Exemplary shown are these two cleavages using the compounds propylbenzene (C_6H_5–CH_2–CH_2–CH_3, figure a) and 1-hexene (b).
>
> a Benzyl and b allyl cleavage
>
> While σ-cleavage can (and will) easily occur again in the saturated propyl or butyl residue, a different cleavage process nevertheless leads to much more abundant cations:
> — The mass spectrum of the propylbenzene (M = 120) is normalised to the peak with m/z = 91.
> — In the mass spectrum of 1-hexene (M = 84), two approximately equally frequent peaks are noticeable:
>   – m/z = 56 (peak height 95 %) *and*
>   – m/z = 41 (100 %) ◀

▶ http://webbook.nist.gov/cgi/cbook.cgi?Name=propylbenzene&Units=SI&cMS=on#Mass-Spec

or

▶ http://webbook.nist.gov/cgi/cbook.cgi?Name=1-hexene&Units=SI&cMS=on#Mass-Spec

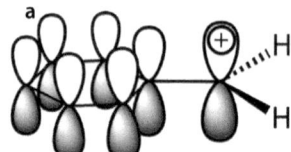
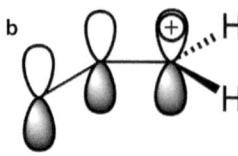

Fig. 2.4 Stabilisation of **a** benzyl and **b** allyl position

The reason for the benzyl and allyl cleavage are the resulting cations:
- Benzyl cleavage produces a benzyl ion whose positive charge is stabilised by the negative charge density of the π-electron system: As shown in Fig. 2.4a, the vacant p-orbital of the sp^2-hybridised benzyl carbon interacts with the aromatic system and thus becomes mesomerically stable (it is resonance-stabilised).
- During allyl cleavage, the positive charge of the resulting ion is stabilised by interaction with the π-electrons of the C=C bond; an allyl cation is present. Analogous to the benzyl cation, the allyl cation is also stabilised by mesomerism/resonance (Fig. 2.4b).

> ► **Example**
> The mass loss in propylbenzene (m/z = 29) can thus be explained by the cleavage of an ethyl radical; the cleavage of the bond marked (a) in the previous example occurs preferentially: The benzyl ion (C_6H_5–CH_2^+) with m/z = 91 is obtained.
> The same applies to the 1-hexene in the right-hand cleavage of the two cleavages marked with (b) in the example figure mentioned: Here the allyl ion (CH_2=CH–CH_2^+) is formed with m/z = 41 (loss of the propyl residue with m/z = 43). ◄

In order to understand where the different peaks within a mass spectrum come from, the fragmentation reactions already described are not sufficient on their own, because in mass spectrometry not only radicals can be split off, but also small *neutral molecules*—provided that their formation is already "inherent" in the structure of the starting compound.

> This is exactly the reason for the peak at m/z = 56 in the spectrum of 1-hexene: In this molecule, the C=C unit can be split off from the "rest" as uncharged ethene (H_2C=CH_2) by transferring a hydrogen atom (because it is uncharged, it will not be detected, i.e. occur in the spectrum); what remains is a butyl radical cation $^{+\bullet}CH$–CH_2–CH_2–CH_3 with m/z = 56. (This can then be fragmented further—keyword: σ-cleavage -, but we do not want to go that far into this topic here.)

The fact that this last type of fragmentation also involves the displacement of individual atoms within the molecular framework—i.e. bonds are broken and others are newly formed—leads us to the next two types of fragmentation, in which **rearrangements** occur.

2.4.2 Rearrangements

The reaction type of rearrangements will doubtlessly be covered in more detail in advanced courses on organic chemistry, but since some reactions of this type are of indispensable importance in mass spectrometry, they will be discussed here in some detail.

2.4 · Fragmentations

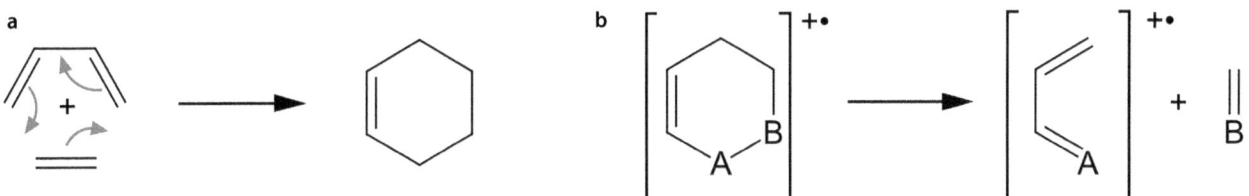

Fig. 2.5 a Diels-Alder and b retro-Diels-Alder reactionIt should be borne in mind that *hetero atoms* can also be part of the resulting ring system

- **Retro-Diels-Alder (RDA)**

The **Diels-Alder reaction** belongs to the **pericyclic reactions,** i.e. to a type of reaction in which new bonds are formed and others broken by the **concerted** displacement of several electrons. A simple example of the Diels-Alder reaction is the reaction of 1,3-butadiene with ethene, which leads to the formation of cyclohexene (Fig. 2.5a):

The *reversal of* this reaction, i.e. the cleavage of a corresponding product into its starting materials, is then referred to as a **retro-Diels-Alder reaction (RDA** for short). This reaction is particularly common in six-membered ring systems containing a double bond. In general, these reactions follow the scheme shown in Fig. 2.5b, in which A and B *can be* hetero atoms, but do not have to be.

> ► **Example**
>
> Assume that atom A is an oxygen atom, while atom B is a nitrogen atom to which a hydrogen atom is also bonded (i.e.: B = N–H). For our analyte with the sum formula C_4H_7NO the following peaks would be expected:
> - the *parent peak* M^+ with m/z = 85
> - the cationic fragment $H_2C=CH–CH=A^{+\bullet}$, specifically: $H_2C=CH–CH=O^{+\bullet}$ with m/z = 56, which is formed via RDA.
> - The neutral molecule split off in the course of the RDA reaction ($H_2C=B$, i.e. in our example $H_2C=NH$), on the other hand, is uncharged and therefore *does not appear* in the mass spectrum.
>
> However, the reaction can also proceed in such a way that the positive charge (and the radical character) remains with the latter molecular fragment, i.e. the ion $H_2C=B^{+\bullet}$ is formed.
> - Then we also get a peak for the ion $H_2C=NH^{+\bullet}$ at m/z = 29.
> - The peak at m/z = 56 is then correspondingly smaller in the associated spectrum than if this alternative fragmentation did not occur. ◄

- **McLafferty Rearrangement**

Analogous to the Diels-Alder reaction, in which new C–C bonds are formed, a hydrogen atom can also change its bonding partner; an example of this is shown in Fig. 2.6.

Again, X stands for a hetero-atom that has at least one free electron pair and can accordingly stabilise the positive charge arising in the course of ionisation quite well. The fragmentation described here is a β-cleavage, in which not the C–C bond in α position to the hetero atom is broken, but its β-C–C bond (one bond further away from the hetero atom). At the same time—and herein lies the parallel to RDA—a hydrogen atom in γ-position changes its bonding partner and is bonded to the hetero atom itself in the course of this reaction, which also proceeds in a concerted manner. (If necessary, look again at Fig. 2.2.) Thus, in addition to β-cleavage, an *H-shift* also occurs. (It should not remain unmentioned that this reaction can also take place in steps, but this aspect will not be discussed here any further.)

☐ **Fig. 2.6** McLafferty rearrangement

This fragmentation, known as the McLafferty rearrangement, occurs quite frequently under the conditions of mass spectrometry, but can also be brought about under other reaction conditions. Then it is called an **En reaction**, and you will also encounter it again in more advanced courses on organic chemistry.

▶ **Example**

Let us again take a concrete example: If X stands for an oxygen atom, our analyte would be the aldehyde butanal ($CH_3-CH_2-CH_2-CHO$, M = 72 g/mol).
- With highest probability, we will obtain the molecular peak with m/z = 72.

The additional peaks resulting from the McLafferty rearrangement depend on which of the two resulting fragments takes on the positive charge:
- If the radical cation of the enol form of acetaldehyde ($CH_2=CH-OH^{+\bullet}$) with m/z = 44 is formed, uncharged ethene is also split off, which explains the mass loss of 28 (72 − 44).
- Alternatively, radical cationic ethene ($CH_2=CH_2^{+\bullet}$) with m/z = 28 is formed, and neutral acetaldehyde (in its enol form) is split off.

Since both processes can occur equally, it is statistically expected that at least three peaks occur in the corresponding mass spectrum of butanal: at m/z = 72 ($M^{+\bullet}$), m/z = 44 ($M^{+\bullet}$ − 28) and m/z = 28 ($CH_2=CH_2^{+\bullet}$; $M^{+\bullet}$ − 44). ◀

2.4.3 And Now: Step by Step

Here, as an exercise, is another spectrum that should (hopefully) finally make the principle clear: An unknown, white plastic was analysed via **GC/MS pyrolysis** (as part of a laboratory course). It turned out (how exactly that was found out shall not be discussed here) that it is the plastic polystyrene (PS) obtained by polymerisation of styrene (vinylbenzene, phenylethene, $C_6H_5-CH=CH_2$) (☐ Fig. 2.7).

▶ http://webbook.nist.gov/cgi/cbook.cgi?Name=toluene&Units=SI&cMS=on#Mass-Spec

During the GC/MS analysis of the various pyrolysis products, among other things a mass spectrum was recorded, which clearly shows that in the process of thermal decomposition of PS toluene (methylbenzene, $C_6H_5-CH_3$) was formed. (Again, a reference spectrum of this compound can be found in the NIST database.)

☐ **Fig. 2.7** Polystyrene

2.4 · Fragmentations

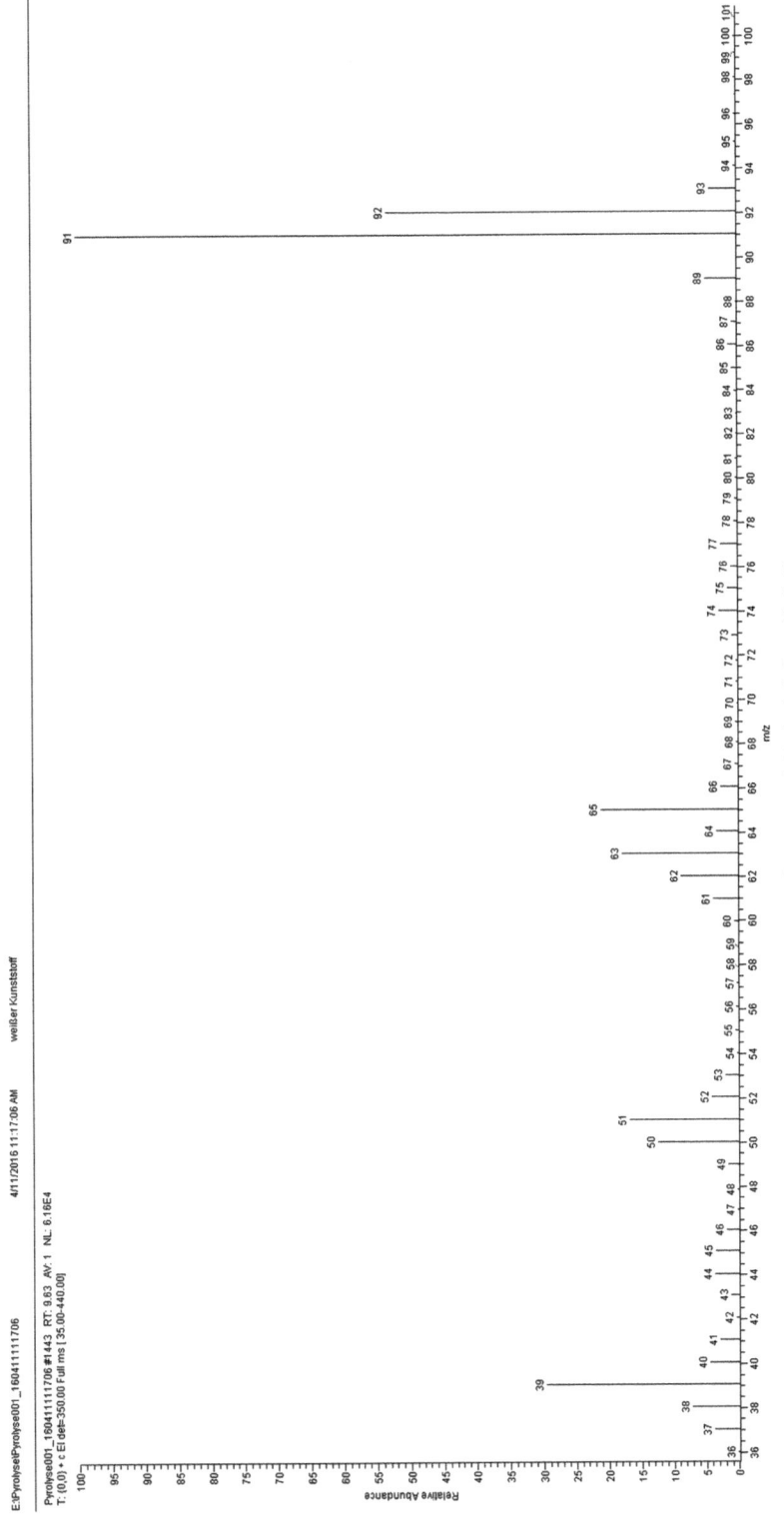

Mass spectrum of toluene ($C_6H_5-CH_3$); kindly provided by P. Kusch, Hochschule Bonn-Rhein-Sieg

Let us first look at the different peaks; the fragmentation of toluene (together with the m/z ratio of the corresponding molecular ions) is shown in ◘ Fig. 2.9:
- At m/z = 92, this is the *parent peak of* the merely ionised but otherwise unfragmented toluene [M⁺] itself.
- m/z = 93 belongs to a likewise unfragmented toluene, which contains a ^{13}C atom. (Since the framework of toluene consists of seven carbon atoms, the probability that at least one of the C atoms contains an additional neutron is about 7%, given the isotopic ratio $^{12}C/^{13}C$; this peak is correspondingly small.)

If you look closely, you will also discover the peak with m/z = 94, which can hardly be distinguished from the background noise; the even larger numerical values belong to impurities, which need not interest us any further here. (Obviously, the gas chromatographic separation did not run quite perfectly—possibly a *memory effect* manifested itself here: Condensation on cooler parts within the ion source space may still leave [minimal] substance residues of older analytes, which then appear in the next spectrum. Often, however, the associated peaks disappear in the background noise.)
- The spectrum is normalised (as explained in ► Sect. 2.4.1) to the most frequently occurring fragment: m/z = 91. This is the benzyl cation (C_6H_5–CH_2^+) formed by splitting off a hydrogen radical (= [M − 1], the mass number of a ^1H atom; the probability that a ^2H atom is present, i.e. a deuterium, may be neglected) *or* the tropylium ion ($C_7H_7^+$) formed by a simple rearrangement, since this ion is very stable due to its aromaticity.
- The next peak with relevant (%) height can be found at m/z = 65 ($C_7H_7^+$ − 26): this ion is formed when the tropylium ion splits off (neutral) ethyne (acetylene, HC≡CH).
- When ethyne is split off again (i.e. again: −26), a molecular ion with m/z = 39 is formed, which is probably the cyclopropenylium cation ($C_3H_3^+$, the smallest possible aromatic compound) (◘ Fig. 2.8a).
- However, for the benzyl cation (C_6H_5–CH_2^+, m/z = 91) a further fragmentation pathway exists:
  - If a methyl radical (CH_3) is split off (C_6H_5–CH_2^+ − 15), we obtain the phenylium cation ($C_6H_5^+$ with m/z = 77). Although this is very typical of aromatics, it is not particularly stable because the positive charge cannot be delocalised by mesomerism due to its localisation in a (vacant) sp^2 hybrid orbital (◘ Fig. 2.8b). The corresponding peak is correspondingly small.
  - This phenylium cation usually decays even further: by splitting off an ethyne molecule (−26), we obtain the cation $C_4H_3^+$ with m/z = 51. This is also characteristic of aromatic analytes; it is probably the cyclobutenylium cation (◘ Fig. 2.8c).

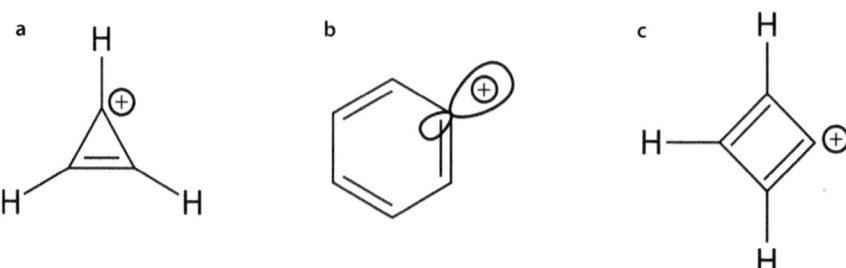

◘ Fig. 2.8 a Cyclopropenylium ($C_3H_3^+$), b phenylium ($C_6H_5^+$) and c cyclobutenylium cation ($C_4H_3^+$).

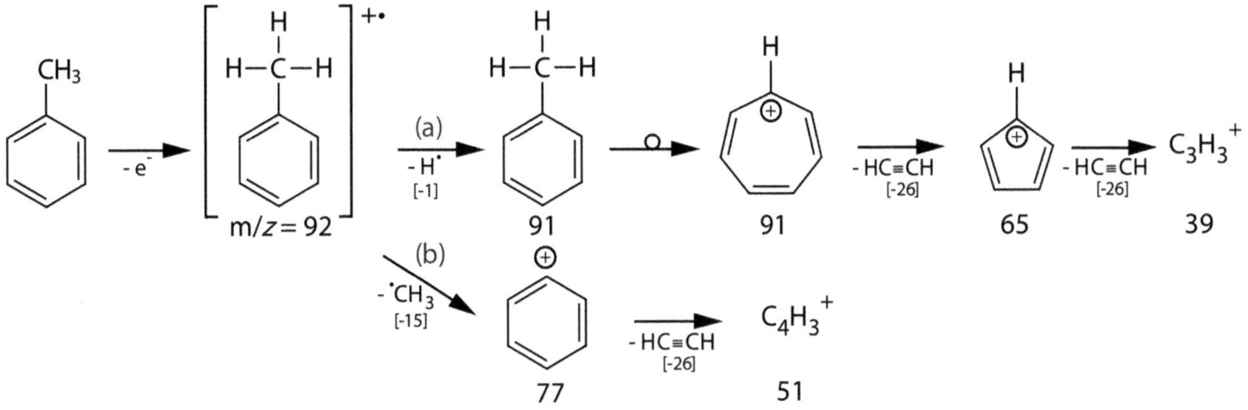

2.4 · Fragmentations

Fig. 2.9 Fragmentation of toluene

A good overview of numerous fragments typical for different types of analytes (such as the benzyl cation, which is almost always found in mass spectra of aromatic analytes) as well as frequently occurring mass losses (such as −26, which is typical for the cleavage of a molecule of ethyne) is provided by the classic of instrumental analysis already mentioned in the introduction: the ▶ Hesse-Meier-Zeeh.

Of course, not all peaks of the spectrum have been discussed here, but the most important ones whose occurrence can be explained by the fragmentation scheme shown in ◘ Fig. 2.9.

Bienz S, Bigler L, Fox T, Meier H (2016) Hesse-Meier-Zeeh: Spektroskopische Methoden in der organischen Chemie

❓ Questions

5. Why do we find in the EI mass spectrum of bromobenzene (C_6H_5Br), next to the molecular peak with m/z = 156, another peak with m/z = 158, the intensity of which corresponds almost exactly to that of the peak at 156?
6. Which peaks do you expect to see in the mass spectrum (obtained after EI) of (a) ethylcyclohexane; (b) chlorobenzene; (c) propylbenzene?
7. Which peak in the EI-MS of 1-iodo-6-methylcyclohexane from the NIST database can be explained by a retro-Diels-Alder reaction? What is the probable origin of the three clearly more prominent peaks at m/z = 222, m/z = 127 and m/z = 95?
   ▶ http://webbook.nist.gov/cgi/cbook.cgi?ID=C40648100&Units=SI&Mask=200

Nuclear Magnetic Resonance Spectroscopy (NMR)

Contents

3.1 Physical Principles – 28
3.1.1 Spin States in the Magnetic Field – 29
3.1.2 Energy Considerations – 30

3.2 First NMR Spectra – 32
3.2.1 Influence of the Electron Density – 33
3.2.2 Multiplet – 36
3.2.3 The Chemical Shift δ, Considered in More Detail – 39
3.2.4 Anisotropy Effects – 44

3.3 ^{1}H-NMR – 46
3.3.1 Chemical Equivalence – 46
3.3.2 Couplings – 48

3.4 ^{13}C-NMR – 57
3.4.1 Shifts, Couplings, Spectra – 57
3.4.2 Two-dimensional NMR – 61

3.5 Other Usable Cores – 63

Further Reading – 77

© Springer-Verlag GmbH Germany, part of Springer Nature 2025
U. Ritgen, *Analytical Chemistry II*, https://doi.org/10.1007/978-3-662-68710-9_3

Nuclear magnetic resonance (NMR) spectroscopy, perhaps the most important method for determining the structure of covalently built (mostly organic) compounds, is a non-destructive method of analysis; here, unlike in MS, the analyte can be recovered afterwards,—which is of course quite desirable, especially in the case of a preparatively labourious substance, which you may have synthesised only on a milligram scale and of which you now want to know whether you have really obtained what you were after. (Whether this is always feasible in practice, however, is another matter, because some analytes can only be dissolved in solvents such as dimethyl sulfoxide (DMSO, $(CH_3)_2SO$), and the attempt to separate this solvent by distillation is quite questionable in view of its boiling point (bp = 189 °C); there are not too many organic analytes that can withstand such thermal stress ....)

SKOOG, Chapter 19: Nuclear magnetic resonance (NMR) spectroscopy

▶ Skoog devotes the whole of Chapter 19 to NMR; in addition to the absolute basics, which are indispensable for evaluating the corresponding spectra, he also deals in detail with the necessary equipment and selected technical details. In some cases, details are discussed which go far beyond the scope of this book. Therefore, this part of the book is only intended to give you a (very brief) overview of the physical basics of nuclear magnetic resonance experiments, in order to move on from there to the *practical* applications of this analysis method, which are clearly more relevant for you. It lies in the nature of things that certain instrumental requirements cannot be addressed to the extent that would be necessary for an in-depth introduction to the fine art of NMR. If you are interested in these aspects to an extent that exceeds even Skoog (and it is really worthwhile to delve into this topic!), please refer to the further reading at the end of this chapter. In particular, Breitmaier's work deals in detail with the importance of circularly polarised radiation and phenomena such as spin-lattice and spin-spin relaxation processes—important aspects for recording clean spectra. However, for the evaluation of spectra *already available*—and this is the topic we want to focus on here—it is not absolutely necessary to go into such detail.

3.1 Physical Principles

The analysis is based on the absorption of electromagnetic radiation from the radio frequency range (with frequencies from 4 to 900 MHz). The absorption takes place through the atomic *nuclei* in the presence of a magnetic field.

> However, not all atomic nuclei can be excited to absorption of radio waves. This is only possible with nuclei in which the number of protons and neutrons is not even.

> This already shows us that the most common carbon isotope (^{12}C) is *not* suitable for nuclear magnetic resonance spectroscopy, because carbon-12 has 6 protons and 6 neutrons in its nucleus. However, this is not the case for the (much rarer) carbon isotope ^{13}C.

If, however, a suitable nucleus absorbs correspondingly, the phenomenon of *resonance* occurs (which is why it is called NMR). In order to indicate which atomic nuclei are brought to resonance in a corresponding experiment, the isotope used is named for eacH-NMR spectrum, i.e. one speaks of a ^1H-NMR or a ^{13}C-NMR spectrum, etc.

3.1 · Physical Principles

■ Resonance

Important for this phenomenon is, that atomic nuclei (even without external excitation!) rotate around their own axis—they have a **nuclear spin.**

Please note that we are looking at *atomic* nuclei on the *microscopic* level here, and this means that this rotational motion is quantised in the same way as you should already know (perhaps even from Part IV—Molecular Spectroscopy—of "Analytical Chemistry I"?): Only certain energy states can be taken—those which can be described by a *half-* or *integer* multiple of the value h/2π. Which and how many different (nuclear) spin states are possible depends on the (atomic-nucleus-specific) spin quantum number I.

Nuclei of particular importance in nuclear magnetic resonance spectroscopy are: ^1H, ^{13}C, ^{19}F and ^{31}P. For all of them, I = ½. Of particular importance here are the ^1H and the ^{13}C spectra. (We will deal with others within the framework of this part at most in the form of side remarks.)

> **A Brief Reminder**
> Of course, the question immediately arises: Why are half-integer values possible at all? After all, it has always been the case with quantum numbers that only integer values were permissible!—What is important here is not the values themselves, but the *differences that are possible* between the individual spin states. For example, if we take the values +½ and −½, the difference between them is exactly 1, which of course is an integer. And you already know this from basic courses on *general chemistry:* It is exactly the same as with electron spin, where the two spin states +½ and −½ are possible: Electrons also have the spin quantum number I = ½.

3.1.1 Spin States in the Magnetic Field

Without a magnetic field, the different spin states cannot be distinguished—you already know this from the p- or d-orbitals of the same principal quantum number, which cannot be distinguished in field-free space—i.e. without the presence of an (electro-)magnetic field.

From *physics it* is known that charged bodies in rotation generate a magnetic field B, and atomic nuclei *are* charged. (Please do not forget that there is always at least one proton p^+ in the atomic nucleus, depending on the atomic number even considerably more.) This results in a magnetic moment μ_0.

In a magnetic field of strength B_0, a rotating atomic nucleus can then align itself in different ways. Depending on I, this results in different possible magnetic quantum states M:

$$M = +I, +(I-1), +(I-2)\ldots \text{ to} -I$$

For the nuclei mentioned above, which are used very often in NMR, only the spin states +½ and −½ are possible, because with (+½ − 1) we already obtain −½. Here, the orientation in the magnetic field can be compared with a compass needle, which can only be oriented
- in parallel or
- antiparallel

to the magnetic field.

> **Looking Over the Brim of the Proverbial Tea Cup**
> For other nuclei, which are much less important in "everyday analytics", the situation regarding the nuclear spin is somewhat different: The lithium isotope ^{6}Li, for example, has a nuclear spin of I = 1, which means that there are not only two but *three* different spin states (+1, 0 and −1), and in the case of lithium isotope ^{7}Li it goes even further: the nuclear spin I(^{7}Li) = 3/2 even allows *four* spin states: +3/2, +½, −½ and −3/2. Particularly creative in terms of nuclear spin is bismuth-209 with I(^{209}Bi) = 9/2, leading to a total of 19 (!) different spin states (everything from +9/2, +7/2, ... to −9/2). Here, the similarity to the compass needle in the magnetic field does not really work, anymore. However, we will not discuss such nuclei further in the context of this introduction: We stick as far as possible to the convenient special case I = ½, leading to the two possibilities "parallel" and "antiparallel".

Skoog, Section 19.1.1: Quantum mechanical description of NMR

These two spin states differ slightly in their potential energy (E_{pot}). Convention states that +½ is slightly more energetically favourable than −½. The relative energy difference of the two spin states is shown graphically in Fig. 19.1 in Skoog

> **For Those Who Want to Know More**
> The variable γ given in the above-mentioned sub-figure 19.1.1b is the *gyromagnetic ratio* specific to the respective atomic species. It tells us something about how pronounced the interaction of the atomic nucleus with the electromagnetic radiation is in each case. For *quantitative* considerations, this value (specific for the respective isotope) is very important; however, since we want to deal with nuclear magnetic resonance spectroscopy primarily *qualitatively*, it is mentioned here only for the sake of completeness.

3.1.2 Energy Considerations

As with other excitations by electromagnetic radiation (UV/VIS, IR, etc., possibly known from "Analytical Chemistry I"), a transition from one quantum state (here: the spin state) to another leads to a change in the energy content of our analyte (in this case: the energy content of the excited atomic nucleus), so in the case of such a state change, energy
- is released when the system changes from an energetically less favourable to an energetically more favourable state (this leads to the **emission** of electromagnetic radiation), or
- is consumed when the system is forced into a less favourable state from an energetically more favourable state (**absorption** of electromagnetic radiation then takes place).

According to Eq. (3.1) (perhaps already known from Part I of "Analytical Chemistry I")

$$E = h \times \nu = h \times \frac{c}{\lambda} \tag{3.1}$$

there is a direct relationship between the energy released or absorbed and the frequency (ν) or wavelength (λ) of the electromagnetic radiation emitted or absorbed.

3.1 · Physical Principles

- In the absence of a specifically applied magnetic field, the different spin states are equally likely, so both spin states (we continue to restrict ourselves here to systems with I = ½) are adopted equally often.
- However, if a (sufficiently powerful) magnetic field is applied, the lower-energy +½ state is understandably preferred.

Since the energy difference between the two states is really very small, the "surplus" of atomic nuclei in the +½ state is not particularly large—but still significant, because if there were *no* surplus at all (i.e. if both states were still occupied equally often), the amount of energy released by emission (during the transition from −½ to +½) would be exactly as large as the amount of absorbed energy that causes the transition from the more favourable +½ state to the less favourable −½ state.

■ The Classical Way of Representation

Since an atomic nucleus rotates (otherwise there would be no magnetic moment μ_0), there is a **precession motion**—analogous to a spinning top, which is not oriented perfectly vertical, but wobbles around its slightly obliquely oriented axis of rotation; you can see this in Fig. 19.2 from Skoog. The rotating atomic nuclei thus *precess* around the vector of the magnetic field. The precession frequency ν, which is also called the **Larmor frequency,** corresponds to the frequency of the absorbed or emitted radiation from ▶ Sect. 3.1.2.

Skoog, Section 19.1.2: Classical description of NMR

Upon emission of the associated frequency (recall once again the quantisation of energy and thus the precisely defined quantum states), the cone from the above figure turns "upside down" if it was previously in the state I = +½; correspondingly for I = −½, exactly the reverse results, so that after emission of a photon of frequency ν (recall Eq. 3.1), the cone "stands" exactly as in Fig. 19.2 from Skoog

It is important to note that the Larmor frequency of an atomic nucleus
(a) depends not only on the gyromagnetic ratio γ (from ▶ Sect. 3.1.1),
(b) but also on the strength of the applied magnetic field B0 (we will encounter this again in ▶ Sect. 3.2),
(c) as well as on the bonding conditions of the respective atoms under consideration.

> This is exactly why chemically different atoms can be distinguished, e.g. the two C-atoms of an ethanol molecule, for one thing the **methyl-C-atom** (to which three H-atoms are bonded) for the other the **methylene-carbon** (which also is connected to a hydroxy group). Of course, the same is true for the three different kinds of H-atoms in this molecule (the three methyl hydrogens, the two methylene hydrogens and the hydrogen atom that belongs to the OH-group of the molecule).

More on this in ▶ Sect. 3.2.

■ A Quick Look at More Details

Since in the magnetic field the number of nuclei in the energetically more favourable +½ state is slightly larger, somewhat more absorption than emission takes place (as mentioned above). However, this poses the problem that due to the absorption the number of nuclei in both spin states within a short period of time becomes exactly the same, so that a *saturated system* is formed, in which absorption and emission are exactly balanced and thus the absorption signal would ultimately drop to zero. Of course, this must be prevented.

At the same time, it has to be considered that a nucleus, which has been put into the energetically less favourable state $-½$ by absorption of a corresponding photon, is quite "anxious" to return to its (energetically more favourable) initial state, i.e. to **relax.**

Thus, it must be ensured by apparatus that the **relaxation of the nuclei** occurs at least as quickly as the change of the spin state due to the interaction with the electromagnetic radiation. (The fact that in reality there are even several different relaxation processes shall be mentioned only briefly, because this goes clearly beyond the scope of this introduction.)

For nuclear magnetic resonance spectroscopy to work at all, the magnetic field of the respective nucleus must be at right angles to the applied magnetic field. Such apparative aspects would also go a bit beyond the scope here; what is decisive is only that it is ultimately possible to determine both Larmor frequencies—and from these frequencies, nuclear magnetic resonance spectra can then be generated.

3.2 First NMR Spectra

Which Larmor frequency belongs to a given atomic nucleus depends not only on the atomic species under consideration (the fact that the isotope-specific gyromagnetic ratio γ also plays a role here has already been mentioned in ▶ Sect. 3.1.1), but also on various other factors which—and this is the good thing about it—are *directly related* to the chemical (bonding) conditions present in the respective analyte. For this reason, the different atoms of a polyatomic analyte can usually be easily distinguished. In which form—and why—the bonding conditions and the like have an effect, you will learn in ▶ Sect. 3.2.1; first, however, let us take a look at the first spectra as an example. It is important to bear in mind that chemically different atomic nuclei of the same type of atom have different Larmor frequencies precisely *because* of different (bonding) conditions, or in other words: can be excited/brought to resonance with different wavelengths.

Quite analogous to the spectroscopic methods you already learned about in Part IV of "Analytical Chemistry I", a certain part of the electromagnetic spectrum (this time radio waves) must also be scanned in NMR. In addition, a strong magnetic field is needed: Several Tesla are desirable. (The stronger the magnetic field, the more detailed the spectra obtained will be—you will find out why in ▶ Sects. 3.3 and 3.4.) And just as in IR spectroscopy, a distinction is made between two different methods:

- **CW NMR (Continuous Wave)**

Here—again as with IR—the radio wave range is gradually scanned. Two different methods are used for this:
(a) It is indeed possible to selectively change the frequency of the excitation radiation in the presence of the magnetic field; this would be the "correct" scanning of the relevant wavelength range.
(b) Since the Larmor frequency of the analyte atoms also depends on the applied magnetic field strength (B_0) (this was mentioned in ▶ Sect. 3.1.2), one can also choose the opposite way and keep the frequency constant and vary the magnetic field strength. Nowadays this is the more common way: One uses monochromatic excitation radiation from the radio wave range and scans "little by little" different magnetic field strengths.

In both cases, relatively "fuzzy" spectra are obtained, in which considerable signal broadening occurs, so that the depth of detail is limited. (In laboratory jargon, sucH-NMR spectrometers are often referred to as "estimators". They may be just about good enough for very, very rough statements, but they can hardly be used for serious analysis.)

Such a CW spectrum is shown (stylised) in Fig. 19.12a from Skoog. It is an anything but well-resolved ^{1}H-NMR of ethanol (C_2H_5OH, CH_3–CH_2–OH).

Skoog, Section 19.1.4: NMR spectra types

Fortunately, the fact that computers are now basically ubiquitous even in the laboratory world can be exploited in NMR:

- **FT-NMR (Fourier Transform NMR)**

Analogous to IR spectroscopy in Part IV of "Analytical Chemistry I", it is also possible to use polyfrequency radiation (this time in the radio wave range) in NMR spectroscopy, thanks to the Fourier transform method, and thus scan all wavelengths simultaneously. This not only has the advantage that the measurement is much faster, but also that the problem of signal broadening does not arise. Accordingly, a—much higher resolution—FT-NMR spectrum also shows details that are (usually) lost in CW; see Fig. 19.12b from Skoog- again, this ^{1}H-NMR spectrum also belongs to ethanol (CH_3–CH_2–OH). (Why the three signals from the spectrum above turn to this seeming multitude of signals is explained in ▶ Sect. 3.3.)

Skoog, Section 19.1.4: NMR spectra types

> **Technical Language Tip**
> In nuclear magnetic resonance spectroscopy, as you may have noticed, we consistently talk about *"signals"*, not "peaks" or "bands". Some examiners place a lot of emphasis on this, so for the sake of your grades, you should conform to this usage in any exam.

3.2.1 Influence of the Electron Density

Let us first look at the not particularly informative ^{1}H-NMR spectrum of Fig. 19.12a from Skoog. Despite the low resolution, it is clear that the total of six hydrogen atoms present by no means all are brought to resonance by the same wavelength (at constant magnetic field) or, if only *one* wavelength is to be used for excitation, at the same magnetic field strength: We observe *three* signals. (The axis labelling "Magnetic field →" tells us that in fact only one excitation wavelength and thus a variable magnetic field was used for generating this spectrum.)

Skoog, Section 19.1.4: NMR spectra types

The fact that, although there are six hydrogen atoms in the ethanol molecule, we only get three signals tells us that some of the H atoms are **chemically equivalent:**

> Atomic nuclei are considered chemically equivalent if they are in an equivalent environment, i.e. if they all have the same type of bonding partner and are otherwise indistinguishable from one another.

We will look at chemical equivalence in more detail in ▶ Sect. 3.3.1.

In addition, it becomes clear that the various signals differ in their height—or better: in the area that is located under the respective signal.

> Because of possible signal broadening, signal heights alone are not meaningful in NMR, but the *areas* lying under the signals *are*, i.e. the **integral** belonging to the respective signal. This contains even highly important information.

Integrate and you get:
1. for the signal marked –OH a relative area of 1
2. for the methylene group (–CH_2–) an area of 2 and
3. for the methyl group (–CH_3) an area of 3.

These differences in area can be attributed to the fact that a different number of *chemically equivalent* hydrogen atoms can be excited at the same time (namely one, two or three—just as the labelling of the signals suggests).

Now, of course, the question arises as to *why* these three types of hydrogen atoms (or better: atomic *nuclei*), as part of the respective groups, can be excited to resonate with different ease. This is where **shielding effects** come into play.

> **Important**
> Please note that the integrals under the respective signals only allow *relative* statements. If one only pays attention to the areas, the compound 2,3-dimethyl-2-butene-1,4-diol—regardless of whether its *(Z)*-form (a) is present or the *(E)*-isomer (b)—has exactly the same area ratio as the ethanol spectrum, although the molecule has exactly twice as many hydroxyl methylene and methyl groups as C_2H_5OH.

2,3-Dimethyl-2-buten-1,4-diol

Nevertheless, the three compounds can be distinguished from each other by nuclear magnetic resonance spectroscopy, because the relative position of the respective signals also holds analytical information—plenty of it, to be precise: The *shielding* of the respective H atoms from the different (functional) groups does indeed differ in the different molecules.

■ **Shielding and Chemical Shift**

The extent to which a corresponding atomic nucleus can be excited to resonance (transferred to spectra: how far "to the left" or " to the right" the corresponding signal appears in the spectrum) is referred to as the **chemical shift.** It depends on various factors. A decisive role is played by the *electron density* at the atom in question.

This can be explained by the fact that this electron density has a **shielding effect**:

> The higher the charge density on an atom (nucleus), the more it is shielded from the effect of the magnetic field. Effectively, this means that the local shielding of an atom causes a correspondingly less pronounced, i.e. weakened, magnetic field to act on its nucleus.

3.2 · First NMR Spectra

As expected, this affects both measurement techniques (changed magnetic field strength when excited by a constant wavelength or constant magnetic field strength when excited by radiation of variable wavelength).
- In the case of excitation with electromagnetic radiation of constant wavelength and variable magnetic field strength, the following applies: If the effect of the magnetic field on the atomic nucleus is weakened by the electron density, a correspondingly somewhat stronger magnetic field is required to bring about resonance. Accordingly, the associated signal is shifted "to the **right**", i.e. to the **high field**; this is then referred to as a **upfield shift**.
- Similarly, at constant magnetic field strength, a corresponding atomic nucleus can only be excited by somewhat *higher-energy* electromagnetic radiation.

> If, on the other hand, the shielding by electrons is reduced, for example because the electron density at the atomic nucleus under consideration is lowered by one or more electronegative bonding partners, resonance can already be achieved somewhat more easily: Such a nucleus is called **deshielded**.

In such a case, you need:
- either (at constant wavelength of the excitation radiation) a not so strong magnetic field
- or (at constant magnetic field) a slightly less energetic excitation radiation.

In this case one speaks of a **downshift**, the corresponding signal is found "further **left**" in the NMR spectrum: It is shifted to **lower field**.

This can even be seen in the rather poor spectrum (Fig. 19.12a from Skoog):

Skoog, Section 19.1.4: NMR spectra types

- The signal of the hydrogen nucleus of the OH group appears at lower field because the significantly more electronegative oxygen atom significantly reduces the electron density at the hydrogen in question, so that this nucleus can already be excited to resonance in the somewhat weaker magnetic field—the signal is downshifted.
- The shift in charge density still has an effect even if the hydrogen atoms are not directly bonded to the more electronegative atom (here: oxygen), as is the case with the methylene group (which carries the methyl group and the OH group): Due to the higher electronegativity of oxygen compared to carbon, the latter is deprived of part of its "own" electron density, and this atom compensates for this by "pulling" at least part of the electron density it now lacks from the two hydrogen atoms, which are also deshielded in this way. Accordingly, the signal of the methylene hydrogens is also shifted to lower field (=downshifted), although not as strongly as in the case of the hydroxy hydrogen.
- Finally, the methyl group is bonded to the slightly "electron-depleted" methylene carbon, which also leads to a slight deshielding of the corresponding methyl hydrogens. However, this is much less pronounced than for the two preceding hydrogens, resulting in a (seemingly) upshifted signal for the methyl group *relative* to the other two moieties of the molecule: It appears in the right part of the spectrum.

❓ Questions

8. In which of the compounds given do you expect an upshift for the C and H atoms respectively, and in which do you expect a downshift? Give reasons for your answer.

a. $CH_3-C(=O)-OH$ b. $CH_3CH_2CH_2-C(=O)-OH$ c. $CH_3CH_2-NH_2$ d. $(CH_3CH_2)_2B-CH_2CH_3$ (triethylborane)

9. For which of the two methyl groups of acetic acid methyl ester (CH$_3$–C(=O)–O–CH$_3$) do you expect a stronger shift to lower field? Again, you are asked to give a reason.

Skoog, Section 19.1.4: NMR spectra types

The influence of the electron density on the position of the respective atomic nuclei is obviously enormous, but this is not the only thing to consider. *Anisotropy effects* also play a role; we will discuss these in ▶ Sect. 3.2.4. Now we shall first be interested in why, in a higher-resolution NMR spectrum (such as the one shown in Fig. 19.12b from Skoog), the three relatively broad signals we saw in Fig. 19.12a are split into **multi-line systems**—so-called **multiplets**. We will also address the question of why the respective relative signal heights of the respective multiplets differ so obviously (and characteristically!).

3.2.2 Multiplet

Skoog, Section 19.1.4: NMR spectra types

If you look at the three signals from ▶ Skoog Fig. 19.12b—now split into multi-line systems—you will notice that the signal belonging to the –OH group remains practically unchanged. (Thanks to the higher resolution, it is now only much less broadened—the area under the signal, i.e. the integral, remains completely unaffected.)

The situation is different for the signals for the (rather downshifted) methylene signal and the (rather upshifted) methyl signal: Here you see four lines (for the methylene signal) and three lines (for the methyl signal), respectively. Such multi-line systems are, as already mentioned, called **multiplets**.

If at this point you think of **multiplicity**, which you (might) already know from Part IV of "Analytical Chemistry I", you are exactly on the right track: In this section you will see that nuclear magnetic resonance spectroscopy is indeed also about multiplicities—which should not be surprising, since you learned in ▶ Sect. 3.1.1 that *spin states* are also of immense importance in NMR, even if this time it is not about the *electron* spin but the *nuclear spin*.

Accordingly, the multi-line systems occurring in a spectrum are designated according to the number of "partial signals" (i.e. lines) present, with the individual lines of a multiplet being exactly equidistant from their respective neighbours:

Skoog, Section 19.1.4: NMR spectra types

- If there is only one single line (as in the case of the signal of the OH group in Fig. 19.12b from Skoog), it is called a **singlet**.
- Two lines, each of which is (approximately) the same height, are called a **doublet**.
- Three lines (whose signal heights are approximately in the ratio 1:2:1) are called a **triplet** (e.g. signal of the CH$_3$ group in Fig. 19.12b).
- Four lines (with relative signal intensities of 1:3:3:1) are called **quartets** (seen, for example, in the signal of the CH$_2$ group in Fig. 19.12b).
- We continue with **quintets, sextets, septets,** etc.

As you can see, the relative signal heights of the individual lines belonging to such a multiplet are not only important, but can even be predicted quite easily: All multiplet splittings *follow Pascal's triangle*. We will deal with the question *why* this is so in ▶ Sect. 3.3.2, but what this means for analysing spectra, you should learn right now:

3.2 · First NMR Spectra

Here you experience the effects of the **coupling** (more precisely: *spin-spin-coupling*) of neighbouring atoms of the same type of atom (in this case: hydrogen atomic nuclei).

The number of resulting lines of a respective multiplet in ^{1}H- and in ^{13}C-NMR spectra depends on the number of (hydrogen) nuclei which are bonded to an *immediately adjacent* carbon atom. For the H atoms of the methyl group, these are the two H atoms of the methylene group, and for the hydrogens of the methylene group, correspondingly, the three H atoms of the adjacent methyl group.

> If hetero atoms such as oxygen or nitrogen are involved, this coupling usually does not occur, which is why the H atom bonded to the oxygen does not also affect the signal of the methylene group: Couplings do not (or better: *rarely*) act across hetero atoms.

The number of lines that appear in each case follows a formula that you have already become familiar with in Part IV of "Analytical Chemistry I". The multiplicity then results according to:

$$M = 2S + 1 \tag{5.2}$$

S is the sum of the spin quantum numbers of all considered particles.

> **Do You Remember?**
> In the aforementioned book, where the multiplicity was determined exclusively for electrons with the spin quantum number ½ (with the possible spin states +½ and −½), we used a simpler formula:
>
> $$M = (\text{Number of unpaired electrons}) + 1 \tag{3.3}$$
>
> In Part IV of "Analytical Chemistry I", this was formula ▶ 3.5: it described the number of possible ways a system with a given number of unpaired electrons has to arrange/orientate them in the magnetic field.

In NMR, however, it is not the *electron* spin but the *nuclear spin* that is important. It makes our lives immensely easier that both ^{1}H and ^{13}C nuclei also have a spin of $m_s = ½$, i.e.—quite analogous to electrons—only the two spin states +½ and −½ can occur.

In this respect, Eq. (3.3) thus can also be extended to atomic nuclei:

$$M = \left(\text{Number of coupling atomic nuclei with } m_s = \tfrac{1}{2}\right) + 1 \tag{3.4}$$

(For nuclei in which other spin states can also occur—in ▶ Sect. 3.1.1 some atomic nuclei were mentioned as examples—other values result, correspondingly. For the time being, however, we will restrict ourselves to the nuclei ^{1}H and ^{13}C).
Meaning:
1. If the nucleus under consideration does *not* interact with any other nucleus, S = 0, and thus, according to Eq. (3.4), M = 1, i.e. a **singlet**—just as we observe in our example spectrum of ethanol with the hydrogen atom of the hydroxyl group.
2. If the C atom immediately adjacent to the carbon atom, to which the currently considered hydrogen atom(s) is/are themselves bonded, carries only *one* hydrogen atom (i.e. if there is a **methine group**), then M = (2 × ½) + 1 = 1 + 1 = 2. A **doublet** (with signal intensity 1:1) is obtained.

3. If the "neighbouring carbon" carries *two* H atoms (such as our methylene group, $-CH_2-$), the corresponding result is M = (2 × (½ + ½)) + 1 = (2 × 1) + 1 = 3, that is, a **triplet** (with intensity 1:3:1).
4. And if a methyl group ($-CH_3$) is attached to the carbon atom carrying our H atom under consideration (i.e., the one responsible for the signal in question), this leads to a **quartet** (with relative intensities 1:3:3:1), because M = (2 × (½ + ½ + ½)) + 1 = (2 × 3 / 2) + 1 = 3 + 1 = 4.

Of course, this can go on—and you will soon see that it does, but for now, let's leave it at that.

Skoog, Section 19.1.4: NMR spectra types

— Thus, in the case of our high-resolution ethanol spectrum (Fig. 19.12b from Skoog), the *neighbouring* methyl group causes the signal of the hydrogen atoms of the methylene group to split into a quartet, while this same methylene group in turn splits the signal of the methyl group hydrogens into a triplet.

❗ Please avoid a (very popular) mistake in thinking: If a signal is split into a triplet, it means that the hydrogen atom (or hydrogen atoms) belonging to this signal couples (or couple) with a neighbour to which two H atoms are bonded. *It does not at all mean that the signal itself belongs to two corresponding atoms or that two H atoms must be bonded to this atom!*

Now we already know how we can describe the individual signals of an NMR spectrum, if they experience a multiplet splitting—just by interaction with neighbouring atoms. For a complete description of a spectrum (and sometimes this has to be possible without mapping the spectrum completely), this is clearly not sufficient, together with statements such as "less strongly down-shifted" or the like.

Tip for the Lab
As a solvent, one prefers to use water, if the analyte dissolves well in it—however, in ^{1}H-NMR, which is all about hydrogen atoms and their nuclei, one cannot simply use "normal" de-ionised H_2O which itself contains covalently bonded hydrogen atoms (obviously), and since the solvent molecules are usually present in vast excess compared to the analytes, the associated signals would far exceed all others in the spectrum (so that the signals that are actually of interest would be lost in the background noise) and/or be masked. So instead of using plain water, where almost all hydrogen atoms correspond to the isotope ^{1}H (you remember the natural abundance of the different hydrogen isotopes from ▶ Sect. 2.1?), one uses the fully **deuterated** counterpart (^2H_2O, or D_2O for short). Analogously, one can also use deuterated methylene chloride (CD_2Cl_2) or deuterated chloroform ($CDCl_3$), etc.

However, since the element symbol D for the hydrogen isotope ^{2}H is not IUPAC-compliant, the degree of deuteration of a solvent is rather indicated by additions to the molecular formula: For fully deuterated water (= all ^{1}H atoms are replaced by ^{2}H atoms) one correctly writes H_2O-d_2, for deuterated methylene chloride CH_2Cl_2-d_2, for corresponding chloroform $CHCl_3$-d_1, etc. If the solvent dimethyl sulfoxide (DMSO) with the molecular formula $(CH_3)_2SO$—mentioned at the beginning of this chapter—is used in NMR experiments, DMSO-d_6 is used accordingly.

10. Which multiplet results for (a) the methylene group and (b) the methyl group of chloroethane (CH_3–CH_2–Cl)?
11. What are the multiplets for (a) the methylene group and (b) the methine group of 1,1,2-trichloroethane?

3.2.3 The Chemical Shift δ, Considered in More Detail

How strongly a signal is upshifted or downshifted, i.e. which Larmor frequency the corresponding atomic nucleus has, depends—as already mentioned in ▶ Sect. 3.1.2—on the strength of the magnetic field used in each case. However, not all NMR spectrometers have magnets of exactly the same strength, so for the sake of comparability of the measurement results, it is indispensable to introduce a *standard* relative to which all other shifts can then be specified exactly.

For this purpose, the compound tetramethylsilane (TMS, $(CH_3)_4Si$) has become established in ^1H and also ^{13}C nuclear magnetic resonance spectroscopy. In this molecule, the carbon and thus also the hydrogen atoms are quite strongly shielded due to the lower electronegativity of silicon. Accordingly, the signals of the hydrogen atoms in the ^1H-NMR spectrum and also the signals of the carbon atom in the ^{13}C-NMR spectrum are clearly upshifted compared to practically all hydrogens occurring in "normal" organic compounds: The signal of the internal standard is usually on the far right-hand side in ^1H- and ^{13}C-NMR spectra. (We will discuss ^1H-NMR spectroscopy in more detail in ▶ Sect. 3.3 and ^{13}C spectroscopy in ▶ Sect. 3.4.)

> **Tip for the Lab**
> Tetramethylsilane is usually added to the samples to be measured as an *internal standard*. Since TMS, with its boiling point of 27 °C, can be easily removed again if necessary, there is practically nothing to prevent recovery of the analyte substance.

Thus, in the case of excitation by a constant electromagnetic wavelength with a varying magnetic field strength (the alternative with a constant magnetic field and a variable excitation radiation will not be considered further here), there is a direct relation between the Larmor frequency ν and the strength of the magnetic field B. In brief:

$$\nu \sim B$$

Furthermore, the local shielding of the nucleus under consideration, discussed in ▶ Sect. 3.2.1, has to be taken into account. So far, we have only dealt with the shielding by the local electron density (especially the *deshielding effect* in case of *lowered* electron density or changed electron density distribution); in ▶ Sect. 3.2.4 you will get to know further effects, which also (significantly) lead to shielding or deshielding.

But wherever the shielding may come from, its effect on the actual locally prevailing magnetic field (B_{local}) can be summarised by a (locally specific) *shielding constant* σ. For an applied magnetic field of strength B_0 we can say:

$$B_{local} = B_0 - (\sigma \cdot B_0) = B_0 \cdot (1-\sigma)$$

B_{local} stands for the resulting magnetic field "on site", related to the respective atomic nucleus with all its shielding and deshielding local conditions. (As already stated: more about this in ▶ Sect. 3.2.4.)

Again assuming constant wavelength of the excitation radiation, the local magnetic field must be stronger for a shielded atomic nucleus (such as the H atoms in TMS) than for a comparatively significantly deshielded (and thus more easily excitable) atomic nucleus (from the analyte). Thus, one can say:

$$\sigma_{TMS} > \sigma_{considered\ analyte\ nucleus}$$

For the sake of simplicity, we will now refer to this $\sigma_{considered\ analyte\ nucleus}$ generally as σ_{sample}, without forgetting that a sample may well have more than one relevant atomic nucleus (to be brought to resonance)—we have already seen this in the ethanol spectra from ▶ Skoog Fig. 19.12. The following considerations apply equally to each individual relevant nucleus, but with different factual (numerical) values for σ_{sample} in each case.

If we now want to represent the connection between the respective Larmor frequency of the TMS standard (ν_{TMS}) and the respective sample atom (ν_{sample}) with the applied magnetic field (B_0), we must briefly fall back once more on the already mentioned (atomic nucleus-specific) gyromagnetic ratio γ (which will not be discussed further here). And because the resonance of an atomic nucleus is more or less a circular motion (keyword: precession!), we must also divide this gyromagnetic ratio by 2π (this is a mathematical necessity). In total, then, the following applies to *each* precessing nucleus (whether TMS standard or sample atom) as a function of its individual shielding constant σ:

$$\nu_0 = (\gamma / 2\pi) \cdot B_0 \cdot (1-\sigma)$$

Since the same atomic species are considered for the TMS standard and for the respective sample (the strongly shielded, upshifted hydrogen atoms of the TMS and the rather deshielded H atoms of the sample in ^{1}H-NMR spectroscopy or the equally well-shielded [because comparatively electron-rich] carbon atoms of the TMS and the carbon atoms of the sample in the ^{13}C-NMR), the same γ-value applies to the standard and the sample. Since both substances (TMS and sample) are also based on the same applied magnetic field B_0, we can summarise the term $(\gamma/2\pi) \cdot B_0$ to a constant k.

Then first of all:

$$\nu_{TMS} = k \cdot (1 - \sigma_{TMS}) \tag{3.5}$$

and

$$\nu_{sample} = k \cdot (1 - \sigma_{sample}) \tag{5.6}$$

> **! Attention**
>
> Here one can easily get confused, because in these considerations we assume—and we hardly did that, so far—a constant magnetic field B_0, although we actually work with *constant excitation radiation* and a *variable magnetic field*. Nevertheless, this is not wrong, because it should be emphasised once again:
> — If we work with a *constant* magnetic field, differently shielded atomic nuclei are excited to resonance with electromagnetic radiation of different energy: In the case of deshielded nuclei, slightly less energetic radiation is sufficient for this (in which case the Larmor frequency directly correlates with the corresponding radiation wavelength according to Eq. 3.1).
> — However, this also works with a *variable* magnetic field: In this case, atoms that are shielded or deshielded to different degrees can be excited at the same wavelength, but this depends on the magnetic field in question: In the case of deshielded nuclei, a less strong magnetic field is suf-

Skoog, Section 19.1.4: NMR spectra types

3.2 · First NMR Spectra

ficient to cause resonance, while the more shielded nuclei are only brought to resonance by a stronger magnetic field.

So previous considerations could also be transferred to variable magnetic field, however this would be much more complex concerning calculations.

Accordingly, it follows:

$$v_{sample} - v_{TMS} = k \cdot (\sigma_{TMS} - \sigma_{sample}) \tag{3.7}$$

If we now divide Eq. (3.7) by Eq. (3.5), we obtain:

$$\frac{v_{sample} - v_{TMS}}{v_{TMS}} = \frac{\sigma_{TMS} - \sigma_{sample}}{1 - \sigma_{TMS}} \tag{3.8}$$

This has two advantages:
- You do not have to bother with the constant k anymore.
- You put everything "relative to TMS" and that's the important thing about here.

It's made even easier by the fact that, generally speaking, we know:

$$\sigma_{TMS} \ll 1$$

This makes the denominator in the right-hand part of Eq. (3.8) practically 1, and leads to:

$$\frac{v_{sample} - v_{TMS}}{v_{TMS}} = \sigma_{TMS} - \sigma_{sample} \tag{3.9}$$

We then define the **chemical shift** denoted by the Greek letter **δ** as:

$$\delta = (\sigma_{TMS} - \sigma_{sample}) \cdot 10^6 \tag{3.10}$$

Here δ is *dimensionless*; this quantity describes the relative shift in **ppm**, i.e. *parts per million*. That way, the resulting shift values are completely independent of the spectrometre used and the corresponding measurement variables (magnetic field strength, frequency of the excitation radiation).

Even though we will go into selected details about the various atomic species used in NMR and the corresponding spectra in some of the coming sections (^{1}H-NMR will be dealt with in ▶ Sect. 3.3, ^{13}C-NMR in ▶ Sect. 3.4), two important orders of magnitude should be mentioned in advance:

> **Important**
> - In ^{1}H-NMR, the shift values usually lie in the range of 0 ppm (shielded) to 12 ppm (strongly deshielded).
> - In ^{13}C-NMR spectra, δ-values ranging from 0 ppm (high field) to 220 ppm (downshifted) are commonly expected.

■ **Extreme Shift Values**

If a covalently bonded hydrogen atom is very strongly positively polarised, the corresponding ^{1}H-NMR spectra show, at first glance, unexpectedly large shift values: The ^{1}H signal of the acid proton of a carboxylic acid may well show values of δ > 13, and even in the case of β-dicarbonyl compounds such as 2,4-pentanedione (◻ Fig. 3.1a), which can also tautomerise to the enol form (b), the hydrogen atom of the enolic OH group, which forms a hydrogen bond with the carbonyl group in the β-position, is extremely strongly deshielded: Depending on which substituents the C_5 chain still has (not shown in the figure), values δ > 17 are possible. Accordingly, it is important to keep in mind

Fig. 3.1 Keto-enol tautomerism on 2,4-pentanedione

that ^{1}H-NMR spectra usually only cover the shift range $12 > \delta > -1$—if necessary, one has to check whether one has not "lost" one or the other H atom.

However, one should not mistakenly assume that it is impossible for a ^{1}H-NMR signal to "appear to the right of TMS": For organometallic complexes in which there is a *metal-hydrogen* bond, the hydrogen is the *more electronegative* bonding partner—this means that it takes up electron density from the metal and is thus even more strongly shielded than the H atoms of the reference substance TMS: For such (latent anionic, i.e. hydridic) hydrogen atoms, shift values of $\delta < -10$ are also not uncommon. (For e.g. the compound tetracarbonyl iron hydride (HFe(CO)$_4$), which one could also treat as a complex and call it pentacarbonylhydridoiron(0), $\delta = -10.5$.) If you do not have this in mind, you might "lose" one or the other hydrogen atom in the spectrum for corresponding analytes, because most standard NMR spectra do not show signals with shift values smaller than $\delta = -1$ (or -2); the spectrum is accordingly "truncated on the right" relatively close to the standard TMS.

Concerning the Resolution

Skoog, Section 19.1.4: NMR spectra types

Figure 19.12a, b from Skoog already showed that a higher resolution spectrum also offers a higher information density. But this still leaves one question unanswered: How *can* the resolution be increased?—While it has already been noted that, in general, FT-NMR spectra lead to a higher resolution than CW spectra, there is another, much more crucial factor:

The resolution of a spectrum depends significantly on the *excitation frequency*.

This in turn is inextricably linked to the applied *magnetic field strength* (our B_0) from this section:

With greater field strength of the magnet used, more energetic excitation radiation may be used. Just a few examples:

- With a magnet of strength 1.41 T (= Tesla, the unit of magnetic flux density, please do not confuse this unit symbol with tritium ...) one uses electromagnetic radiation with a frequency of 60 MHz; accordingly, one speaks of a 60 MHz spectrometer.
- If the strength of the magnetic field is increased to 2.35 T, the excitation takes place with radiation of 100 MHz.
- With a field strength of 9.4 T, resonance can be achieved with radiation of the frequency 400 MHz.
- The most powerful magnets currently used in NMR (field strength: 28.2 T) (as of April 2023) enable the first 1.21 GHz spectrometers—with which previously unimagined resolutions can be achieved, although even more powerful magnets are already being worked on experimentally: In November 2016, the first magnet with a field strength of 36 T was presented, with which a 1.5 GHz spectrometer could be constructed accordingly. However, since that magnet weighs 33 t, it is probably a little too bulky for everyday use.

3.2 · First NMR Spectra

As you can see, there is a linear relationship between the magnetic field strength and the excitation frequency: Each 0.0235 T corresponds to a 1 MHz increase in excitation frequency.

■ Influence of Magnetic Field Strength on Multiplets

In the previous section, you learned that the higher the field strength (and thus the excitation frequency), the higher the resolution of the resulting spectra. However, thanks to the dimensionless chemical shift, there is still *no change* in the δ-values associated with the respective signals at increased resolution. Compare the 60-MHz ^{1}H-NMR spectrum of ethanol (Fig. 19.13 from Skoog above) with the corresponding 100-MHz spectrum (below).

As you can see, the centres of the individual multiplets are clearly further apart. The singlet of the OH group is missing here in each case; as you can see, the upper of the two spectra is truncated somewhat above δ = 6 ppm, the lower already at δ = 4 ppm. However, the quartet of the methylene group and the triplet of the methyl group are still completely recognisable.

At the same time, however, something else is noticeable: The distance of the "partial signals" belonging to a multiplet remains unchanged: Since the respective Larmor frequencies are also plotted on the respective x-axis of the spectra (in Hz), one could also describe the position of the respective signals via the corresponding *frequency* instead of using the (device-independent) δ-value (note that the middle of the upper as well as the middle of the lower methyl triplet each belong to a chemical shift of about 1.2 ppm)—which, however, is *device-dependent,* so that the comparability with the measured values obtained on other devices seems to be limited. However, on the basis of this Hertz scale, one can also determine the distance of the individual partial signals of a multiplet (i.e. the **line distance**), and here it shows:

> The distances of the individual partial signals of a multiplet are independent of the spectrometer used or the selected measurement conditions.

Skoog, Section 19.2: Influence of the chemical environment on NMR spectra

For this reason, these line spacings can also be given in *absolute numbers* and with a *unit*: For the two triplets studied from Skoog's Fig. 19.13, the line spacing is 7 Hz in each case. In contrast to the position of the respective signals in NMR spectra recorded under different conditions, the *multiplets* resulting from the aforementioned couplings are therefore *independent of the measurement conditions* chosen in each case—both in their shape and in the spacing of their individual lines. This leads to a term, which is introduced already right now and will be explained further in ▶ Sect. 3.3.2: **coupling constants**.

Skoog, Section 19.2: Influence of the chemical environment on NMR spectra

> **Generally Speaking:**
> — Absolute resonance frequencies (in MHz) are measurement-dependent.
> — Resonance frequencies given relative to the reference substance (TMS) are measurement-dependent.
> — The dimensionless δ-values of the chemical shift, related to the respective reference substance (the internal standard, usually TMS) are *independent* of the measurement conditions.
> — Line distances within a multiplet are *independent* of the measurement conditions.

We will return to all these points again when we look a little more closely at such couplings in ▶ Sect. 3.3.2.

However, the influence of the electron density (distribution) described in ▶ Sect. 3.2.1 is not the only thing that affects the chemical shift of the respective atomic nuclei.

3.2.4 Anisotropy Effects

Of equal importance are **anisotropy effects** occurring in the respective analyte molecule: the consequences of the fact that any magnetic effects are *dependent on the spatial direction*. Let us compare—again on the basis of ^{1}H-NMR spectroscopy—the chemical shifts of the hydrogen nuclei of ethane, ethene and ethyne:
- For ethane (CH_3–CH_3), the result is $\delta = 0.9$.
- The ^{1}H signal at ethene (CH_2=CH_2) is clearly downshifted with $\delta = 5.8$, so the methylene hydrogens are clearly deshielded.
- For ethyne (CH≡CH), on the other hand, the shift lies in the order of magnitude of $\delta = 2.9$; the methyne hydrogens are obviously shielded in comparison to those in ethyne (although not as strongly as in ethane).

However, this phenomenon is not limited to "normal" double or triple bonds:
- In the case of benzene (C_6H_6), the ^{1}H-NMR spectrum again shows a signal with a distinct downshift: $\delta = 5.86$. (It should be obvious that all six hydrogen atoms of benzene are indistinguishable, i.e. chemically equivalent.) **Aromatic systems** in general are therefore also of special importance.
- The same applies to **multiple bonds containing hetero atoms,** such as the carbonyl group: The ^{1}H-NMR signal of the carbonyl H-atom of acetaldehyde (CH_3–CHO), for example, shows a shift of $\delta = 9.80$ ppm.

Because the underlying phenomenon can be illustrated particularly clearly in aromatic systems, we will turn to these first.

■ Aromatics

Skoog, Section 19.2.2: Chemical shift theory

The reason for the downshift of the NMR signals of hydrogen atoms directly bonded to an aromatic ring is the interaction of the electron density of said aromatic system (i.e.: the conjugated double bonds) with the applied magnetic field (B_0). Let's start with benzene: The applied magnetic field generates a **ring current** in the aromatic system (illustrated in Fig. 19.15 from Skoog). This induces a *secondary magnetic field* (similar to a wire loop), so that inside the ring the resulting magnetic field strength is *lower* than B_0 (which, however, has no effect in aromatics because, for purely steric reasons, no other atomic nuclei can be present *inside* this ring), while the same induced magnetic field ensures that a *stronger* magnetic field is induced on hydrogen atoms (or their nuclei), so these nuclei can be excited more easily (i.e. at lower field—leading to a downshift).

> **For Those Who Want to Know More**
> Strictly speaking, there *are two different forms* of ring current:
> - If an aromatic system is present, i.e. if the number of electrons at the delocalised π-electron system obeys the Hückel rule (number of π-electrons = $4N + 2$, where N can be any integer, including 0—certainly known from the basics of *organic chemistry*), a **diamagnetic ring current** results, which causes exactly what is described above: deshielding of the H atoms bonded to this system and thus *downshifting* of the associated signals.
> - However, if a cyclically conjugated π-electron system does *not* obey the Hückel rule (i.e., if the number of delocalised π-electrons can be described as $4N$, leading to *anti-aromaticity*), the resulting secondary magnetic field has an inverted orientation, so that the nuclei bonded to such a system experience a *upfield shift*. Here a **paramagnetic ring current** occurs.

3.2 · First NMR Spectra

This ring current model can be demonstrated particularly well via ^{1}H-NMR on annulenes:

Annulenes

- For the aromatic [18]-annulene (with $N = 4$, a), a shift of 9.3 ppm is obtained for the signals of the hydrogen atoms lying outside the ring—a real downfield shift.
- In contrast, for the anti-aromatic [16]-annulene (b) with $\delta = 5.2$, this downfield shift is much less pronounced; this value is more consistent with "normal" double bonds, which will be discussed in the immediately following subsection.

This ring current model is supported by the shift values of the hydrogen atoms located in the *interior* of these larger rings (in the case of annulenes, unlike in the case of "ordinary aromatics", H atoms can also be located in the interior of the ring system):
- For the six inner H atoms of the (aromatic) [18]-annulene (a), there is a remarkable upfield shift: $\delta(H_{inner}) = -3.00$.
- The four hydrogen atoms in the "interior" of the (anti-aromatic) [16]-annulene (b) undergo remarkable downshifting: $\delta = 10.3$.

This exactly fits the model: In the case of aromatic compounds, the magnetic field is *increased* by the diamagnetic ring current in the interior of the system, in agreement with Fig. 19.15 from Skoog, so the H atoms located there are much more difficult to excite (and thus are *upshifted*), while the paramagnetic ring current resulting from anti-aromatic compounds is reversed in its direction, so that the hydrogen nuclei located inside experience a significantly weakened magnetic field. The effect is equivalent to shielding, and thus these nuclei are easier to excite to resonance: they are *downshifted*.

Skoog, Section 19.2: Influence of the chemical environment on NMR spectra

■ Multiple Bonds
A similar effect occurs with the non-aromatic ("ordinary") double and triple bonds (even if it is not a true *ring* current, because a ring system is not present here).
- The similarity is particularly strong for ethene and other analytes with **double bonds**: As Fig. 16.16a from ▶ Skoog shows, just as in aromatic systems, the induced magnetic field leads to the hydrogen atoms bonded to these sp^2 carbon atoms being surrounded by an effectively weakened magnetic field, which similarly causes *deshielding* of the H nuclei.

Skoog, Section 19.2.2: Chemical shift theory

- Of course, the same is true for hetero-double bonds (especially C=O): Here, the shielding effect is even more pronounced due to the higher electronegativity of oxygen; accordingly, the ^{1}H-NMR signals of aldehyde hydrogens are also downshifted.
- In the case of ethyne and other compounds with **triple bonds,** it should be noted that—because of the approximately radially symmetric charge distribution of the corresponding π-electrons around the central σ-bond—the induced field is not at right angles to the π-orbitals involved, but instead surrounds the C–C axis *toroidally* (Fig. 19.16b from ▶ Skoog; this is not possible in the case of the double bond due to the nodal plane of the single π-bond). This results in any H atoms bonded to sp-hybridised carbons being *shielded* again: An *upfield*
- Shift occurs (compared to the hydrogen nuclei on a double bond).

   - For the signals of which hydrogen atom (or which carbon atom) of each of the following compounds is the strongest downshift to be expected: (a) *(E)*-2-butene; (b) 2-butenal; (c) toluene?

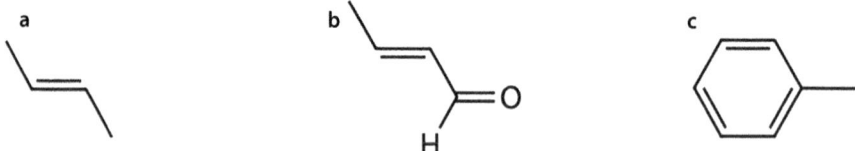

Now that you have become acquainted with the general principles of NMR (especially with regard to nuclei with a spin of ½), we will turn to the two applications that are by far the most common in "everyday analytics": The nuclear magnetic resonance spectroscopic analysis of (organic) analytes via ^{1}H- and ^{13}C-NMR. Conveniently, it is precisely these two types of atoms that are practically ubiquitous, especially in *organic chemistry.*

3.3 ^{1}H-NMR

Skoog, Section 19.1.4: NMR spectra types

The resonance spectroscopy of nuclei of the hydrogen isotope ^{1}H is not only important because almost every organic compound has at least one hydrogen atom somewhere, but especially because the isotope ^{1}H, which has the nuclear spin ½ and can as such be excited and be used in spectroscopy, is by far the most common of the three hydrogen isotopes. (This has already been mentioned in ▶ Sect. 2.17 .) In addition, the couplings in the corresponding ^{1}H-NMR spectra, which have already been briefly touched upon in ▶ Sect. 3.2.2, are particularly informative. *Chemical equivalence* is often important here: If you recall Fig. 19.12 from Skoog, you will remember that the three (chemically equivalent) H atoms of the methyl group of ethanol gave only one joint signal; the same was true for the two (also chemically equivalent) hydrogen atoms of this molecule's methylene group. The chemical equivalence, which was already briefly mentioned in ▶ Sect. 3.2.1, will be discussed again in this section.

3.3.1 Chemical Equivalence

> Atomic nuclei are considered chemically equivalent if they are in an *equivalent environment*, i.e. if they all have the same type of bonding partner and cannot be distinguished from each other in any other way, e.g. with respect to anisotropy effects, etc.

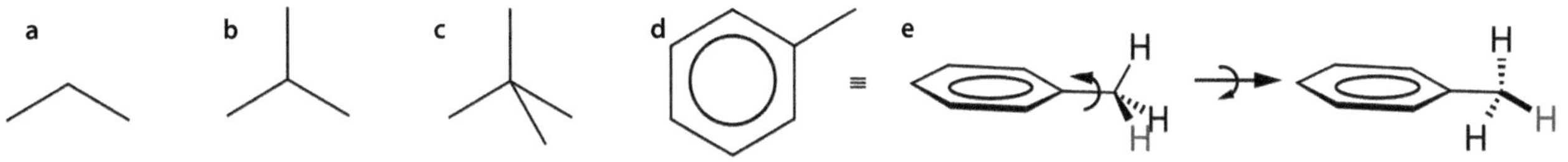

3.3 · ¹H-NMR

Fig. 3.2 Hydrocarbons and aromatic compounds

Simple examples of chemical equivalence are the six methyl hydrogens of propane (Fig. 3.2a), the nine methyl hydrogens of 2-methylpropane (b), the twelve methyl hydrogens of 2,2-dimethylpropane (neopentane, c) and the three H atoms of the methyl group of toluene (toluene, C_6H_5–CH_3, Fig. 3.2d).

One of the main reasons for the chemical equivalence of some atoms is the free rotation around the single bond: The three H atoms of the methyl group of toluene, for example, are positioned differently from the aromatic ring system, when being observed in a "snapshot" (Fig. 3.2e), but since there is free rotation around the $C^{\text{aromatic ring}}$–C^{methyl} bond, *on average* all three methyl hydrogens experience the same interaction, for example with the secondary magnetic field of the aromatic produced by the ring current.

If this free rotation *is not* given, for example in the case of *Z/E isomerism* on double bonds, it is well possible that atoms that appear identical at first sight are *not* chemically equivalent. This applies, for example, to the two methyl groups of the compound 1-chloro-2-methylpropene (Fig. 3.3a): One of the two groups is on the same side of the double bond as the chlorine atom, the other on the opposite side.

For some compounds, especially for single bonds with a certain double-bond character, such as the amide bond –C(=O)–N–, the temperature dependence of the rotation can be seen very clearly: If one records a ¹H-NMR spectrum at room temperature of N,N-dimethylformamide (Fig. 3.3b), one obtains, in addition to the signal for the aldehyde hydrogen (with $\delta = 8.0$ ppm, integral = 1), two further signals: one at 3.1 ppm, the other at 2.9 ppm (each with integral = 3). However, in a spectrum of this compound prepared at temperatures >120 °C, the two signals merge into one singlet (with $\delta = 2.0$ ppm and integral = 6). The reason for this is that above this temperature the thermal excitation is sufficient to cause free rotation around the C–N bond, so that the two methyl groups become indistinguishable. At lower temperatures, however, the double bond character of the C–N bond, illustrated by the resonance structure in Fig. 3.3c, is large enough to make the two CH_3 groups distinguishable.

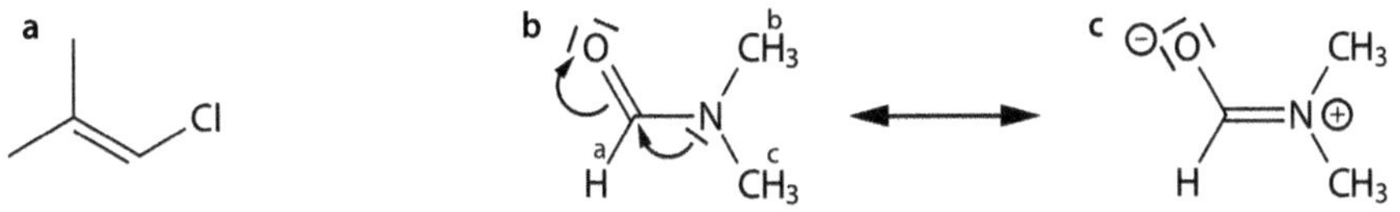

Fig. 3.3 Limited rotation at double bonds

Special Case of Configuration Isomerism

Since chiral factors are not used in NMR experiments (except in the application of *very* fancy methods), mirror-image environments cannot be distinguished from each other; accordingly, the two enantiomers of 2-bromobutane (a/b), for example, show exactly the same NMR spectrum.

In diastereomers, on the other hand, the environments of the different nuclei are *not* equivalent, which is why the two diastereomers (2R,3R)-2-bromo-3-chlorobutane (c) and (2S,3R)-2-bromo-3-chlorobutane (d) can be distinguished very well.

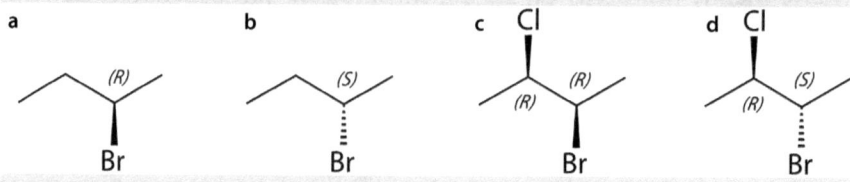

❓ Questions

13. Which hydrogen atoms in the molecule 2,2,3,5-tetramethylhexane (a) are chemically equivalent in each case, and which are not? What are the answers for this question regarding the molecule 2-methylpropanoic acid-2′-aminoethyl ester (b)?
14. Why are the two hydrogen atoms of the methylene bridges in the molecule norbornene (c) not chemically equivalent? Which of the two would you expect to be down-shifteld? (Again, the *reasoning is* important.)

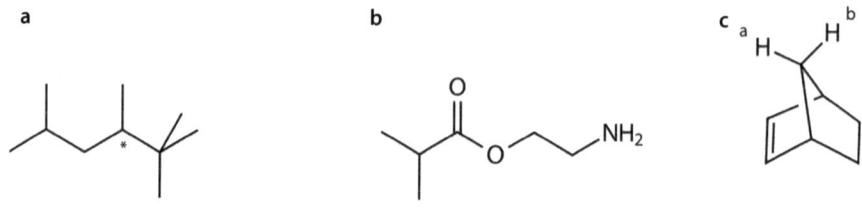

3.3.2 Couplings

As already mentioned in ▶ Sect. 3.2.2, the signal of a hydrogen atom bonded to a carbon atom whose neighbouring carbon also has one or more hydrogen atoms attached to it is split into a multiplet via *coupling*. (The same applies accordingly to the signal of several chemically equivalent H atoms, see ▶ Sect. 3.3.1.)

Two important rules apply:
1. The number of lines of the respective multiplet depends on the number of H atoms bonded to said neighbouring C atom.
2. The relative intensity of the multiplet lines follows Pascal's triangle.

Now, of course, two questions arise: Why does Pascal's triangle play a role here? And what is perhaps even more important: What are these couplings all about?

Let us start small and consider a molecule in which only two different, chemically non-equivalent nuclei play a role: ◘ Fig. 3.4 shows the ¹H-NMR

3.3 · ¹H-NMR

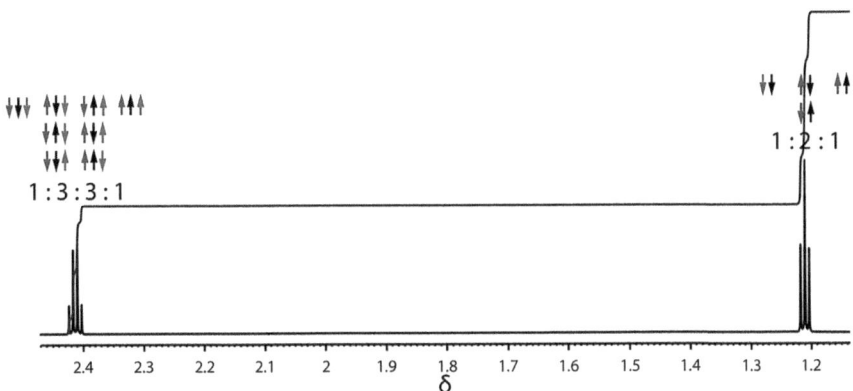

Fig. 3.4 Section of the ¹H-NMR spectrum of propionic acid. (Ortanderl S, Ritgen U: Chemie - das Lehrbuch für Dummies, 1076. Copyright Wiley-VCH Verlag GmbH & Co. KGaA. Reproduced with permission)

spectrum of propionic acid (propanoic acid, CH_3–CH_2–COOH); not shown is the strongly downshifted carboxylic hydrogen (δ = 11.8 ppm). And since there is a heteroatom between it and the next carbon carrying a hydrogen atom anyway (oxygen), no coupling is observed, as already mentioned in ▶ Sect. 3.2.27, so we can simply ignore it here.

As expected, the methyl group splits the signal of the methylene hydrogens (δ = 2.4 ppm) into a quartet, while the methylene group splits the signal of the methyl hydrogens (with δ = 1.2 ppm) into a triplet.

This is due to the fact that the spin state of *all* immediately neighbouring hydrogen atoms affects the nucleus under consideration (i.e. the nucleus whose signal is currently being split): *spin-spin coupling* occurs (which would once again go beyond the scope of an introduction to explain in more detail ...). As you probably already know (e.g. from Part IV of "Analytical Chemistry I"), different spin states differ slightly in their energy content, so we have to take a closer look at the *possible* spin states here as well:

If the considered nucleus is in interaction with only one single "neighbour hydrogen", the nucleus of this coupling neighbour atom can have the spin $+\frac{1}{2}$ or also $-\frac{1}{2}$.

> Please keep in mind that in NMR spectroscopy we do *not* only excite a single analyte molecule to resonance, but a huge number of them.

Accordingly, it can be assumed—purely statistically—that in half of these molecules the coupling nucleus is in the spin state $+\frac{1}{2}$, the other in the state $-\frac{1}{2}$. If these two energetically slightly different nuclei interact with the "nucleus under consideration", two different Larmor frequencies result, i.e. we get two slightly different "shift values", which differ only by a few Hertz, but still *are* distinguishable. Accordingly, we get two lines. And because both spin states of the coupling nucleus ($+\frac{1}{2}$ or $-\frac{1}{2}$) occur with approximately the same probability, both lines also show the same intensity (i.e. an intensity ratio of 1:1).

However, if there are *several* hydrogen atoms in the immediate vicinity, there are naturally also several different energy levels, because there are different spin state combinations: With two neighbours, both could be in the spin state $+\frac{1}{2}$, or one could be at $+\frac{1}{2}$, the other at $-\frac{1}{2}$, or both could be in the $-\frac{1}{2}$ state. This is where multiplicity comes into play: If a nucleus couples with *two* neighbours, the result is a *triplet* (i.e., a three-line system), etc. This is exactly the case for the signal of the methyl group (δ = 1.2): it is split accordingly by the two H atoms of the methylene group. The reason why the middle of the three lines is higher than the two outer lines is that there is only *one* possibility of spin orientation for the two spin-parallel states "2 × (+½)" (i.e.: ↑↑) and "2 × (−½)" (corresponding to: ↓↓), but two for the spin-antiparallel orientation: ↑↓ and ↓↑. Of course, the latter< two have exactly the same energy content, so their NMR signals coincide exactly, resulting in increased signal intensity.

- **Multiplets, Multiplicities, Total Spin and Multiplicity States**

If you remember "Analytical Chemistry I", the rules for the determination of the total spin S_{tot} and the corresponding multiplet states (M_S) might have crossed your mind now—and you would be right: This is exactly what lies behind the formation of the multiplet states in NMR as well, so you are cordially invited to use the corresponding formulas again. However, at least in ^{1}H-NMR (and fortunately also in ^{13}C-NMR, which we will deal with in ▶ Sect. 3.4), one can simplify things even further:

> In agreement with Eq. (3.4) (and since we know spin(^{1}H) = ½), the following immensely important formula applies to ^{1}H-NMR spectra:
>
> Multiplicity of the signal = Number of coupling neighbors + 1
>
> Coupling the methylene signal with the three hydrogen nuclei of the methyl group (δ = 2.4 ppm) to form a quartet results in four different (nuclear) spin states, marked in various shades of grey in ◘ Fig. 3.4. (Their respective total spin S_{tot} and the corresponding multiplicity state M_S could be calculated using the formulas 18.6 and 18.7 from Part IV of "Analytical Chemistry I", but in NMR this is not necessary.)

Since, as mentioned in ▶ Sect. 3.2.3, the spacing of the individual lines of the respective multiplets is independent of the measurement conditions, it is irrelevant here whether this spectrum was recorded with a 60 MHz spectrometer or an instrument with higher resolution, and if you measure the spacing of the individual lines of the triplet and the quartet to their respective immediate neighbours, you will find that they all have exactly the same value, which is then usually given in Hertz. (For the time being, the specific value is irrelevant here, it is of the order of about 10 Hz.)

- **Beyond the Quartet**

The more nuclei are involved in the coupling, the more—energetically minor—differences arise: With four identical nuclei, as already mentioned in ▶ Sect. 3.2.2, the result is a quintet with the relative signal intensities 1:4:6:4:1 (this multiplet would e.g. be obtained for the middle methylene group of the pentane molecule (CH_3–CH_2–**CH_2**–CH_2–CH_3), because the other two methylene groups are chemically equivalent), with five coupling nuclei the result is a sextet (with the intensities 1:5:10:10:5:1), and so it goes on and on—according to Pascal's triangle.

This can be seen nicely in the case of 2-chloropropane, where the signal of the two (chemically equivalent) methyl groups (leading to an integral of 6) is split into a doublet by the methine hydrogen in the middle of the molecule, while conversely the total of six equivalent hydrogen nuclei of the two methyl groups split the methine signal into a *septet* (with an integral of 1), as you can see in ◘ Fig. 3.5:

If you measure the respective distances of the individual lines of this septet (whose relative intensities are in the relative ratio of 1:6:15:20:15:6:1, but this is admittedly not very easy to see here), you will find that it again corresponds exactly to the distance of the two lines of the doublet: the two **coupling con-**

3.3 · ¹H-NMR

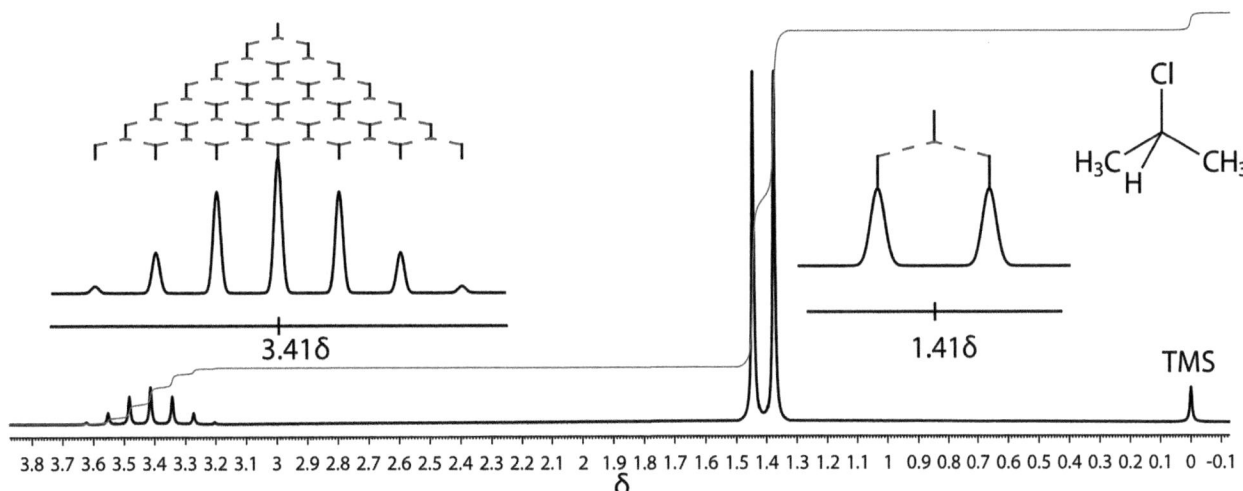

Fig. 3.5 ¹H-NMR spectrum of 2-chloropropane. (Ortanderl S, Ritgen U: Chemie - das Lehrbuch für Dummies, 1077. Copyright Wiley-VCH Verlag GmbH & Co. KGaA. Reproduced with permission)

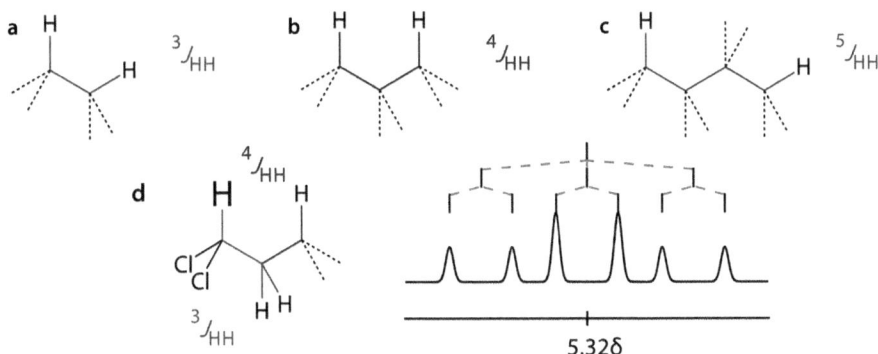

Fig. 3.6 Near and far couplings in ¹H-NMR

stants are identical. Since two H atoms bonded to immediately adjacent carbon atoms are effectively separated by three bonds (◘ Fig. 3.6a), these kind of coupling is called a $^3J_{HH}$ coupling. (The J is the generally accepted symbol in NMR for any form of coupling.)

> You have already seen the same effect in the spectrum of ethanol: The distance between the individual lines of the (methylene) quartet was also exactly as large as the distance between the individual lines of the (methyl) triplet.

The fact that nuclei coupling with each other **always have the same coupling constant** becomes particularly interesting once we have to consider several *not* chemically equivalent atoms capable of coupling.

> Please note: The term "constant" here means that the line distances of the multiplet signals of *two nuclei coupling with each other* have the same value, not that all couplings, no matter between which atoms, would always have the same value. If there are several possibilities of coupling within an analyte, these may well differ in their value—which in most cases is highly informative.

> **Coupling of Chemically Non-equivalent Nuclei**
> Two nuclei that couple with each other *always* have the same coupling constant. If multiplets occur in a (more complex) spectrum that do *not* have the same coupling constant, you know immediately that they are *not* mutually responsible for each other.

- **Multiplets of Multiplets—and How to Describe Them, Part 1**

If a methylene or methine fragment is in the vicinity of two or even three carbon atoms to which one or more hydrogen atoms are also bonded, coupling naturally occurs with *all* the nuclei in question, although not all couplings will have the same coupling constant:

— The methylene group of the molecule propanal (CH_3–CH_2–CHO), for example, will form a $^3J_{HH}$ coupling with both the methyl group and the aldehyde hydrogen:
  – Accordingly, the methylene group will split the (downshifted) signal of the aldehyde hydrogen into a triplet with a coupling constant $^3J_{HH}$ = X Hz.
  – The signal of this methylene group in turn becomes a doublet through the aldehyde hydrogen, also with the coupling constant $^3J_{HH}$ = X Hz.
— At the same time, however, the methylene group of propanal will also form a $^3J_{HH}$ coupling with the methyl group, the coupling constant of which may well have a different value:
  – Thus, the methylene group would also split the signal of the methyl group (whose H atoms are chemically equivalent) into a triplet. In this case, $^3J_{HH}$ = Y Hz.
  – In the reverse move, the methyl group then splits the signal of the methylene group into a quartet with $^3J_{HH}$ = Y Hz.

The signal of the methylene group is thus split twice: Once (with $^3J_{HH}$ = X Hz) into a doublet, once (with $^3J_{HH}$ = Y Hz) into a quartet.

Depending on which coupling constant is larger (here: X or Y), one then speaks of a doublet of quartets (DQ, for X > Y) or of a quartet of doublets (QD, for X < Y). The multiplet to which the larger coupling constant belongs is always mentioned first.

Let's try to deduce what the ^{1}H-NMR spectrum of 1-propanol (CH_3–CH_2–CH_2–OH) should look like:
— The methyl group (which is certainly the most upshifted one due to the electron present on the methyl carbon density alone) is undoubtedly split by the neighbouring methylene group to form a triplet (abbreviated: T). (It should be obvious by now that this multiplet results in an integral of 3, since three chemically equivalent H-atoms together result in only *one* signal, namely this triplet.)
— Three things can be said about the methylene group, to which the OH group is also attached:
  – First, the associated signal will certainly experience the strongest downfield shift due to the higher electronegativity of oxygen compared to all other signals, leading to deshielding of the carbon and as such the hydrogen atoms.
  – Second, this signal will also be split into a triplet (abbreviated: T) because of the neighbouring other methylene group.
  – Third, the integral over this triplet will tell us that there are again a total of *two* hydrogen nuclei associated with this multiline signal.
— And what about the signal from the methylene group that is *not* bonded to the oxygen? This is where things get a little more complex:

3.3 · ¹H-NMR

- On the one hand, it has to be split into a quartet because of the three H atoms of the neighbouring methyl group, and the distance between these four lines will correspond exactly to the distance between the lines of the signal of that methyl group—they have the same $^3J_{HH}$ coupling constant.
- On the other hand, this signal is also split by the two H atoms of the neighbouring (OH-)methylene group: into a triplet. Here again a $^3J_{HH}$ coupling comes into play, whose coupling constant (i.e. the line spacing) corresponds exactly to that which we have already observed for the triplet of the (OH-)methylene group signal. But, and this is the important thing here: This $^3J_{HH}$ coupling does *not necessarily* have the same value as the $^3J_{HH}$ coupling with the other methylene group:
  (a) Assuming that the value of the $^3J_{HH}$ coupling CH_2–CH_2–(OH) is slightly smaller than the $^3J_{HH}$ coupling CH_3–CH_2–(CH_2)- (the difference can be less than 1 Hz, in this example it is 2 Hz), this results in a quartet of triplets (QT for short), shown in (a).
  (b) *If,* however, the value of the $^3J_{HH}$ coupling CH_2–CH_2–(OH) *were* slightly larger than the $^3J_{HH}$ coupling CH_3–CH_2–(CH_2), the result would be a triplet of quartets (TQ for short), shown in brackets in (b). (This does not apply to the coupling constants given in this example—hence the parentheses—but theoretically such a constellation would be possible.) In this figure, the coupling to the "right" methylene group has the value $^3J_{HH}$ = 12 Hz, while the coupling to the "left" methyl group has the value $^3J_{HH}$ = 10 Hz. Note that although the individual lines of this multiplet are located at exactly the same positions in the spectrum, the relative signal intensities of the individual lines in the TQ are different from those in the QT case.

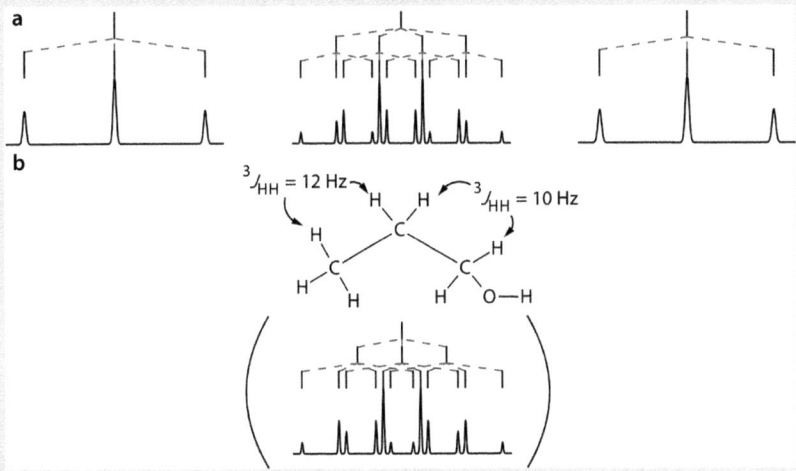

$^3J_{HH}$ multiplet tree for 1-propanol

For Those Who Want to Know *more*
On the one hand, it is very well possible that two different couplings *do* have the same coupling constant (approximately, i.e. within the limits of measurement accuracy). Then we are confronted with an unpleasant problem, because instead of a clear multiplet structure (quartet of triplets—QT—or doublet of triplets—DT—etc.) we get **pseudo-multiplets**, which seriously complicate the interpretation of corresponding spectra—among other things because then the relative

signal intensities of this multiline system no longer necessarily follow Pascal's triangle. This is exactly the problem with the "real existing spectrum" of 1-propanol, which is why the corresponding spectrum is not shown here. From now on the topic becomes too complex for an introduction like this part of the book; here we refer to the further literature (especially to Hesse-Meier-Zeeh).

On the other hand, depending on the structure of the analyte in question, the shifts of individual signals may differ so little that their respective multiplets partially overlap, which of course also complicates interpretation. For this reason, the real spectrum of **pentane**, for example, is not presented in this text. Instead, let us have a look at what is to be expected in a "perfect" spectrum:

- The signal of the two methyl groups, which are chemically equivalent, is split into a triplet by the neighbouring methylene groups; the integral of this multiplet should then tell us that a total of six H atoms are responsible for this signal.
- The signal of the "middle" methylene group should be split into a quintet by the two H atoms of each of the two neighbouring (and chemically equivalent!) methylene groups altogether (with an integral of 2 in total).
- The two (chemically equivalent) "non-middle" methylene groups should be split into a quartet by the methyl group and into a triplet by the middle methyl group. Whether this is called a QT or a TQ depends—as already mentioned—on the respective coupling constants ... which in this case, however, are unfortunately almost identical, so that a pseudo-multiplet is more likely to be expected. In addition, this simple molecule has no functional groups or heteroatoms that would provide serious shielding or screening, and this in turn leads to the fact that although the methyl groups are easily recognisable (they are still highly upshifted compared to all other hydrogen nuclei), the signals of the two methylene groups experience almost the same shift and thus overlap.

▶ https://web.chemdoodle.com/demos/simulate-nmr-and-ms/

It is very helpful and instructive to "play" a little witH-NMR spectra for yourself. Useful for this are appropriate spectra simulators, and the Internet offers quite a lot of them. Two (free of charge) simulators are mentioned here as examples:

- The simulator, which is part of the chemistry drawing program ChemDoodle, allows the simulation of ^{1}H and ^{13}C-NMR spectra as well as mass spectra (as from Chapter ▶ 2).
- On the website nmrdb.org of the *Institute of Chemical Sciences and Engineering* (ISIC), various "special forms" of two-dimensional NMR can be simulated in addition to the "ordinary" ^{1}H and ^{13}C-NMR spectra mentioned above. (Multidimensional NMR is briefly discussed in ▶ Sect. 3.4.2.)

▶ http://www.nmrdb.org/new_predictor/index.shtml

This is not to say that other simulators are not just as usable, but a more comprehensive list would probably not make much sense.

■ **Long-Range Couplings: $^4J_{HH}$, $^5J_{HH}$—and Beyond**

So far, we have only considered couplings of *immediately neighbouring* H atoms. But we cannot (and should not!) leave it at that, because in sufficiently well-resolved NMR spectra, we can also detect **long-range couplings** and occasionally even use them for a more precise structural elucidation.

If you look at the fragment of the molecule in ◘ Fig. 3.6b, you can see the four bonds separating two H atoms that form $^4J_{HH}$ coupling, and ◘ Fig. 3.6

3.3 · ¹H-NMR

shows the fragment required for $^5J_{HH}$ coupling. Again, this results in coupling constants, but the larger the distance between the coupling nuclei, the smaller these constants become, and torsion angles may also play a role. (We will return to this—briefly—in a moment.)

- **Multiplets of Multiplets—and How to Describe Them, Part 2**

For a brief description of such more complex multiplets of multiplets, it has become customary to denote the multiplets resulting from $^3J_{HH}$ couplings with capital letters, while the multiplets resulting from long-range couplings are indicated with *lowercase letters*. This way, e.g. Qt or—for more complex analytes—TQd and the like can be obtained.

> How this affects the resulting multilineage systems is best shown again by an example where we deduce for ourselves what multiplets of multiplets to expect—consider the molecule part of Fig. ▶ 3.6d: Here you can see that the H atom bonded to the doubly chlorinated carbon can form both a $^3J_{HH}$ coupling (with the adjacent methylene group) and a $^4J_{HH}$ coupling, the latter with the methine-H of the carbon adjacent to it, whose other bonding partners should not interest us now, hence the dashed bonds. Accordingly, it is to be expected that the signal is split into a triplet by the immediately adjacent methylene group ($^3J_{HH}$ coupling with a coupling constant in the order of 10 Hz), and each of these three lines (with a relative intensity of 1:2:1) then undergoes the splitting into a doublet with a significantly smaller coupling constant by the $^4J_{HH}$ coupling with the H atom at the "next but one C atom". Altogether, this gives a six-line system, a triplet of doublets (Td) with the relative signal intensities 1:1:2:2:1:1.

The fact that the coupling constants of nuclei that couple with each other are always identical makes it easier for us to evaluate spectra. However, since in the two preceding spectra only two atomic nuclei appeared which couple with each other at all, one might assume this to be purely coincidental here. But this is not the case, as you will see for yourself when you look at the spectra of more complex compounds. Here, however, deeper covering of the matter would once again go beyond the much-cited scope.

> **For Advanced Students: the Influence of the Torsion Angle**
> As is well known, there is in principle free rotation around any single bond. Accordingly, neighbouring nuclei which are in principle capable of coupling are not always equidistant from one another or do not always form the same angle with one another, and different **torsion angles** (= angles of twisting) lead to different coupling constants; this becomes particularly clear in the case of $^3J_{HH}$ couplings. The dependence of torsion angle and coupling constant is described by the **Karplus-Conroy relation** (in some works only called **Karplus relation**), which however again goes beyond this introduction. Here it is advisable to consult more extensive literature. If one wants to make precise statements about the conformation of the analyte under consideration in the case of more complex molecules (e.g. carbohydrate derivatives), the relationship between torsion angles and coupling constants is an extremely useful tool.

? Questions

15. Taking into account $^3J_{HH}$ and $^4J_{HH}$ couplings, what multiplets of multiplets can be expected, at least theoretically, in the ^{1}H-NMR spectra of the compounds (a) through (c)? (The problem of possible pseudomultiplets or coinciding signals is not to be taken into account here.)

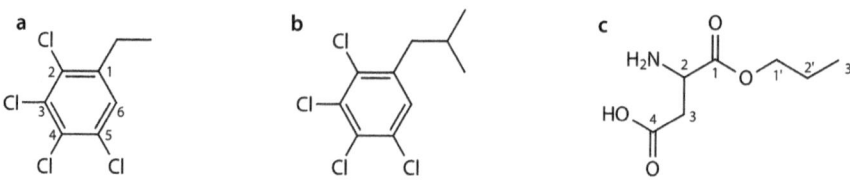

16. Which multiplets are shown in each of the following figures? To which molecular fragments do they each belong?

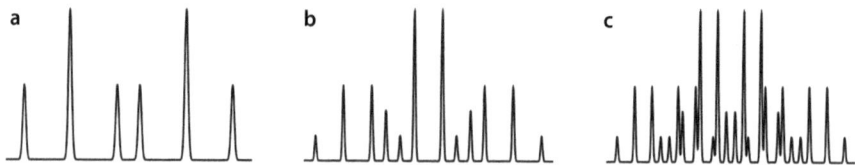

17. Does the ^{1}H-NMR spectrum belong to (a) (E)-1-chloro-2-pentene, (b) (E)-2-chloro-2-pentene, or (c) (E)-5-chloro-2-pentene? Give reasons for your decision.

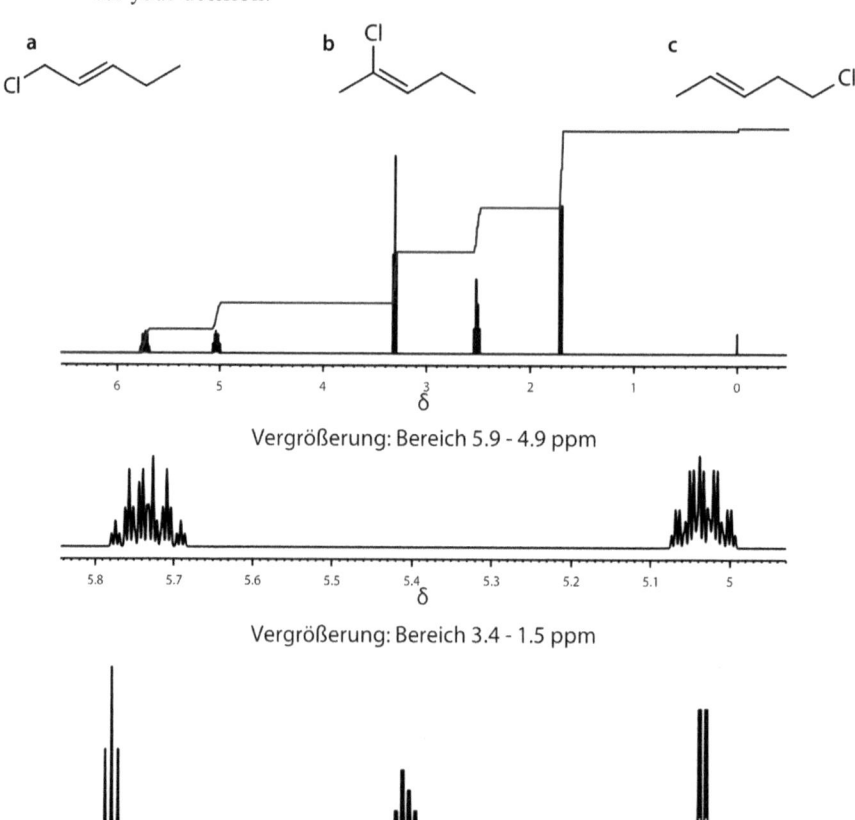

Vergrößerung: Bereich 5.9 - 4.9 ppm

Vergrößerung: Bereich 3.4 - 1.5 ppm

3.4 ^{13}C-NMR

While the most common hydrogen isotope on Earth (^{1}H) conveniently has a spin and can therefore be used in nuclear magnetic resonance spectroscopy, this is unfortunately not the case for the most common carbon isotope (^{12}C, as already mentioned in ▶ Sect. 3.1): Here we have to rely on the isotope ^{13}C, which however only occurs with a relative abundance of 1.1%. If we apply the same statistical considerations as in ▶ Sect. 2.1, we come to the conclusion that on average we need, for example, fifty molecules of ethanol in order to find a single ^{13}C atom in it. (It is "only" fifty, because each ethanol molecule contains *two* carbon atoms; for larger molecules with more carbon atoms, of course, the situation is somewhat different, but here, too, only about 1% of all carbon atoms present can be detected by NMR.) In addition, the gyromagnetic constant of this isotope is much smaller. Accordingly, the sensitivity of the ^{13}C-NMR is significantly lower than that of its hydrogen counterpart: It is only about one six-thousandth.

3.4.1 Shifts, Couplings, Spectra

Because the spin of this NMR-active carbon isotope is also ½, exactly the same rules apply to ^{13}C-NMR spectroscopy as to ^{1}H-NMR:
- The shifts are again given in ppm, but the associated δ-values are much larger: A ^{13}C-NMR spectrum commonly covers a range of about 200 ppm.
- The shifts of the individual carbon atoms are in turn dependent on their shielding (or deshielding) by the local electron density: Carbon atoms with fewer electrons due to more electronegative bonding partners are downshifted, while particularly electron-rich C atoms experience an upfield shift.
- Multiple bonds and aromatic systems show the same anisotropic effect as in ^{1}H-NMR.
- Tetramethylsilane (TMS) again serves as reference substance, because, like the hydrogens, the carbon atoms of this compound are relatively electron-rich (after all, silicon is less electronegative than carbon) and thus well shielded (i.e. shifted to high field compared to most "normal" C atoms). The reference value is δ(TMS) = 0 ppm.

■ **Couplings in ^{13}C-NMR**

Concerning couplings, the isotope carbon-13 again shows remarkable similarity with the ^{1}H nucleus, because—precisely because of the spin of ½—here, too, the interaction with a neighbouring (and also NMR-active) carbon nucleus leads to a doublet, two (chemically equivalent!) neighbouring C atoms would split the signal of the atom under consideration into a triplet, and so on. However, precisely because of the relative rarity of the NMR-active carbon isotope, the phenomenon of coupling occurs quite rarely, since the probability that *both* carbons in an ethanol molecule, for example, are each ^{13}C nuclei is 1% of 1%, i.e., only 0.0001. Accordingly, couplings of directly adjacent carbon atoms (i.e., $^1J_{CC}$ couplings) are of negligible importance analytically (unless one works with analytes in which an excess of ^{13}C atoms has been deliberately incorporated synthetically, but this is very, very costly). Any statistically required CC-coupling "mini-multiplet signals" often disappear in the background noise, but they can make the evaluation of a corresponding spectrum difficult; therefore, it is sometimes helpful to *prevent* these CC-couplings completely by **decoupling techniques**. (If you are interested in exactly how this works, please refer to further technical literature, such as ▶ Skoog or Hesse.)

Skoog, Section 19.5: ^{13}C-NMR spectroscopy

Of particular importance here is the **broadband decoupling,** which prevents *any* coupling of the carbon atoms (after all, there are other couplings than just the $^1J_{CC}$ case, in which a ^{13}C atom couples with another ^{13}C atom; we will discuss these in the following paragraph): Accordingly, for each chemically non-equivalent carbon atom, *one singlet* is obtained; the individual signals then differ accordingly (only) in their chemical shift.

- As an example, consider Fig. 19.27 from Skoog: Part a shows the broadband-decoupled ^{13}C-NMR spectrum of n-butyl vinyl ether (CH_3–$(CH_2)_3$–O–CH=CH_2): The six carbon atoms (all chemically non-equivalent) correspondingly lead to six signals, each with a different chemical shift.
- This spectrum once again shows the influence of the electron (density) and the bonding conditions on the resulting shift values very clearly:
- It can be seen that the methyl group is relatively strongly shielded. The associated signal (marked "4" in Fig. 19.27a), is highly upshifted compared to all others.
- The two signals of the carbon atoms involved in a C–C double bond are lowfield-shifted, whereby the terminal methylene carbon ("6") does not appear as far to the left of the spectrum as the oxygen-bonded methine carbon ("5") because of the two electron-donating hydrogen atoms (C is more electronegative than H).
- As expected, of the three methylene carbons, the signal of the one directly bonded to the (more electronegative) oxygen ("1") experiences the largest downshift.

Skoog, Section 19.5: ^{13}C-NMR spectroscopy

At this point, it must be pointed out that in ^{13}C-NMR—due to technical measurement conditions, which would once again go beyond the scope—the integration of corresponding signals is *not* possible (or meaningful); accordingly, the number of possible chemically equivalent carbon nuclei cannot be taken from these spectra "just like that". This becomes particularly clear in the case of broadband decoupling: The most strongly downshifted signal in Fig. 19.27a ("5") clearly has an oversised signal height compared to the much smaller signal of the methyl group ("4"), although it also represents only *one* C-atom.

▶ **Example**

For example, the broadband decoupled ^{13}C-NMR spectrum of propane would have only *two* signals:
- a signal shifted to a lower field for the methylene group
- a comparatively upshifted signal for the two chemically equivalent (!) methyl groups (in which the C-atoms are somewhat richer in electrons because of the hydrogens bonded to them)

The relative magnitude of the resulting signals does not allow *any* conclusion as to how many chemically equivalent nuclei the signal originates from. ◀

A Brief Outlook

The fact that broadband decoupling leads to signal intensification can certainly be exploited:
- On the one hand, it increases the sensitivity of ^{13}C-NMR.
- On the other hand, it is the basis of a very own NMR variant, which will only be briefly touched upon here, but which should not remain unmentioned.

3.4 · ^{13}C-NMR

> Behind this change in signal intensity is a phenomenon called *nuclear Overhauser effect* (NOE), which is, however, extremely special, so we will not explain it any further. (Even ▶ Skoog says that this topic is "beyond the scope of this textbook"; Hesse or Breitmaier are more detailed.) Nevertheless, it should at least be mentioned, because there is a two-dimensional NMR technique (more on this in ▶ Sect. 3.4.2) which is based precisely on this and which is of considerable importance in—really advanced!—structure elucidation. Here, only the corresponding abbreviation will be mentioned by name: With the help of **NOESY** experiments *(Nuclear Overhauser Enhancement and exchange SpectroskopY)*, it is possible (with a little skill in evaluating the corresponding spectra) to find out which carbon atoms are directly bonded to each other in each case, without having to consider $^1J_{CC}$ couplings. Once one has this information, it is of course much easier to determine the C–C structural framework of even a previously unknown analyte.

However, $^1J_{CC}$ couplings are not the only interaction that ^{13}C nuclei can undergo with neighbouring atoms: Please recall that by far the majority of hydrogen atoms will be of the isotope ^{1}H, which is indeed NMR-active due to its spin of ½—and CH bonds are, after all, hardly uncommon, especially in organic chemistry. Accordingly, there are also $^1J_{CH}$ couplings, and the signal of a carbon atom to which a hydrogen atom is bonded is split into a doublet by its spin (which allows the two energetically slightly different spin states: +½ and −½). A methylene group correspondingly leads to a triplet (again with the relative intensities 1:2:1) and a methyl group correspondingly to a quartet (1:3:3:1). The corresponding coupling constants, however, are much larger than we are used to from ^{1}H-NMR. The signals of carbon atoms that do *not* carry their own H atom (e.g. quaternary carbons) naturally remain singlets.

But there are also $^2J_{CH}$ couplings, i.e. the interaction of an NMR-active carbon atom (nucleus) with the H atoms bonded to an *adjacent* carbon atom. (Please remember that these hydrogens will be ^{1}H atoms with a probability of almost 99.99% [isotope abundance!]; for these $^2J_{CH}$ couplings it is by no means necessary that the neighbouring carbon is also a ^{13}C atom.) Accordingly, this would lead to a further splitting of the present multiplets, and this would again complicate the evaluation of the resulting spectrum. (In addition, $^3J_{CH}$ and other long-range couplings can also occur, which does not exactly make the whole thing easier.)

However, since at least the $^1J_{CH}$ couplings provide useful information, which can be particularly helpful in structural elucidation, but any CC couplings and CH long-range couplings ($^nJ_{CH}$ with n > 1) tend to interfere, **selective decoupling** is often performed, which "hides" both the coupling of neighbouring ^{13}C nuclei and the CH long-range couplings, while the multiplets resulting from the $^1J_{CH}$ couplings are clearly visible in the spectrum.

The corresponding spectrum of the example compound n-butyl vinyl ether is shown in Fig. 19.27b from Skoog (you will certainly already have noticed that this decoupling technique leads to a significant signal broadening):

— It can be seen that the downshifted methine-carbon signal ("5") is split into a doublet.
— The signal from the sp^2-hybridised methylene carbon ("6") undergoes splitting to form a triplet.
— The three sp^3 methylene groups ("1"–"3") also appear as triplets
— Finally, the methyl group ("4") is split into a quartet.

Skoog, Section 19.5: ^{13}C-NMR spectroscopy

And there you can already see a problem arising with selectively decoupled ^{13}C-NMR spectra: The coupling constants of the different multiplets are of the order of small shift differences, so that the multiplets of chemically similar (but not equivalent!) atoms can overlap—which, of course, can make the unambiguous assignment of individual lines immensely difficult. For this reason, both broadband and selectively decoupled spectra of the same analyte are often recorded and then matched.

> The selectively decoupled ^{13}C-NMR spectrum of propane (think back to the last example) would again show only *two* signals, but each would be split into multiplets:
> — The signal of the methylene group shifted to lower field would be present as a triplet due to the two hydrogens bonded to this C atom.
> — In comparison, the upshifted signal of the two chemically equivalent methyl groups would result in a quartet because of the three H atoms bonded to them.

? Questions

18. To which compound of the molecular formula C_4H_9Cl do the following ^{13}C- and ^{1}H-NMR spectra belong?

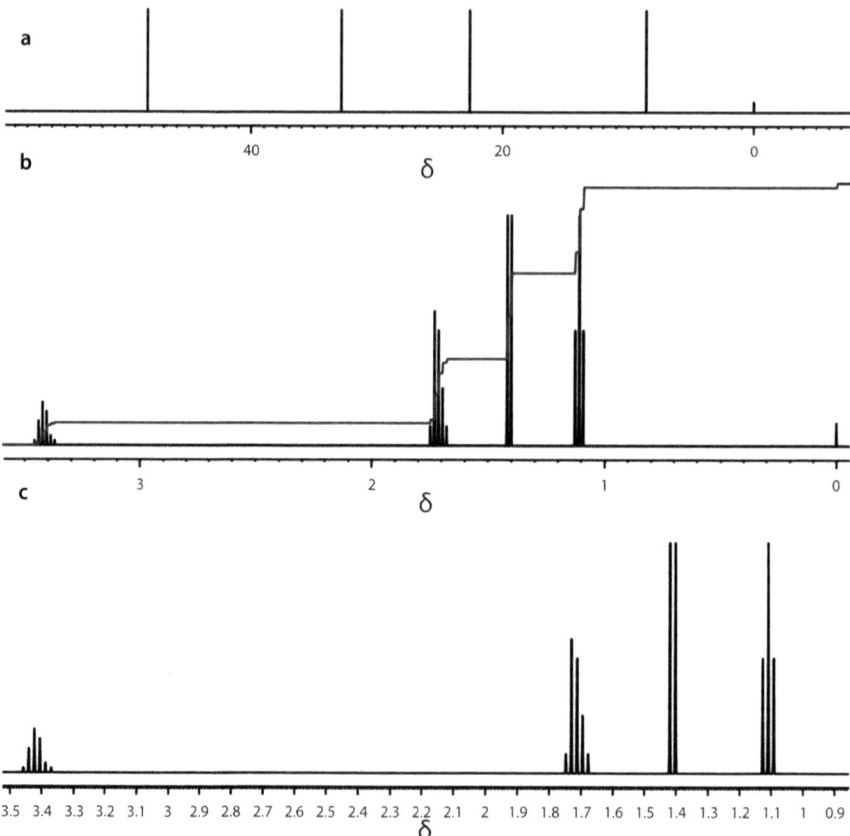

3.4.2 Two-dimensional NMR

You will have noticed by now that nuclear magnetic resonance spectroscopy is not only an extremely helpful and versatile tool: It also offers an almost inexhaustible potential of possibilities (and occasionally: necessities) for further studies. An introductory text such as this section can therefore only provide the absolute basics; should you ever become active in this field yourself, it will be unavoidable to delve into more specialised literature. In addition, ChemgaPedia provides a wealth of information, illustrations and even (extremely helpful) animations on 2D-NMR experiments and the respective technical measurement principles.

Some "more advanced" NMR techniques should be mentioned by name and outlined at least briefly, because in structural elucidation, especially for more complex analytes, much more specialised methods than "only" ^{1}H- and ^{13}C-NMR are of importance. As an example, some 2D-NMR variants should be mentioned here, where corresponding analytical findings (= signals) are not only plotted "in one spatial direction", i.e. along *one* axis of the Cartesian coordinate system, but on two axes (oriented perpendicular to each other), so that the signals' intensities then project "upwards", i.e. into the third dimension. Graphically, this may be quite difficult to represent (and may seem confusing at first glance), but the information content of such 2D-NMR spectra is enormous. Moreover, their evaluation is often easier than one might think (as you will notice in a moment).

▶ http://www.chemgapedia.de/vsengine/vlu/vsc/de/ch/3/anc/nmr_spek/zweidimensionale_nmr.vlu.html

■ **Correlation Spectroscopy** (*COrrelation SpectroscopY*, **COSY**)

With this two-dimensional method, it is quite easy to find out which atomic nuclei couple with each other. A distinction is made between two different variants:
— In *homonuclear* correlation spectroscopy, NMR spectra of the same atomic species are considered in the two dimensions; the most common variants are the HH-COSY and the CC-COSY.
— In *heteronuclear* correlation spectroscopy, spectra of different atomic species are combined. Here, above all, the CH-COSY is to be mentioned.

As an example, Fig. 19.33 from ▶ Skoog shows the CC-COSY of 1,3-butanediol (the configuration of the chirality center present in this molecule—labeled "2" in the figure—is left out of consideration here). If you first look at the *selectively decoupled* ^{13}C-NMR spectrum of this compound (Fig. 19.33b), you will notice that the multiplets resulting from CH coupling at least partially overlap each other, so that assigning them to specific carbon atoms might seem a bit difficult at first. Although one could, of course, determine the coupling constants by the respective line distances and thus find out which line actually belongs to one or the other multiplet, it is much easier to do this using a CC-COSY: By selective irradiation of various pulses (we really do not want to go into such technical aspects here!), the individual signals are decoupled from each other, so that—separated from each other in time—the individual multiplets appear separately from each other in the 2D spectrum. The rather confusing multi-line system from Fig. 19.33b is thus broken down into a doublet, two triplets and a quartet—viewed in the 2D spectrum (Fig. 19.33a) from "bottom left" to "top right".

Skoog, Section 19.7.2: Two-dimensional NMR

One can even go further: If the broadband-decoupled ^{13}C spectrum were plotted along the horizontal of this two-dimensional spectrum (which, unfortunately, is not the case in this figure), we would observe four singlets there, since the analyte contains four chemically non-equivalent carbon atoms. The shift values of the respective signals would then coincide in each case with the centroid of the associated multiplet from Fig. 19.3b. If one were to draw a diagonal line through the 2D spectrum and then draw a perpendicular line to the broadband-decoupled spectrum at the centroids of the respective multiplets (i.e. always their midpoints, which are exactly intersected by said diagonal line—in the case of the two triplets through the middle signal, in the case of the doublet and the quartet centrally between the two or the two middle signals), we would immediately have assigned the respective multiplets from the 2D spectrum to the individual singlets.

An analogous procedure can also be used for a HH-COSY—this is particularly useful if the different multiplets overlap or even superpose in an "ordinary" one-dimensional ^{1}H-NMR, and such difficulties can also be avoided with (heteronuclear) CH-COSY spectra by introducing the second dimension.

(If you are interested in the technical aspects of the respective techniques, please refer to further literature; for us it is only relevant here *that* the relevant information can be obtained.)

- **CH–COLOC** (*COrrelation through LOng-range Coupling*)

While in the COSY just presented any long-range couplings (i.e. $^nJ_{CH}$ with n > 1) are eliminated by selective decoupling, CH–COLOC *is* based exactly on these couplings: Here it is mainly $^2J_{CH}$ and $^3J_{CH}$ that are used for structural elucidation.

- **CC-INADEQUATE**

Even the $^1J_{CC}$ coupling, which is quite rare because of the relative rarity of the isotope carbon-13, serves as the basis of a 2D-NMR technique: the INADEQUATE experiment, which could also be called ^{13}C–^{13}C–COSY. (The lovely acronym INADEQUATE stands for *Incredible Natural Abundance DoublE QUAntum Transfer Experiment*, but the quantum mechanical aspects clearly take us too far here.) This technique, precisely because the ^{13}C content of the analytes is quite small, is anything but sensitive, but if one has enough sample substance available (and the analyte can also be dissolved in the chosen solvent ...), there is probably no method more elegant to determine the **connectivity** of an analyte of unknown structure. Within the scope of an introduction, the complete evaluation of an appropriate 2D spectrum cannot be discussed in adequate detail, hence only the end result: based on coupling constants and shift values, one obtains a spectrum in which one effectively has to connect the individual signals, resembled by dots, just like in the popular *"connect the dots"* images, and at the end one directly obtains the carbon framework of the analyte. Even though INADEQUATE spectra rarely occur in everyday laboratory work, they should be mentioned here because there is probably no type of spectrum whose evaluation is more fun!

- **Other Techniques**

▶ http://www.chem.ox.ac.uk/spectroscopy/nmr/acropage.htm

You may have already noticed the tendency of NMR spectroscopists to use ... creative abbreviations or acronyms. The few examples mentioned in this introduction are only the tip of the iceberg: You can find several collections of interesting (and really informative) NMR techniques on the Internet, often with extensive information on the technical aspects.

■ 3D-NMR and Even Higher Dimensional Spectra

Now it becomes decidedly too special for an introductory text. It should only be mentioned that, especially in the case of highly complex analytes, the combination of several NMR experiments reveals correlations that remain undetected in a single (1D-)NMR spectrum or would simply not be detectable due to signal superposition. A prime example of the immense power of these combination techniques is the structural elucidation of proteins in their native folding state (you have certainly already learned something about this in introductory courses on *biochemistry*), which until a few years ago could be carried out almost exclusively via X-ray structure analysis—which of course was only possible if the protein in question could also be crystallised with sufficient purity (and by no means all proteins do us this favour). Nowadays, such analyses are often carried out by three- or even more-dimensional NMR experiments (which is extremely difficult to represent graphically).

Incidentally, it is very possible that you have already come into contact (directly or indirectly) with 3D-NMR spectroscopy in your life outside the laboratory: The increasingly widespread imaging technique in medical diagnostics known as *magnetic resonance imaging* (or *magnetic resonance tomography*, MRI for short) is based on exactly that.

3.5 Other Usable Cores

In ► Sect. 3.1.1 you already learned that not only the nuclei of the isotopes ^{1}H and ^{13}C can be excited to Larmor precession in the magnetic field: In principle, this is possible with *all* atomic nuclei if they do not have an even number of protons and neutrons (which is why carbon-12 with six protons and neutrons each cannot be used for NMR). However, not all theoretically conceivable atomic nuclei are used for NMR experiments. The scope of the previous text on nuclear magnetic resonance spectroscopy will probably already give you an idea that ^{1}H- and ^{13}C-NMR are by far the most important techniques, especially, of course, because they are practically ubiquitous in organic chemistry.

Nevertheless, occasionally also other atomic nuclei are considered in NMR—for more special cases. The most important of these "exotic cases" are briefly (!) outlined here.

■ ^{19}F NMR

On Earth, only the isotope fluorine-19 naturally occurs, so fluorine is a **pure element** (although other isotopes are accessible synthetically). For this reason, this nucleus, whose spin = ½ corresponds to that of ^{1}H and ^{13}C, is very well suited for the characterisation of fluorine-containing compounds: ^{19}F nuclear magnetic resonance spectroscopy is similarly sensitive as ^{1}H-NMR. The chemical shifts of the associated signals span a range of about 400 ppm, with both downshifted (with $\delta > 0$) and upshifted signals (with $\delta < 0$) appearing in corresponding spectra. However, due to the high electronegativity of fluorine alone, the electron density there is usually quite high, so that shielding (and thus upfield shift) tends to occur; accordingly, negative shift values are anything but rare. The reference substance is usually trichlorofluoromethane ($CFCl_3$; for obvious reasons, the usual TMS would make little sense here); because of the likewise considerable electronegativity of the chlorine atoms, which accordingly also draw electron density from the carbon to themselves, the comparatively electron-poor (because hardly negatively polarised) fluorine is assigned the shift value $\delta = 0$ ppm here.

Due to the natural abundance of this isotope (100 %, it *is* a pure element), couplings can also be observed in ^{19}F NMR spectra; $^3J_{FF}$ coupling constants, such as occur in **vicinally** difluorinated compounds, are of the order of 20 Hz.

∎ ^{31}P NMR

Phosphorus is also a pure element: only the isotope ^{31}P is stable. This nucleus also has spin = ½; however, its sensitivity is much lower than that of hydrogen—less than 10 %. Nevertheless, this method is quite common, especially in *organoelement chemistry/organometallics*, mostly used to support findings already obtained via ^{1}H- and ^{13}C-NMR.

As with fluorine, the range of chemical shifts here is quite large: a ^{31}P NMR spectrum usually covers more than 500 ppm, in extreme cases 700 ppm. Phosphoric acid (H_3PO_4) serves as a reference substance with a defined shift value of δ = 0 ppm; depending on the bonding state and the type of bonding partners, most signals lie between +200 (downshifted) and −250 ppm (upshifted). Because of the relative abundance of the isotope ^{31}P, coupling constants can often be determined, but these cover a wide range of values: In compounds with P–P bonds, the associated $^1J_{PP}$ couplings can have values from 15 to 600 Hz.

∎ ^{15}N, ^{29}Si and ^{207}Pb—and nuclei with spin ≠ ½

The isotope nitrogen-15, as well as the isotopes silicon-29 and lead-207 also have a nuclear spin of ½, but these NMR variants are to be regarded as quite special; only ^{15}N-NMR, employed predominantly in the analysis of organic amines (to which amino acids and thus peptides and proteins also belong), can be said to be used "more than only in exceptional cases".

In ▶ Sect. 3.1.1 ▶ it has already been stated that there are also atomic nuclei with a spin > ½; In addition to the examples mentioned in the section above, the hydrogen isotope ^{2}H (deuterium) must be mentioned, whose nuclear spin is 1 and which therefore—just like lithium-7—has *three* possibilities of aligning itself in the magnetic field (which makes the evaluation of any multiplets in the associated spectra a little more difficult), as well as the oxygen isotope ^{17}O (which is very rare, with a relative abundance of less than 0.04%), which can even assume *six* different spin states thanks to its spin of 5/2. But you can already see: Things are getting *very* specific again. Let's leave it at this, ands consider this just a rough overview.

Harris, Section 21.2: Atomisation: Flames, Furnaces and Plasmas

✓ Answers

1. According to Table 21.1 from Harris, the following isotopes are of concern: natural abundance ^{12}C = 98.93%; ^{13}C = 1.07%; ^{35}Cl = 75.78%, ^{37}Cl = 24.22%; ^{79}Br = 50.69%, ^{81}Br = 49.31%.

   (a) The relative abundance ratio of the two chlorine isotopes is particularly important here: Ignoring—at first—the probability that every one hundredth molecule has the isotope ^{13}C as its central C atom, we obtain the nominal masses 50 (3 × ^{1}H, 1 × ^{12}C, 1 × ^{35}Cl) and 52 (3 × ^{1}H, 1 × ^{12}C, 1 × ^{37}Cl). Since the isotope ^{35}Cl occurs three times as often as the isotope ^{37}Cl, the corresponding peak is also three times as high. Thus, one observes two peaks, which have a height ratio of 3:1 and which differ by the mass number 2—this is the characteristic **isotope pattern** for a singly chlorinated compound. In addition, the possible nominal masses 51 (3 × ^{1}H, 1 × ^{13}C, 1 × ^{35}Cl) and 53 (3 × ^{1}H, 1 × ^{13}C, 1 × ^{37}Cl) are also obtained, but these will hardly be discernible. (Isotope pattern ^{12}C/^{13}C: at a distance of 1 mass number, there is a height difference of 99:1.) You will still recognise the peak at 51 in

3.5 · Other Usable Cores

spectrum (a) (but the one at 50 is 99 × higher), the 53 peak is lost in the background noise (it is only one third as high as the one at 52, because of the chlorine isotope pattern).

(b) If we again ignore the carbon isotopes for the time being, we get the nominal masses 94 (3 × ^{1}H, 1 × ^{12}C, 1 × ^{79}Br) and 96 (3 × ^{1}H, 1 × ^{12}C, 1 × ^{81}Br). Again, the **isotope pattern** is important: tthe two bromine isotopes are nearly equally abundant and differ in mass number by 2—so if you see two peaks of virtually the same height 2 u apart, that strongly suggests a singly brominated compound. (Look again at Table 21.1 from Harris : no other of the common elements brings such a 50:50 ratio!) Because of the isotope ^{13}C, there are also tiny peaks at 95 (3 × ^{1}H, 1 × ^{13}C, 1 × ^{79}Br) and 97 (3 × ^{1}H, 1 × ^{13}C, 1 × ^{81}Br), but their heights are each only 1% of that of their immediate neighbour to the left.

(c) Here, two isotope patterns overlap: the 3:1 ratio of ^{35}Cl/^{37}Cl and the 1:1 of ^{79}Br/^{81}Br. Even if we (once again) initially ignore ^{13}C, a somewhat more complex spectrum results accordingly:

- The peak at 128 can be attributed to 2 × ^{1}H, 1 × ^{12}C, 1 × ^{35}Cl, 1 × ^{79}Br,
- at 130, the peaks of 2 × ^{1}H, 1 × ^{12}C, 1 × ^{37}Cl, 1 × ^{79}Br and 2 × ^{1}H, 1 × ^{12}C, 1 × ^{35}Cl, 1 × ^{81}Br overlap (in a high-resolution spectrum, separation of the two peaks would probably be possible, but that is going too far here),
- at peak 132, 2 × ^{1}H, 1 × ^{12}C, 1 × ^{37}Cl, 1 × ^{81}Br are present.

For each of the three peaks in this spectrum discussed so far, there is also the peak of the corresponding molecules with a central ^{13}C atom, shifted by 1 u to higher masses (i.e.: "to the right"); because of the relative rarity of this isotope, however, the peaks in question are again only vanishingly small. (But please do not forget that analytes can easily have *significantly more than one* C-atom, and then the additional peaks resulting from the ^{13}C-atoms simply cannot be ignored!)

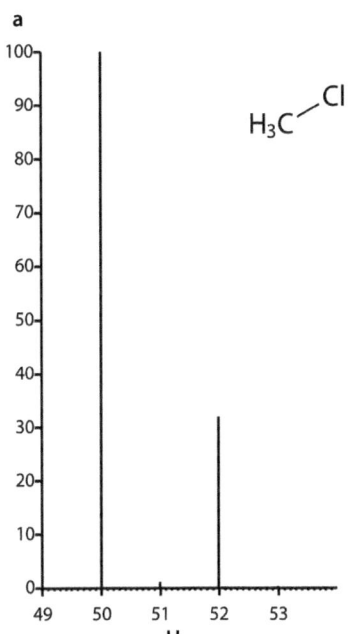
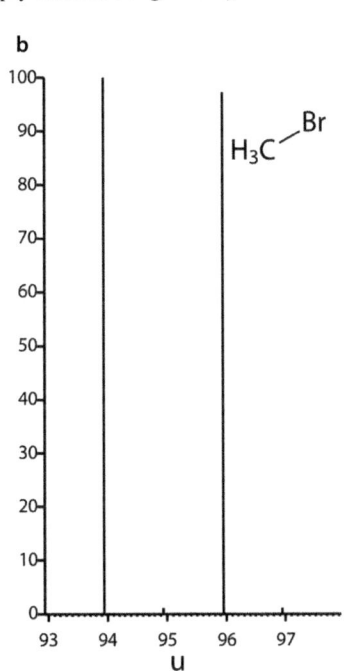
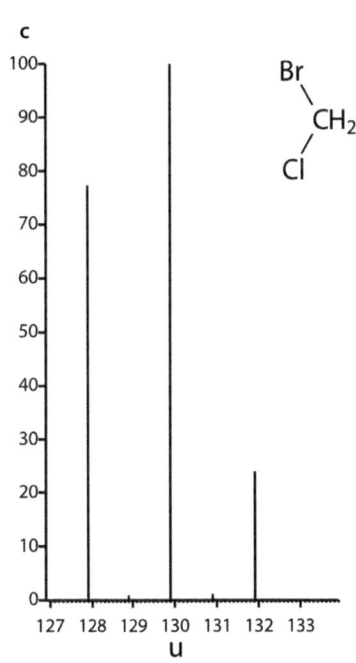

2. With m(H) = 1, m(C) = 12 and m(O) = 16, the electron impact ionisation, which leads to the loss of one electron (or, in the case of multiple ionisation, correspondingly several electrons), results in (a) with single ionisation to m/z = 60/1 = 60, (b) with double ionisation correspondingly to m/z = 60/2 = 30. (c) With m(D) = 2, single ionisation leads to m/z = 61. (d) Here, an analyte which has been deliberately deuterated several times by synthetic means is considered: $m(CD_3COOH) = 64$, single ionisation gives m/z = 64, (e) double ionisation thus 64/2 = 32.

3. In the case of chemical ionisation, ionisation takes place through the uptake of an additional H^+ ion, correspondingly the mass number of the analytes per charge increases by 1, thus resulting in: (a) m/z = (60 + 1)/1 = 61; (b) m/z = (60 + 2)/2 = 31 (*double* ionisation means: uptake of *two* hydrogen cations!); (c) m/z = (61 + 1)/1 = 62; (d) m/z = (64 + 1)/1 = 65; (e) m/z = (64 + 2)/2 = 33.

4. In thermospray ionisation, the positive charge(s) is/are transferred by the cations of the buffer system attaching to the analyte. With $m(Na^+) = 23$ and $m(NH_4^+) = 18$, this gives: (a) m/z = (1442 + 23)/1 = 1465; (b) (1442 + (2 × 23))/2 = 1488/2 = 744; (c) m/z = (1442 + 18)/1 = 1460; (d) (1442 + (2 × 18))/2 = 1478/2 = 739.

5. It should be noted that the two bromine isotopes ^{79}Br and ^{81}Br occur with almost the same abundance. While the molecular peak with m/z = 156 belongs to the molecule $C_6H_5^{79}Br$, the almost equally frequent molecule $C_6H_5^{81}Br$ provides the peak with m/z = 158. By the way, the spectrum, which you can take from the NIST database, is normalised to the peak with m/z = 77, which results when the C–Br bond is cleaved, forming the phenyl radical cation $C_6H_5^{+\bullet}$.

▶ http://webbook.nist.gov/cgi/cbook.cgi?Formula=C6H5Br&NoIon=on&Units=SI&cMS=on#Mass-Spec

(a) ▶ http://webbook.nist.gov/cgi/cbook.cgi?ID=C1678917&Units=SI&Mask=200#Mass-Spec

(b) ▶ http://webbook.nist.gov/cgi/cbook.cgi?Formula=C6H5Cl&NoIon=on&Units=SI&cMS=on#Mass-Spec

(c) ▶ http://webbook.nist.gov/cgi/cbook.cgi?ID=C104518&Units=SI&Mask=200#Mass-Spec

6. (a) In the case of ethylcyclohexane (C_6H_{11}–C_2H_5), which contains no heteroatoms of any kind, the peaks of the two ions resulting from σ-cleavage are to be expected in addition to the molecular peak M^+ with m/z = 112: Either a neutral ethyl radical $C_2H_5^\bullet$ can be split off, so that the resulting cyclohexyl radical cation $C_6H_{11}^{+\bullet}$ with m/z = 83 (M – 29) is detected, or the charge remains with the ethyl radical, so that the signal of the ethyl radical cation $C_2H_5^{+\bullet}$ with m/z = 29 appears in the spectrum. However, since this radical cation has a disproportionately larger charge density, it will not appear nearly as often (peak height!) as the cyclohexyl cation. You can look up the spectrum yourself in the NIST database—by the way, the other peaks occurring in this spectrum can be explained by the further fragmentation of the cyclohexyl ring, but that would again go beyond the scope of this introduction. b) In the case of the chlorobenzene (C_6H_5Cl), a *double* parent peak is to be expected because of the isotope pattern of chlorine ($^{35}Cl/^{37}Cl$ = 75/25): In addition to the molecular ion $C_6H_5^{35}Cl$ with m/z = 112, the peak of the molecular ion $C_6H_5^{37}Cl$ with m/z = 114 is also to be expected, but its peak height reaches only one third of that of 112. Furthermore, the crucial fragmentation reaction will again be σ-cleavage, so that the phenyl radical cation ($C_6H_5^{+\bullet}$) formed by losing the chlorine atom leads to a peak at m/z = 77. Incidentally, the fact that in the corresponding spectrum from the NIST database the height of the peaks at m/z = 35 and 37 is practically zero tells us something else: the C–Cl-σ cleavage evidently only very rarely proceeds in such a way that the charge remains with the halogen atom. (c) For the butylbenzene (C_6H_5–CH_2–$(CH_2)_2$–CH_3), too, the molecular peak with m/z = 134 is to be expected; since this molecule contains ten carbon atoms after all, the probability that at least one of these C atoms belongs to the isotope ^{13}C is thus 10%, so we can also expect a (much smaller) peak at with m/z = 135. The hitherto familiar σ-cleavage leading to the phenyl radical cation ($C_6H_5^{+\bullet}$) plays practically no role in the

3.5 · Other Usable Cores

fragmentation of this analyte, and the peak at m/z = 77 practically disappears in the background noise of the associated NIST spectrum. Much more important here is the benzyl cleavage, giving the benzyl radical cation (C_6H_5–$CH_2^{+•}$) with m/z = 91—the spectrum is even normalised to this peak. (The additional propyl radical [with m = 43] formed during *benzyl cleavage* is not detected; the possibility that the charge remains with this other fragmentation product and a propyl radical cation CH_3–CH_2–$CH_2^{+•}$ with m/z = 43 is theoretically possible, but since it, unlike the benzyl cation, is not resonance-stabilised, its formation can be safely neglected; it plays no role in the associated spectrum.) More interesting in this spectrum is the also quite strong peak at m/z = 92: This is a protonated benzyl cation ($C_7H_8^+$); a hydrogen atom has been shifted from the other fragment, resulting in neutral (and therefore undetectable) propene (CH_3-CH=CH_2). This type of hydrogen rearrangement is very characteristic for alkylbenzene derivatives.

7. To answer this question, it is first necessary to clarify which products *could* be formed in an RDA. Here the figure helps: On the path marked (a), the radical cation $C_2H_4^{+•}$ with m/z = 28 (M − 194) is formed from the (already ionised) molecular radical cation $C_7H_{11}I^{+•}$ with m/z = 222, in addition to a (non-detectable) neutral molecule; if the positive charge ultimately lies with the larger molecular fragment, as path b) describes, a cation C_5H_7I $C_5H_7I^{+•}$ with m/z = 194 (M − 28) is formed in addition to (neutral) ethene. We do indeed find a corresponding peak, but it is hardly worth mentioning: Obviously, the fragmentation of this analyte takes place only to an insignificant extent via RDA. Clearly more significant are two of the three peaks mentioned with m/z = 222, m/z = 127 and m/z = 95: The former is the molecular ion $M^{+•}$ itself, 127 is the iodine radical cation $I^{+•}$ formed by cleavage of the I–C bond. (Iodine is a pure element: only the isotope ^{127}I occurs in nature, other isotopes are accessible by synthetic means but do *not* commonly lead to an isotope pattern in MS.) Now the peak at m/z = 95 becomes understandable, as well: It belongs to the molecular ion where the iodine atom has been split off: $C_7H_{11}^{+•}$ (M − 127).

8. In the case of acetic acid (a), the signal of the carboxyl hydrogen (–COO**H**) will be shifted to the lower field for two reasons: On the one hand, the electron density will be significantly lowered by the more electronegative bonding partner—the oxygen atom (H: δ+, O: δ−); on the other hand, this effect is further enhanced by the fact that the other bonding partner of this O atom (the carboxyl carbon –**C**OOH) is already strongly positively polarised by the doubly bonded, also more electronegative bonding partner (=O) (C: δ+), correspondingly the oxygen atom to which this hydrogen is bonded cannot attract too much charge density from this C atom, which results in its electron-pull towards the hydrogen being further enhanced. (This strong positive polarisation is, after all, the reason for the comparatively large acid strength of carboxy compounds: Compare the pK_A values of acetic acid ($pK_A(CH_3COOH)$ = 4.75) with that of ethanol ($pK_A(CH_3CH_2OH$ = 15.9: ethanol is more than ten orders of magnitude less acidic and does *not* react as an acid towards water!). The strong positive polarisation of the carboxylic carbon then also causes its signal in the ^{13}C-NMR spectrum to experience a considerable downshift as well. And

the strong positive polarisation of the carboxyl carbon will also have at least a slight deshielding effect on the methyl group of this molecule: The corresponding ^{1}H and ^{13}C signals will not appear "as far to the left" in the corresponding spectrum as the respective other ones of this compound, but they will still be recognisably downshifted for methyl signals. In the case of butyric acid b), this effect will be even more pronounced: In both ^{1}H- and ^{13}C-NMR spectra, the carboxyl group will be strongly shifted to lower field, and this will also affect the ^{1}H and ^{13}C signals of the immediately adjacent methylene group (–**CH2**–COOH), while the other methylene group (CH$_3$–**CH2**–CH$_2$–) will already be found much less far to the left in the spectrum due to the decreasing "electron-withdrawing effect" of the carboxyl group. Finally, the terminal methyl group will be located bonth in the ^{1}H- and in the ^{13}C-NMR spectrum furthest "to the right" of all signals. Also for propylamine (c) the higher electronegativity of the hetero atom (here: N) has a corresponding effect: The methylene group directly bonded to the nitrogen will be shifted further to the lower field in both spectra than the CH$_2$ group adjacent to it, and the terminal methyl group is found furthest "to the right" in both spectra. In triethylborane d), the conditions are reversed: Since boron is *less* electronegative than carbon is, a *negative* polarisation results for each of the two methylene groups (C: δ–). The corresponding signals in the ^{13}C-NMR spectrum are noticeably upshifted for methylene signals, while the methyl signals are found "farthest to the right" as usual. The increased electron density of the methylene carbons also affects the associated H atoms: The signal of the methylene groups is upshifted compared to that of the methyl groups to such an extent that they are found—which is rather unusual for CH$_2$ groups—"further to the right" in the ^{1}H-NMR spectrum than the signal for the methyl hydrogens.

- Exemplary are the simulators of nmrdr.org:
- for ^{1}H-NMR spectra: ▶ http://www.nmrdb.org/new_predictor/index.shtml?v=v2.56.0
- for ^{13}C-NMR spectra: ▶ http://www.nmrdb.org/13c/index.shtml

- **For those who want to try it for themselves**
- The internet offers several ^{1}H and ^{13}C-NMR simulators that calculate ^{1}H and ^{13}C-NMR spectra for practically any compound imaginable—the structure of which can be entered via a Java mask. The deviation of the simulated spectra based on theoretical considerations from real, "self-recorded" spectra may occasionally be of importance in everyday laboratory/analytical work, but for practice these spectra simulators are worth their weight in gold.

9. In the case of the methyl group, which is bonded to the carboxyl carbon, "only" the strong positive polarisation of the carboxyl carbon has an effect, whereas in the case of the methoxy group, both this C atom and the hydrogen atoms bonded to it are clearly de-shielded by the direct linkage of the carbon to the oxygen. Accordingly, the associated signal in both the ^{13}C- and ^{1}H-NMR spectra is shifted to lower field. (Of course, the ^{13}C signal of the carboxyl carbon, whose electron density is by far the most lowered one, is even more strongly downshifted.)

10. (a) The signal of the methylene group (–CH$_2$–) is split into a quartet by the three hydrogen atoms of the adjacent methyl group (–CH$_3$); (b) The signal of the methyl group is split into a triplet by the two H atoms of the methylene group.

11. (a) The methylene signal is split into a doublet by the adjacent methine group; (b) The signal from the methine group is again split into a triplet by the methylene group.

12. (a) The two hydrogen atoms attached to the sp^2-carbon atoms and thus also the hydrogen atoms attached to them are chemically equivalent, so they produce only a single common signal (with an integral of 2); the two methyl groups are also equivalent. Because of the anisotropy effect of double

bonds, the signals of the two sp² carbons and the corresponding hydrogens experience a downfield shift: for the methyl group, $\delta(^1H) = 1.6$ ppm and $\delta(^{13}C) = 17$ ppm; for the sp²-CH groups, however, $\delta(^1H) = 5.5$ ppm and $\delta(^{13}C) = 125$ ppm. (b) The anisotropy effect of the C=O double bond is further enhanced by the higher electronegativity of the oxygen atom: for the CH group at C^2 (starding to count in accordance with IUPAC recommendations with the carbonyl carbon): $\delta(^1H) = 6.3$ ppm and $\delta(^{13}C) = 134$ ppm, and for that of C^3: $\delta(^1H) = 6.9$ ppm and $\delta(^{13}C) = 154$ ppm. The signals of the aldehyde carbon and its associated H atom are also even more significantly downshifted: $\delta(^1H) = 9.5$ ppm and $\delta(^{13}C) = 193$ ppm. The terminal methyl group with $\delta(^1H) = 2.0$ ppm and $\delta(^{13}C) = 18$ ppm is noticeably upshifted compared to the other three signals in the spectrum of this compound. (c) Because of the ring current of this aromatic compound, the H atoms immediately attached to the ring will be downshifted compared to the methyl group: $\delta(^1H)$(methyl) = 2.3 ppm, $\delta(^1H)$(ring) = 7.2 ppm. Please note: Although here all H atoms bonded to the ring itself undergo exactly the same chemical shift, these five hydrogen atoms are by no means chemically equivalent: chemically equivalent are only the two H atoms in *ortho* and in *meta* position to the methyl substituent, respectively. The fact that the signals of all three—chemically different—hydrogen atoms just coincide is to be regarded as a coincidence. The chemical non-equivalence of the different ring members becomes clear in the ^{13}C-NMR spectrum: While the methyl group does not undergo any remarkable shift here either—$\delta(^1H) = 1.6$ ppm; $\delta(^{13}C)$(methyl) = 21 ppm -, the four chemically different (!) carbon atoms are similar but not identical in their shift values: $\delta(^{13}C)(C^1$, to which the methyl group is attached) = 137 ppm, $\delta(^{13}C)(C^2$, *ortho-position*) = 129 ppm, $\delta(^{13}C)(C^3$, *meta-position*) = 128 ppm, and $\delta(^{13}C)(C^4$, *para-position*) = 126 ppm.

13. (a) In 2,2,3,5-tetramethylhexane, the C and H atoms of the three methyl groups at C^2 are chemically equivalent. The same applies to the C and H atoms of the two methyl groups at C^5. Furthermore, thanks to the free rotation around the single bonds, the two hydrogen atoms at C^4 are also chemically equivalent (and would thus split the ¹H-NMR signal of the hydrogens at C^3 and C^5 into a triplet each, but that's just in passing). One more remark: Please note that C^3 is a chirality center. (Although this is not important here, it should not go unmentioned.) (b) There are also 2 chemically equivalent methyl groups in the 2-methylpropanoic acid-2′-aminoethyl ester: at C^2. Moreover, in the case of the two methylene groups of the 2′-aminoethyl radical, the two hydrogen atoms are each chemically equivalent to one another.

14. Due to the rigid backbone of the system bicyclo[2.2.1]hept-2-ene, the hydrogen atom marked H^a is always closer to the C=C double bond of the six-membered ring, while the hydrogen atom marked H^b points in the other direction. Unfortunately, the classical Lewis notation gives the impression that atoms linked by covalent bonds are actually "quite far apart"; in this case, it would be more informative (albeit perhaps a bit confusing) to depict the molecule using a space-filling model: Then it became obvious that the atom marked H^a actually comes *very* close to the C=C double bond—close enough to enter the sphere of influence of the anisotropy resulting from the double bond. For this reason—i.e., the corresponding secondary magnetic field—this hydrogen atom undergoes minor shielding and is somewhat less easily excited to resonance than H^b: $\delta(H^a) = 1.3$ ppm, $\delta(H^b) = 1.1$ ppm.

15. In molecule (a), whose ring atoms have been numbered consecutively starting from the ethyl substituent in accordance to IUPAC, only two couplings have to be considered: The hydrogen atoms of the methylene group are split into a quartet via a $^3J_{HH}$ coupling with the H atoms of the methyl group (Q, again with a coupling constant of the order of 10 Hz), making the signal of the

methyl group a triplet (T) of the same $^3J_{HH}$ coupling constant. In addition, however, there is also a $^4J_{HH}$ coupling with the H atom attached to the aromatic ring (at C^6). This coupling, with a coupling constant of the order of 5 Hz, leads to the signal of the aromatic H atom (which will therefore be recognisably downshifted) splitting into a triplet (t), while the signal of the two methylene hydrogens as a whole splits into a Qd.

It gets funnier with molecule (b): The methylene and methine groups are of particular importance here. Let us first look at who couples with whom—and which coupling constants have to be considered in each case:

— The two (chemically equivalent) methyl groups couple via a $^3J_{HH}$ coupling with the adjacent methine hydrogen and via a $^4J_{HH}$ coupling with the methylene group.
  – The methine group is split via $^3J_{HH}$ = A to form a septet (sept), with A lying in the order of magnitude of 10 Hz.
  – The methylene group is also split into a sept via the $^4J_{HH}$ coupling. The associated coupling constant $^4J_{HH}$ = B will be of the order of 5 Hz.
— The methine group couples
  – via the (already mentioned) $^3J_{HH}$ coupling with the two equivalent methyl groups, so that their signal is split to a doublet with $^3J_{HH}$ = A.
  – with the methylene group: This $^3J_{HH}$ coupling causes its signal to be split into a doublet. The associated coupling constant $^3J_{HH}$ = C will also be of the order of 10 Hz, but need not be identical with A.
— In the case of the methylene group, the following is noticeable:
  – the (already mentioned) $^3J_{HH}$ coupling with the neighbouring methine group (with $^3J_{HH}$ = C), which splits its signal into a triplet.
  – the $^4J_{HH}$ coupling (also mentioned above) with the two equivalent methyl groups (with $^4J_{HH}$ = B), whose signal is thus split into a triplet.
  – the $^4J_{HH}$ coupling with the H atom on the aromatic system, which splits its signal—just as in molecule (a)—into a triplet with the coupling constant $^4J_{HH}$ = D.
  – The hydrogen atom belonging to the aromatic compound cannot enter into a $^3J_{HH}$ coupling, but it can enter into the $^4J_{HH}$ coupling (as just mentioned) to the methylene group, whose signal is therefore split with $^4J_{HH}$ = D to form a doublet.
  – Those were all the couplings. And now you still have to find out which multiplets result:
  – For the two (chemically equivalent) methyl groups, the result is a doublet (with $^3J_{HH}$ = A) of triplets ($^4J_{HH}$ = B), thus: Dt.
  – The methine group becomes a septet ($^3J_{HH}$ = A) of triplets ($^3J_{HH}$ = C), SeptT for short. But if C > A holds, it would be a triplet of septets, TSept.
  – The signal of the methylene group becomes a doublet ($^3J_{HH}$ = C) of septets ($^4J_{HH}$ = B) of doublets ($^4J_{HH}$ = D). Again, if D > B holds, it would rather be a doublet of doublets of septets.
  – The signal of the aromatic H atom undergoes the splitting into a triplet (with $^4J_{HH}$ = D).
  – Molecule (c) looks more complicated than it is, because we already said that, at least in a first approximation, couplings across hetero-atoms can be neglected. (If one approaches it more exactly, this is not unrestrictedly true, but this text here is about the *basics*.) The signal of the methine group labelled 2 is split into a triplet via the $^3J_{HH}$ coupling with the methylene group at C^3 (with $^3J_{HH}$ = A), and in turn it splits the signal of the methylene group labeled 3 into a doublet (of course with the same coupling constant):

3.5 · Other Usable Cores

- 1: Triplet ($^3J_{HH}$ = A); T.
- 2: Doublet ($^3J_{HH}$ = A); D.
- With that, we are already done on this side of the molecule. On the other side of the carboxyl group, however, a little more happens:
- The methylene group labeled 1' is split into a triplet via $^3J_{HH}$ coupling with the adjacent methine group (2') (with $^3J_{HH}$ = B), and in addition, further splitting of the signal occurs due to $^4J_{HH}$ coupling with the methyl group ($^4J_{HH}$ = C).
- On the one hand, the methylene group marked 2' couples with the already mentioned neighbouring methylene group 1', resulting in a triplet; we already know the associated coupling constant: $^3J_{HH}$ = B. At the same time, there is also a $^3J_{HH}$ coupling with the methyl group, resulting in a quartet with the coupling constant $^3J_{HH}$ = D.
- Finally, the methyl group couples with the adjacent methylene group (triplet with $^3J_{HH}$ = D) and with the 1' methylene group, although we already know this coupling constant as well: $^4J_{HH}$ = C.
- Overall, for the right molecule part:
  - for 1': a triplet ($^3J_{HH}$ = B) of quartets ($^4J_{HH}$ = C), Tq for short.
  - for 2': if B > D, we get a triplet of quartets (TQ); if B < D, we get a quartet of triplets (QT).
  - for the 3' methyl group: a triplet ($^3J_{HH}$ = D) of triplets ($^4J_{HH}$ = C): Tt.

16. (a) shows a doublet of triplets: As the corresponding multiplet tree in subfigure (a) below clearly shows, the coupling constant leading to the doublet is significantly larger than that of the triplet. Shown below is the corresponding molecular section responsible for such a multiplet splitting. (The bonds "going nowhere", i.e. where no binding partners are indicated, mean that the binding partners there are not involved in the coupling, perhaps because it is a heteroatom or similar.) Note the relative signal intensities of the "sub-multiplet": as expected (and in agreement with Pascal's triangle), the relative intensities of the doublet are 1:1, while the triplet has the intensity ratio 1:2:1. (b) represents a triplet of quartets, where—as the multiplet tree in the figure below shows—again the first coupling constant is recognisably larger than the second. Again, the responsible molecular section is shown below. (c) finally, is a bit trickier (if you figured this one out: Kudos!): there is a doublet of triplets of quartets, where the coupling constant of the doublet is only slightly larger than that of the triplet, while the quartet is of a completely different order of magnitude. Try to follow this reasoning using the multiplet tree below (c)!

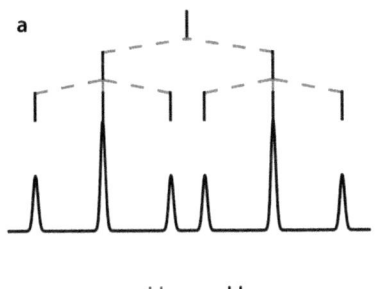
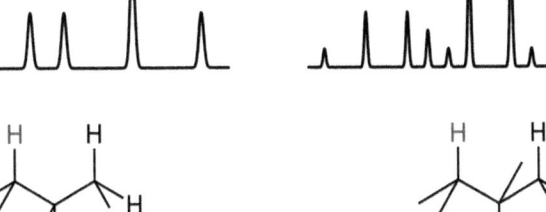
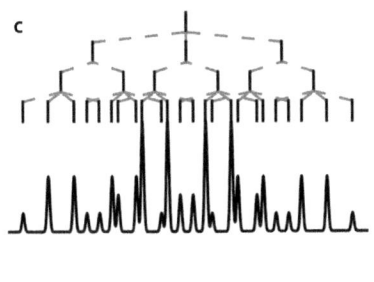

17. The elegant way to solve this problem would be, of course, to take a closer look at the multiplets, which are reasonably well recognisable in the detail magnifications of the spectrum, to find out which signals have the same $^3J_{HH}$ or $^4J_{HH}$ coupling constant. For this, however, you would need much more experience in estimating in which orders of magnitude corresponding couplings actually lie. Nevertheless, this task can be solved, and quite easily at that: Let's first look at what we can say about the spectrum: It consists of five signals, each split into multiplets, whose integrals can be easily determined in the spectrum (by measuring and comparing them with each other):

- $\delta = 5.73$ ppm—a rather obscure multiplet even in magnified view; integral I = 1.
- $\delta = 5.03$ ppm—this multiplet can also not be decoded immediately; I = 1
- $\delta = 3.31$ ppm—one triplet; I = 2
- $\delta = 2.51$ ppm—a quartet which is then split even further—obviously by a long-range coupling: Qd; I = 2
- $\delta = 1.71$ ppm—a doublet that has undergone further splitting—also by via long-range coupling: Dd; I = 3
- The small signal at $\delta = 0.00$ ppm stems from the internal standard TMS.

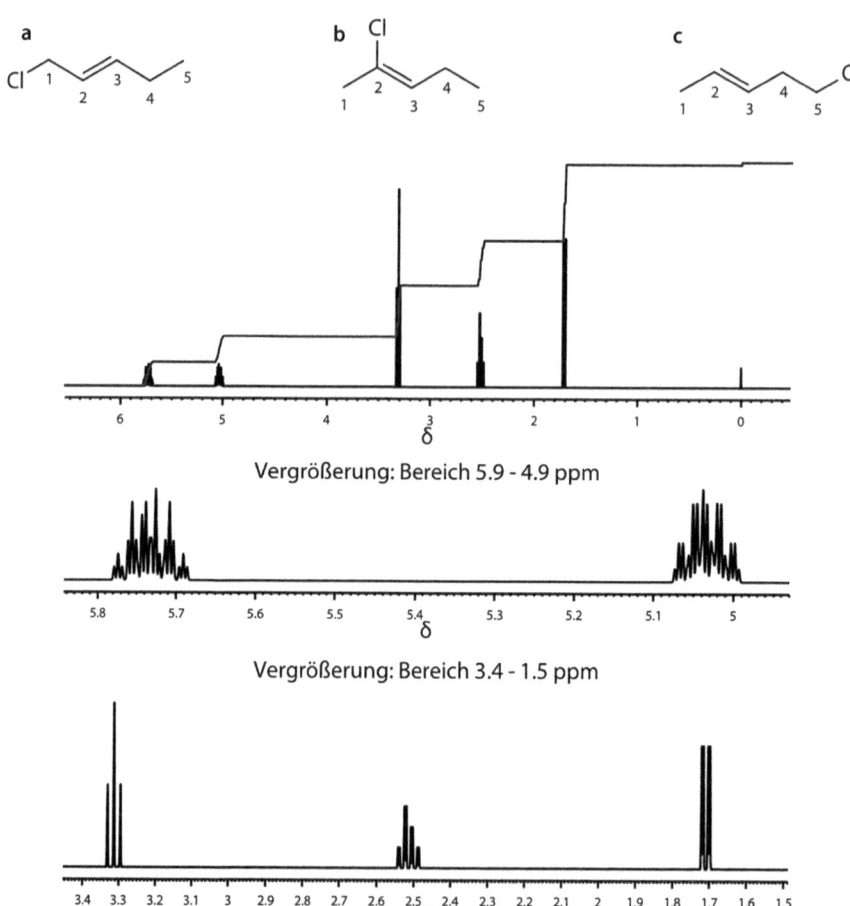

^{1}H-NMR spectrum to be interpreted

And now let's go through what can be predicted about the molecules (a)–(c) and how they differ from each other. For the sake of comparability, the carbon atoms of the chain (and thus also the corresponding hydrogen atoms) are arbitrarily

3.5 · Other Usable Cores

numbered from left to right, as in the figure above—we accept that this numbering is not IUPAC-compliant for molecule (c), the 1-chloro-3-pentene (after all, according to the nomenclature rules, the chlorine-bearing carbon is assigned the "1"):

— In 1-chloro-pent-2-ene (a) there are five chemically non-equivalent hydrogen atoms or groups, accordingly five signals can be expected in the spectrum.
  – the methylene group of C^1 (whose signal in the ^{1}H-NMR would have to have an integral of 2 and would be shifted to lower field due to the high electronegativity of chlorine),
  – the methine group of C^2 (integral: 1),
  – the methine group of C^3 (integral: also 1),
  – the methylene group of C^4 (integral: 2; compared to the methylene group of C^1, this signal should be upshifted) and
  – the methyl group of C^5 (integral: 3).
— In 2-chloro-2-pentene (b) there are:
  – the methyl group of C^1 (integral: 3)
  – the methine group of C^3 (integral: 1),
  – the methylene group of C^4 (integral: 2) and
  – the methyl group of C^5 (integral: 3)
  – Since no hydrogen is bonded to the carbon atom marked "2", there are only *four* chemically non-equivalent hydrogens, so this compound can already be ruled out: The present spectrum *cannot possibly* belong to this compound.
— That leaves the 1-chloro-3-pentene (c). There is:
  – the methyl group of C^1 (integral: 3),
  – the methine group of C^2 (integral: 1),
  – the methine group of C^3 (integral: 1),
  – the methylene group of C^4 (integral: 2) and
  – the methylene group of C^5 (integral: 2). Because of the higher electronegativity of chlorine, this signal would be expected to be shifted to lower field compared to the signal of the methylene group of C^4.
  Let us now consider what multiplet would be expected for each of these signals: As usual, we assume $^3J_{HH}$ and $^4J_{HH}$ couplings, using uppercase letters for the former and lowercase letters for the latter to describe the resulting multiplet. If there is more than one coupling of the same magnitude, the order in which they are mentioned in the table is meaningless for the time being. Moreover, we already know that the signals belonging to the hydrogens bonded to the sp^2 carbons (to C^2 and C^3) must be downshifted because of the anisotropy effect of the C=C double bond.

◘ Expected multiplets for the ^{1}H-NMR signals of compounds (a)–(c)

	(a)	(b)	(c)
H to C^1	Dda (I = 2)	Sd (I = 3)	Dd (I = 3)
H to C^2	TDt (I = 1)	–	QDt (I = 1)
H to C^3	DTq (I = 1)	Tq (I = 1)	DTqt (I = 1)
H to C^4	QDd (I = 2)	QD - none $^4J_{HH}$ (I = 2)	DTd (I = 2)
H to C^5	Td (I = 3)	Td (I = 3)	Tda (I = 2)

[a]Shift of the signal to the lower field due to the electronegativity of the chlorine

In order to decide whether our analyte is (a) or (c), we should look for an easily recognisable signal in the spectrum—the one with the largest integral: δ = 1.71 ppm; I = 3. This is undoubtedly a methyl group. This signal is split into a doublet of doublets—which is exactly what was expected for compound (c), while the signal of the methyl group of compound (a) would be split into a triplet (or a triplet of doublets). Thus, the analyte must be 1-chloro-3-pentene. Try to unambiguously assign the other four signals via the present (or expected) multiplets! (Two hints for this: Firstly, not every theoretically expected long-distance coupling ($^4J_{HH}$) will necessarily be recognisable in the spectrum, and secondly, you should bear in mind that the resulting chemical shift is not *exclusively* determined by the electronegativity of any binding partners.)

To, here are the ^{1}H-NMR spectra of the other two compounds: (a) shows the ^{1}H-NMR spectrum of 1-chloro-2-pentene, (b) shows the ^{1}H-NMR spectrum of 2-chloro-2-pentene.

Antworten

a

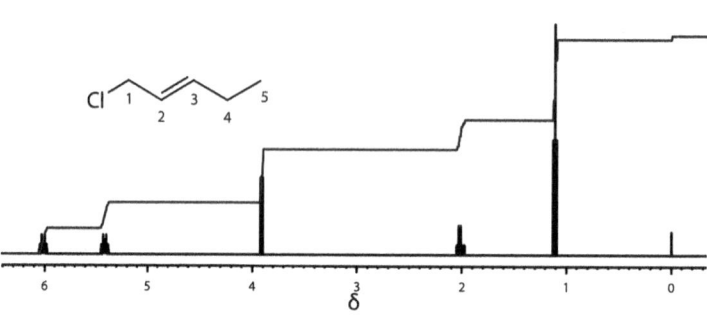

Vergrößerung: Bereich 6.1 - 5.3 ppm

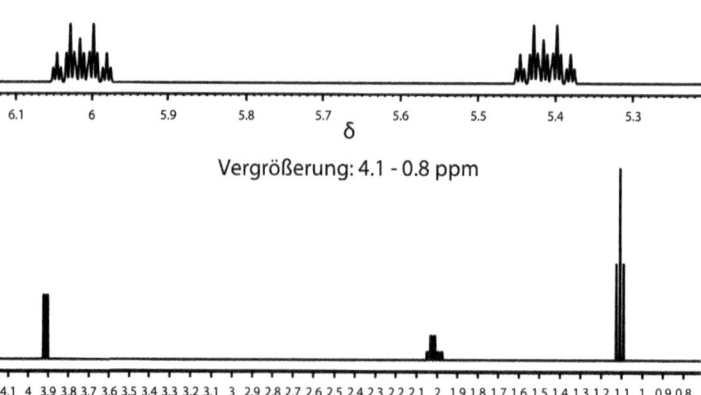

Vergrößerung: 4.1 - 0.8 ppm

3.5 · Other Usable Cores

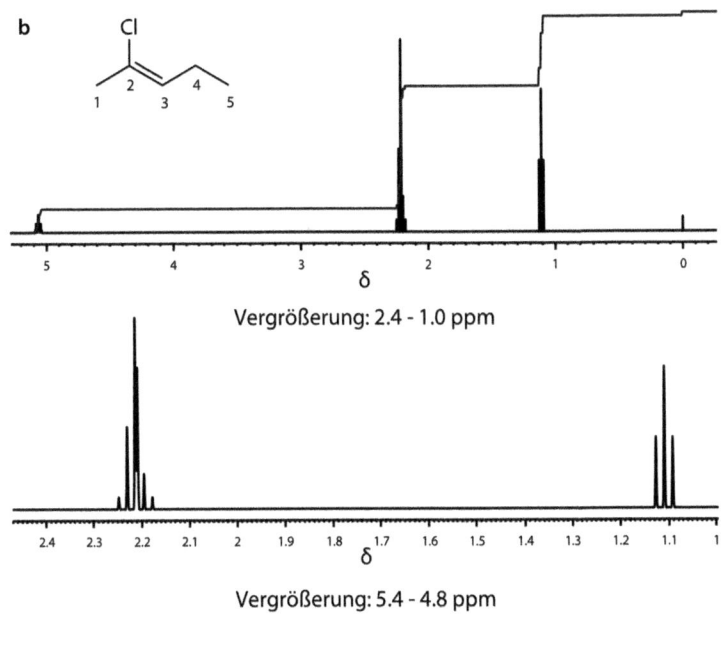

Vergrößerung: 2.4 - 1.0 ppm

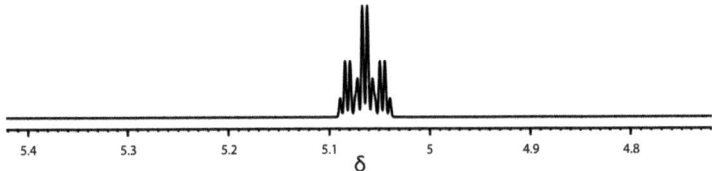

Vergrößerung: 5.4 - 4.8 ppm

Again, try to assign the signals based on their chemical shift and the multiplet obtained.

18. Again, we can simplify our lives immensely: Assuming that carbon always satisfies the octet rule (which it almost always does) and that chlorine only forms a single bond (which is commonly not the case only after reaction with strong oxidising agents), the sum formula C_4H_9Cl can belong to only four possible compounds:
    - 1-chlorobutane (a)
    - 2-chlorobutane (b)
    - 1-chloro-2-metylpropane (c) or
    - 2-chloro-2-methylpropane (d)

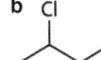

Let us first look at the ^{13}C-NMR spectrum—so that you do not have to constantly flip (or scroll) back and forth, it is given here as (a) once again: There are four signals (please disregard the small TMS signal at $\delta = 0.00$ ppm). This already eliminates compounds (c) and (d), because the two (c) and three (d) methyl groups of these compounds, respectively, are chemically equivalent and therefore yield only one signal. (Try to assign the four signals to the respective C-atoms!)

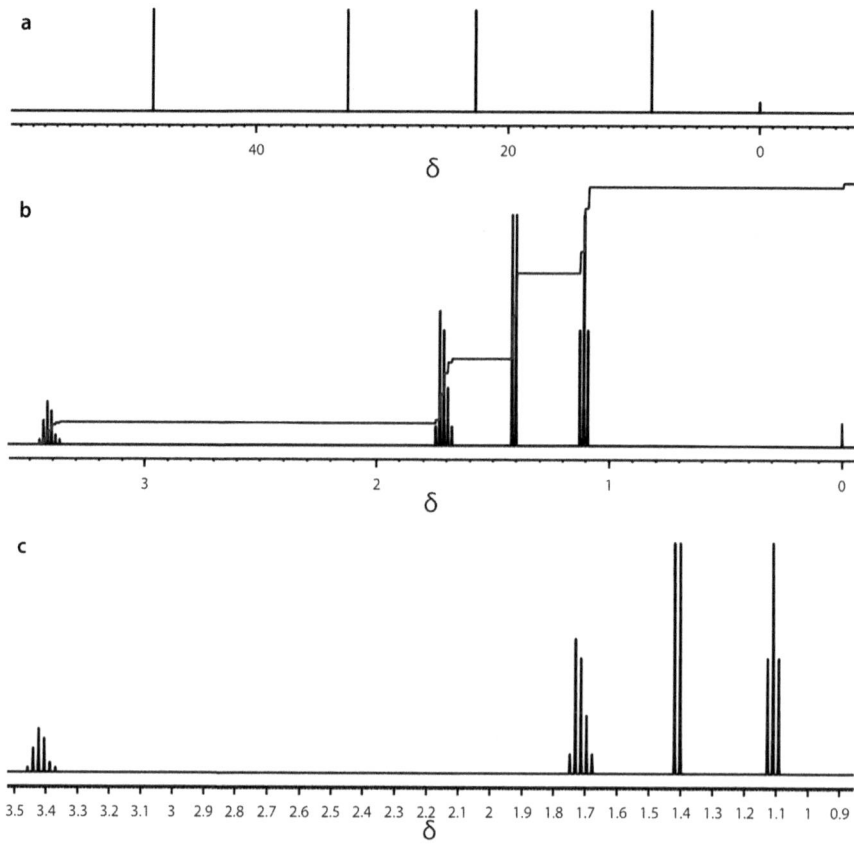

^{13}C and ^{1}H spectra of C_4H_9Cl

Turning now to the ^{1}H-NMR spectrum (b), we see that, according to the integrals, there are two chemically non-equivalent methyl groups, leaving only compound (b) to be considered. This finding is confirmed by the signal at $\delta = 3.41$ ppm, whose integral is only I = 1: Such a signal is incompatible with the structure of 1-chlorobutane. But we can cite even more evidence: The signal from the methyl group of compound (a), which is closer to the chlorine atom (and therefore is slightly deshielded and thus downshifted), must be split into a doublet by the $^3J_{HH}$ coupling with the methine hydrogen at carbon-2—and this is exactly what applies for the signal at $\delta = 1,41$ ppm (as can be seen more clearly in (c), the spectra magnification), while the other methyl group at $\delta = 1.11$ ppm is split into a triplet by the adjacent methylene group, as expected. (Also try to assign the other ^{1}H signals and explain the multiplets present!)

Summary
- **Mass Spectrometry (MS)**

In mass spectrometry, individual analyte molecules are fragmented by interaction with an electron beam or with strongly excited atoms/molecules in the presence of a magnetic field, and the fragments obtained are separated on the basis of their mass-to-charge ratio (m/z). (Depending on the type of ionisation, multiply charged fragments may also be produced.) Which bonds of the respective analytes are preferentially cleaved depends on the type of bonds and, to a large extent, also on any heteroatoms (i.e. everything except carbon and hydrogen) present. The question of whether the respective intramolecular conditions permit rearrangements also plays a major role.

Since mass spectrometry actually detects individual molecule fragments, the individual fragments differ recognisably in their mass/charge ratio if different isotopes are present. The different natural abundance of the individual isotopes usually leads to characteristic isotope patterns, which—provided a little practice—allow certain statements about the analyte at first glance.

- **Nuclear Magnetic Resonance (NMR) Spectroscopy**

In nuclear magnetic resonance spectroscopy, suitable atomic nuclei (especially ^1H and ^{13}C, but certain isotopes of some other atoms qualify for this type of analysis) are excited to precession by resonance with low-energy electromagnetic radiation in a strong magnetic field. Which wavelength range (from the radio waves spectrum) is suitable for this in each case depends on various factors. Particularly important are:

— The shielding of the atomic nuclei by increased (or de-shielding of the atomic nuclei by decreased) electron density (here especially hetero-atoms play an important role). Shielded nuclei can be excited by less energetic radiation (or in a weaker magnetic field); in the NMR spectra, the signals of the corresponding nuclei are downshifted (to be found more on the left-hand side of the spectrum), while nuclei shielded by increased electron density lead to a upshifted signal.
— Anisotropy effects, which can induce a secondary magnetic field and in turn lead to de-shielding or shielding.

In both cases, it should be noted that the different shift values are given relative to a reference substance; in ^1H- and ^{13}C-NMR, tetramethylsilane (TMS, $(CH_3)_4Si$) is used in each case; for other nuclei (i.e. in the hetero-nuclear NMR), other reference substances are utilised.

All NMR experiments have in common that atomic nuclei that are spatially sufficiently close to each other can couple with each other, which leads to the splitting of the respective signals into multi-line systems, where the number of resulting lines of a signal depends on the number of nuclei coupling with the atom under consideration. Such multiplets exhibit characteristic relative signal intensities/heights (they follow Pascal's triangle); the relative difference of the individual shift values depends on the respective coupling constants, whereby evaluation of the spectra is immensely simplified by the fact that nuclei coupling with each other inevitably show exactly the same coupling constant.

If the atomic nuclei involved in coupling are chemically equivalent, they lead to simple multiplets; chemically non-equivalent nuclei, on the other hand, split the signal of the respective atomic nucleus under consideration into multiplets of multiplets, whereby one traditionally states the multiplet with the larger coupling constant first in each case.

Several different NMR experiments can be combined to form multidimensional NMR spectra. These then allow—on the basis of coupling constants and/or due to other intramolecular interactions—much more detailed statements about the connectivity/the molecular framework or also—in even more complex experiments—about bond distances or bond angles and the like.

Further Reading

Auch hier gehen manche der erwähnten Werke hinsichtlich ausgewählter Gebiete der Analytik weit über "den Skoog" und "den Harris" hinaus; gerade beim Thema "Kernresonanzspektroskopie" sei ausdrücklich auf das Werk von Breitmaier und auf den Hesse – Meier – Zeeh verwiesen.

Bienz S, Bigler L, Fox T, Meier H (2016) Hesse – Meier – Zeeh: Spektroskopische Methoden in der organischen Chemie. Thieme, Stuttgart

Binnewies M, Jäckel M, Willner, H, Rayner-Canham, G (2016) Allgemeine und Anorganische Chemie. Springer, Heidelberg

Breitmaier E, Vom NMR-Spektrum zur Strukturformel organischer Verbindungen (2005) Wiley-VCH, Weinheim

Breitmaier E, Jung G (2012) Organische Chemie. Thieme, Stuttgart

Budzikiewicz H, Schäfer M (2013) Massenspektrometrie: Eine Einführung. VCH Weinheim

Cammann K (2010) Instrumentelle Analytische Chemie. Spektrum, Heidelberg

Christen HR, Vögtle F (1988) Organische Chemie – Von den Grundlagen zur Forschung Band I. Salle, Frankfurt

Hage DS, Carr JD (2011) Analytical Chemistry and Quantitative Analysis. Prentice Hall, Boston

Harris DC (2014) Lehrbuch der Quantitativen Analyse. Springer, Heidelberg

Lambert JB, Gronert S, Shurvell HF, Lightner DA (2012) Spektroskopie. Pearson, München

Lottspeich F, Engels JW (2012) Bioanalytik. Springer, Heidelberg

Ortanderl S, Ritgen U (2019) Chemie – das Lehrbuch für Dummies. Wiley, Weinheim

Ortanderl S, Ritgen U (2015) Chemielexikon kompakt für Dummies. Wiley, Weinheim

Reichenbächer M, Popp J (2007) Strukturanalytik organischer und anorganischer Verbindungen. Teubner, Wiesbaden

Schröder B, Rudolph J (1985) Physikalische Methoden in der Chemie. VCH, Weinheim

Schwister K (2010) Taschenbuch der Chemie. Hanser, München

Skoog DA, Holler FJ, Crouch SR (2013) Instrumentelle Analytik. Springer, Heidelberg

Skrabal PM (2015) Spektroskopie. VDF, Zürich

Williams D, Fleming I (2008) Spectroscopic methods in organic chemistry. McGraw-Hill, Berkshire

Electroanalytical Methods

Requirements

The title of this part, "Electroanalytical methods", already reveals where the focus lies this time: This part of the book deals primarily with electrochemical processes. Accordingly, you should be familiar with the associated basics:
- Redox reactions
- Oxidation numbers
- Oxidising/reducing agent
- Voltage series
- Normal hydrogen electrode
- Standard potentials
- Nernst equation
- Electric charge
- Voltage
- Current intensity
- Current flow

The last points of this (certainly not complete) list already show you: Once again, the boundaries between chemistry and physics are blurring; it cannot hurt to keep the teaching material of any introductory courses on the subject of physical phenomena and their description handy.

In addition, there are the "general principles" of chemistry, for example
- the law of mass action
- acid/base reactions
- the pH-value
- esterifications

and the like.

Once again, in this part, various aspects that you actually already know are linked in a new way, so that new possibilities arise to find out something or the other about the analyte.

Learning Objectives

In "electroanalytical methods", the analytes to be quantified are modified electrochemically. The electrochemical reactions cause measurable changes in chemical potentials or in the current flow resulting from potential differences.

This part of the book is intended to present to you a selection of methods that are of immense importance in routine and/or trace analysis today. You will learn to understand the associated (electro-)chemical processes as well as the resulting measurement methodologies, so that the technical aspects of the respective experimental setups are also addressed, so it becomes not only understandable what is actually measured in each case, but also how it can be technically carried out and which (technical) limits are set for one or the other method. Differences between ideal conditions and the real conditions that prevail in each case—and also how these differences have an effect—should also be taken into account.

The chemical basics of the processes dealt with in this part are certainly familiar to you from General and Inorganic Chemistry (they are, however, briefly repeated again); thus, we again deal with a topic in which there is "actually" nothing new to learn, but in which, once again, the interaction of various topics dealt with separately in previous teaching units is important:

- In the different variants of potentiometry, we will refer to Nernst's equation again and again; this is of indispensable importance for the potentials of the analyte systems as well as for the potentials of the reference systems used in the measurements. In order to understand how ion-selective electrodes work, recourse to solid-state chemistry is indispensable, and you may already be familiar with diffusion phenomena from Part III of "Analytical Chemistry I":
- The fundamentals of redox reactions dominate this entire section; in coulometry, they are supplemented by acid/base and esterification reactions.
- Redox processes, in particular electrolysis, are also important in the topics of amperometry and voltammetry; recourse to certain aspects of physics is also necessary here—Faraday's laws in particular play a major role here.

Only the interaction of these different parts of chemistry (or better: of the natural sciences in general) allows not only to apply the techniques considered in this part, but also to understand them.

Contents

Chapter 4   General Aspects – 81

Chapter 5   Potentiometry – 93

Chapter 6   Coulometry – 113

Chapter 7   Amperometry – 117

Chapter 8   Voltammetry – 121

General Aspects

Skoog, Section 22.6: Electroanalytical methods

This part is about the applications (and thus the applicability) of electrochemistry, the basics of which you have already learned in introductory courses or textbooks on general and inorganic chemistry (which were then probably deepened in physical chemistry). An overview of the different electroanalytical methods is given in ▶ Skoog; the *absolute* basics are briefly (!) summarised here.

Harris, Section 13.1: Basic Concepts (Fundamentals of Electrochemistry)

Electrochemistry is about the transfer of electrons from one reactant to another. Some technical terms, which you are certainly already familiar with, should be explicitly mentioned here once again:

- The reactant that gives up the electron (or electrons) is **oxidised;** in the process, the **oxidation number** of the respective atom **increases. Oxidation** takes place here.
- The reactant that accepts the electron(s) is **reduced;** the *oxidation number* of the respective atom is **decreased** in the process. In this partial step of the reaction, the **reduction** takes place.
- The *electric charge* is given in coulombs (C).
    - A single electron has the charge 1.602×10^{-19} C.
    - The charge of one mole of electrons is then calculated to be $(1.602 \times 10^{-19} \text{ C}) (6.022 \times 10^{23}) = 9.649 \times 10^4$ C.
    - This value, the charge of 1 mole of electrons, is commonly called **Faraday constant (F).** It appears again and again in the topic of electrochemistry, usually stated with the value **96,485 C/mol**.

If a given amount of charge flows through an electric circuit, the resulting *current intensity* can be specified. The associated unit is the ampere (A), where: 1 ampere = 1 coulomb per second.

> **Roman or Arabic?**
> In older language textbooks, oxidation numbers were given exclusively with *Roman* numerals; in complex chemistry, this is often still done today, and in the nomenclature of compounds like iron(II) oxide (FeO) and iron(III) oxide (Fe_2O_3), Roman numerals are still indispensable. In the last years, however, IUPAC has advocated the use of *Arabic numerals,* as these are less likely to lead to confusion. (IV and VI are clearly easier to confuse than 4 and 6.) Nevertheless, it is not "wrong" to decide to use Roman numerals (for whatever reason). However, you should *know both* variants: Some textbooks ask you to opt for one, others for the other.
> Please note that there are two basic differences between the oxidation numbers and the charge numbers:
> - With oxidation numbers, you *first* state the *charge sign,* then the *number*. For example, the calcium atom of the Ca^{2+} cation (with the charge 2+) has the oxidation number +2 (or +II, if you prefer).
> - In contrast to monovalent charges, where one limits oneself to charge signs (one writes Na^+, not Na^{1+}), the 1 is also to be mentioned for oxidation numbers: The sodium cation Na^+ thus has the oxidation number +1 (or +I).

Just as an acid/base reaction—in which a transfer of hydrogen cations occurs—can only take place if, in addition to the substance that can *give up the* corresponding H^+ ion, there is also a reaction partner that then *accepts* this hydrogen cation*,* a redox reaction can also only take place if, parallel to the *oxidation* (i.e.: electron donation), a *reduction* (accepting an electron) takes place. The electrons must never "just disappear into the void" (or "emerge from the void").

Chapter 4 · General Aspects

There are further parallels between acid/base and redox reactions:

(a) In acid/base reactions, the pK_A value tells us how eager an acid is to donate one proton. (If it is a polyprotic acid, there is an individual pK_A value for each deprotonation step; for the sake of this simile, let's stick to monoprotic acids.) Similarly, the pK_B value describes a compound's willingness to accept one proton.
(b) In redox reactions, the effort to donate (or accept) one (or more) electron(s) is described by the **standard potential.**
(c) In acid/base reactions, we have an acid and its conjugate base (HA/A^-) or a base and its conjugate acid (B/BH^+).
(d) In redox reactions, we have redox couples: the oxidised form and the reduced form, often represented $Me^{x+} + n\ e^-/Me^{x-n+}$. (The fact that Me here stands for "metal" does not mean non-metals or semi-metals could not be involved in redox reactions, as well.

■ The Standard Potential

For the standard potential (E^0), the following applies:

(a) Substances with a *negative* potential ($E^0 < 0$) are called *base (metals)*; they react with 1-molar hydrochloric acid (pH = 0) to release elemental hydrogen:

$$M + xH^+ \rightarrow M^{x+} + \frac{x}{2} H_2 \uparrow$$

(b) For substances with a *positive* potential ($E^0 > 0$) this reaction does not occur; they often are called "noble".
(c) Please note that "noble" is a relative term: When comparing standard potentials of two different redox couples, the one with the higher potential is said to be "more noble" than the other.

The reference for the standard potential is the **normal hydrogen electrode (NHE),** which you certainly already know from General Chemistry (or maybe Part II of "Analytical Chemistry I"?). The condition that pH = 0 applies also comes from this experimental setup. Accordingly, the standard potentials of all substances are "standardised" to this reference value; they form the **electrochemical series.**

> According to the electrochemical series, a *more noble* substance present in *oxidised* form will oxidise a *less noble* substance present in *reduced* form (and itself be *reduced* in the process).

▶ Example

A redox reaction occurs when you dip a rod of elemental zinc into a solution of a copper(II) salt: The (less noble) zinc is oxidised to form zinc(II) ions, while the electrons given off in the process reduce the copper(II) ions to elemental copper:

OX: $\quad Zn \rightarrow Zn^{2+} + 2e^-$

RED: $\quad Cu^{2+} + 2e^- \rightarrow Cu$

The overall equation of this reaction is then:

$\quad Zn + Cu^{2+} \rightarrow Zn^{2+} + Cu \quad$ ◀

If the corresponding experiments are not carried out as a "short-circuited battery" (you already know this from Part II of "Analytical Chemistry I"), but as a **galvanic element with half-**cells that are separated from each other but conductively connected—each containing the oxidised form of one or the other substance (as a dissolved salt) *and* the reduced form (as an elementary metal) -, the electron flow can be measured as a voltage: It results from the *potential difference of* the two half-cells. This experiment is shown graphically in ▶ Harris at the end of Sect. 13.2 (unnumbered figure); the fundamentals of galvanic cells per se are also summarised in this section.

Harris, Section 13.2: Galvanic cells

■ The Galvani Voltage

The decisive point here is that you *cannot measure the potential of a single half-cell directly*. It is quite true that a potential *is formed* within each of the two half-cells, but it is only verifiable (i.e.: measurable) in interaction with the other half-cell.

The potential that is formed comes from the **Galvani voltage**, resulting from the *internal electrical potential* of the **electrode** (the solid conductor, in our example the elementary metal) and the internal electrical potential of the **electrolyte** (the substance that is present in the form of dissolved ions). We call it the *inner* potential because we refer to the potential *within the respective phase*, not about the potential that arises at the interface to another phase: If, for example, the zinc rod from our example were to be dipped into the corresponding (zinc ion) solution, a certain (very limited) number of individual zinc atoms would go into solution as ions (Zn^{2+}), while the (former) valence electrons of this atom would remain in the zinc rod. Accordingly, an electrical potential already forms between the positive charge carriers and the now slightly negatively charged metal rod. This potential, for the time being, prevents further dissolution of the zinc rod: An equilibrium is established and a (precisely defined, but *not measurable*) potential is obtained. (It is not measurable because in any conceivable experimental arrangement new phases and phase boundaries would arise—for example between a measuring electrode and the zinc solution—which would then lead to *new* potential differences). But it is precisely this excess of electrons in the slightly negatively charged zinc electrode that ultimately leads to these electrons migrating to the easily reducible copper(II) ions, which causes the actual electrical current flow in this galvanic element.

■ Concentration Dependence of Potentials

From General Chemistry (or from Part II of "Analytical Chemistry I", perhaps?) you know that standard potentials are not only based on standardised reaction conditions, but also on standardised concentrations (1.00 mol/L). If other concentrations are involved, there are two things to consider:

Harris, Section 13.4: The Nernst equation

1. On the one hand, the potential of the half-cells in question can be determined using the Nernst equation, which of course you already know from General Chemistry (or from Part II of "Analytical Chemistry I"). For the sake of completeness, it is mentioned here once again:

$$E = E^0 + \frac{RT}{zF} \cdot \lg \frac{[Ox]}{[Red]} \tag{4.1}$$

In this general notation, R is the gas constant (with the value 8314 J · mol^{-1} · K^{-1}), T the absolute temperature (in K). Furthermore, *z* stands for the number of electrons exchanged in the course of the associated redox reaction, and F stands for the Faraday constant (F = 96,485 C/mol).

The concentration dependence is shown in the term to be logarithmised: [Ox] stands for the concentration of the *oxidised* form of the redox couple

Chapter 4 · General Aspects

in question, and [Red] for the concentration of the *reduced* form. (It would be better to use the formulation "of the *not so highly* oxidised form", after all, this equation is also suitable for determining the potential of a half-cell in which *both* reaction partners are already present in a [more or less strongly] oxidised form).

> An example would be a redox system in which iron(II) ions are oxidised to iron(III) ions: Then [Ox] = [Fe^{3+}] and [Red] = [Fe^{2+}]. Compared to elemental iron, these iron(II) ions of course already represent iron in an oxidised form.

❯ The reduced form thus does not have to be elementary; it just needs to be not as highly oxidised as the form described as "oxidised".

Under standard conditions, T = 298 K, so the RT term represents a *constant*, and that F is a *constant* was mentioned (again) in the paragraph directly above. If one then takes into account the conversion factor from the natural to the decadic logarithm (ln x = 2.303 × lg x, so once again we deal with a *constant*), the result is the much more familiar (and easier to handle) form of Nernst's equation:

$$E = E^0 + \frac{0{,}059 \text{ V}}{z} \cdot \lg\frac{[\text{Ox}]}{[\text{Red}]} \quad (4.2)$$

Equations (4.1) and (4.2) also show that it is possible to construct galvanic cells which contain the same electrodes and electrolytes, i.e. which do not differ in their E^0 values, but in their concentrations, leading to differing actual potentials which can be calculated with the aid of these equations: Such a galvanic cell is known as a **concentration chain**.

The concentration dependence of electroanalytical cells and also the potentials that arise—also as a function of concentration—are summarised again briefly in ▶ Skoog.

Skoog, Section 22.2: Potentials in electroanalytical cells

Skoog, Section 22.3: Electrode potentials

■ **Concentration and Activity**

We generally assume that all the ions present in a solution behave *ideally*, i.e. that they do not interact with each other in any way. Strictly speaking, however, this assumption is only valid if the concentration of the solution in question is sufficiently small that the individual ions can actually be regarded as *isolated from one another*. The reason for this is that oppositely charged particles attract each other, i.e. several anions will group around each cation in solution and vice versa: an *ionic atmosphere* is formed (graphically represented in ▶ Harris in Fig. 7.2) which leads to a changed **ionic strength**. (We will return to ionic strength in ▶ Sect. 5.1.1.)

Harris, Section 7.1: The influence of ionic strength on the solubility of salts

If the concentration of the dissolved ions is too high, the interactions of the particles lead to a (slightly or also massively) reduced "reactivity", so that one must consider these ions' **activity**, because here an *ideal solution is no longer present*.

The relationship between concentration and activity can be quantified using the *concentration-dependent (!)* **activity coefficient γ,** which must be determined experimentally for each ion (and also for the concentration present in each case); that is is why it is usually stated as γ_X for an ion X; the notation γ(X) can also be found:

$$A_X = \gamma_X \cdot c(X) \quad \text{or} \quad A_X = \gamma_X [X] \quad (4.3)$$

Harris, Section 7.2: Activity coefficients

Skoog, Appendix B (activity coefficients)

Very dilute solutions, in which (almost) no interactions of the dissolved particles occur, behave almost ideally, i.e. here the activity coefficient approaches the value $\gamma = 1$, but with increasing concentrations it can drop even quite drastically—this effect also depends on the *charge density* of the ion under consideration. The relationship between activity and ionic strength is shown abstractly in Fig. 7.4 from ▶ Harris for ions of different charge with the same ionic radius; ▶ Skoog compares various concrete monovalent and polyvalent ions in this respect in Fig. B.1.

> The exact point at which one can speak of a "very dilute" solution is not clearly defined; different textbooks contain correspondingly deviating key figures. In the context of this part, we consider solutions to be "very dilute" when their concentration does not exceed the value 10^{-3} mol/L. But this value is by no means "set in stone".

■ **The Law of Mass Action, Viewed in a More Physico-chemical Way**

When it comes to activity, the units suddenly do not seem quite coherent: Activity is in fact, as surprising as it may seem at first glance, *dimensionless*. The reason for this is that the corresponding calculations actually use the *molar ratios of* the substances X under consideration to the amount of substance of *all* components present. This may also be surprising, but please think back to the considerations on the pH value of aqueous solutions of acids and bases from General Chemistry, so that the underlying logic becomes clear. It was assumed that the amount of solvent (in this case: water) is not changed significantly, even if one or the other water molecule is lost (or: "consumed") in the reaction.

$$HA + H_2O \rightarrow H_3O^+ + A^-$$

Accordingly, a constant amount of substance of the water—$n(H_2O)$—is assumed, and thus all "concentration data", from a physico-chemical point of view, are simply *ratios of amounts of substance* or proportions of amounts of substance x_i, as we already had them in Part I of "Analytical Chemistry I" (and know from DIN 1310):

$$x_i = \frac{n_i}{n_{tot}} \tag{4.4}$$

However, such mass fractions (logically) always have the unit mol/mol, i.e. they are actually dimensionless.

– For *non-ideal solutions,* where the respective activity coefficient of each component is $\gamma_i < 1$, the "ordinary" law of mass action for the general reaction

$$aA + bB \rightleftarrows cC + dD$$

then must be reformulated accordingly on the basis of Eqs. (4.3) and (4.4):

$$K = \frac{c(C)^c \cdot c(D)^d}{c(A)^a \cdot c(B)^b} \text{ then becomes } K_a = \frac{\gamma_C \cdot x_C^c \cdot \gamma_D \cdot x_D^d}{\gamma_A \cdot x_A^a \cdot \gamma_B \cdot x_B^b} = \frac{a_C^c \cdot a_D^d}{a_A^a \cdot a_B^b} \tag{4.5}$$

(The subscript "a" of the equilibrium constant K_a tells us right away that we are not working with concentrations here, but with *activities* and thus with *proportions of substances*. This is also referred to as the *thermodynamic equilibrium constant*; for more information, refer to books or lectures on *Fundamentals of Physical Chemistry*).

- For ideal solutions—with $\gamma_i \cong 1$—the whole thing simplifies considerably:

Here, one just approximates: $K_c = \dfrac{c(C)^c \cdot c(D)^d}{c(A)^a \cdot c(B)^b} \approx K_a$ (4.6)

So in the case of ideal solutions, one may, if one wishes, also work with the "correct concentrations", but then, depending on the stoichiometric factors, one should not be bothered by supposedly nonsensical units.

Of course, there are also corresponding tables on this subject. However, since the activity coefficient also depends on the selected reaction conditions (and is *concentration-dependent*—this cannot be emphasised strongly enough!), *standard values* (under standard conditions) are usually tabulated. (The theory behind this and further information about the handling of activity coefficients can be found again in ▶ Skoog and ▶ Harris.)

Skoog, Appendix B (activity coefficients)

Harris, Section 7.2: Activity coefficients

> **If You Want to Be More Precise**
> Strictly speaking, one would *always and everywhere* have to use the activity where the concentration plays a role—especially when the concentration is too large to speak of an "ideal solution". Of course, this also means that—just as strictly speaking—whenever we are dealing with the law of mass action (LMA), we would not have to use the *concentrations* of the atoms/molecules/ions involved, but the corresponding *activities*. Also, when dealing with the Nernst equation, we would always have to replace the term "concentration" with "activity". For the sake of simplicity, we will refrain from doing this and continue with the usual "concentrations" (except when it is really important to emphasise the activity again specifically).
> However, two things should be noted about the law of mass action:
> - When using Eq. (4.5) (for non-ideal solutions) and Eq. (4.6) (for ideal solutions), of course the concentrations that are *present at equilibrium* are to be used, not the initial concentrations. (The latter would not make any sense act all, because at the beginning of any reaction no products have formed yet, so what would you want to put into the numerator?).
>   *In other words:* When equilibrium has been reached, the quotient of the (mathematical) product of the concentrations/activities of the products (if necessary with their stoichiometric factors, which enter the calculation as exponents) and the (mathematical) product of the concentrations of the reactants (for which any stoichiometric factors also enter as exponents) always results in a *constant value* for the respective equilibrium under consideration.
> - The term *law of mass action* may be surprising, because it raises the question of how the respective *masses* of the individual components (reactants and products) actually affect each other. In fact, this designation has purely historical reasons: When the concept of the law of mass action was first published in 1867, the technical term for what was later called "activity" was *"active mass"*. The LMA does not actually have much to do with the actual mass of the individual constituents of the equilibrium, except that, according to the well-known formula n = m/M, there is of course a direct relationship between amount of substance and mass.

For those who want it *very* precisely
Even though the two notations—c(X) and [X]—are almost universally regarded as synonymous, there are many lecturers (and also some textbooks) who strictly reject this (especially if they deal with *physical chemistry*). There, [X] stands *exclusively* for a *concentration present at equilibrium*. Now, when dealing with equilibrium constants, it is usually just those equilibrium concentrations that are meant, but there are certainly situations in which one wants to describe systems in which equilibrium has *not yet* been established. In this case—especially if the lecturer puts great stress on it—one should refrain from using the square-brackets shorthand notation.

The Law of Mass Action and the Units— Again, Because This Is Prone to Cause Problems
In the vast majority of cases, you have considered solutions and thus also their concentrations in the law of mass action. Accordingly, the stoichiometric factors a, b, c and d always result in purely mathematical terms as the unit for the LMA:

$$\left(\frac{mol}{L}\right)^{(c+d)-(a+b)}$$

In *gas-phase reactions,* where gaseous reactants and products are to be considered, instead of the concentrations of the individual components i, one uses their respective pressures p_i in the equilibrium state (with a corresponding unit for the pressure, i.e., for example, mbar, Pa, or torr). Let us again consider the general reaction equation:

$$aA_{(g)} + bB_{(g)} \rightleftarrows cC_{(g)} + dD_{(g)}$$

Of course, a law of mass action can also be established for this, but statements about "concentrations" would be a bit out of place when dealing with gases. That is why it is even easier to understand why one quickly arrives at a *dimensionless* final result for reactions that take place in the gas phase:

What is being considered here is not the *concentration* of the individual components (c_i), but the **partial pressures pi** (i.e. the pressures each of them individually contribute to the total pressure)—and for each of the respective components (i), those are nothing other than their respective mass fraction x_i, multiplied by the total pressure prevailing in the system:

$$p_i = \frac{n_i}{n_{tot}} \cdot p_{tot} = x_i \cdot p_{tot} \qquad (4.7)$$

This clearly again involves the substance amount fraction x_i from DIN 1310 (and Eq. 4.4) with the unit mol/mol—already mentioned above: They are dimensionless, thus we are dealing with a *pressure ratio.*

Accordingly, the expression for the associated law of mass action for general gas phase reactions can be summarised as:

$$K_p = \frac{p(C)^c \cdot p(D)^d}{p(A)^a \cdot p(B)^b} \qquad (4.8)$$

K_p then has a "pressure unit" only if the sums of the exponents in the numerator and in the denominator differ.

> Please keep in mind that, in the end, statements about c as well as about p always represent statements about a "number of particles per volume".

An Overview

If you really want to describe equilibria "cleanly", you should first consider what you are dealing with:

- If it is a homogeneous solution equilibrium in which the *concentrations* of the substances involved are relevant, and if the concentrations are low enough in each case to treat the mixture as an "ideal solution", then the associated equilibrium constant is given the index c (for *concentration*) in the international technical literature, i.e. one writes K_c. Here

$$K_c = \frac{c(C)^c \cdot c(D)^d}{c(A)^a \cdot c(B)^b} \qquad (4.9)$$

- If the concentration of the components involved in the solutions is so high that one *cannot* consider it an "ideal solution", one must take into account the activity coefficients γ or the respective activity a_i of the individual components, dealing with the constant K_a. The corresponding equation is:

$$K_a = \frac{a_c^c \cdot a_d^d}{a_a^a \cdot a_b^b} \qquad (4.10)$$

- For (necessarily homogeneous) *gas* equilibria, the constant K_p (for *pressure*) is used, since the individual partial pressures p_i are involved here. Here is the corresponding equation again:

$$K_p = \frac{p(C)^c \cdot p(D)^d}{p(A)^a \cdot p(B)^b} \qquad (4.11)$$

The relationships between the various equilibrium constants should also be obvious now:

It follows from Eqs. (4.3), (4.9) and (4.10):

$$K_a = K_c \cdot \frac{\gamma_c^c \cdot \gamma_d^d}{\gamma_a^a \cdot \gamma_b^b} \qquad (4.12)$$

And for gases:

$$K_p = K_c \cdot (RT)^{(c+d)-(a+b)} \qquad (4.13)$$

Especially when dealing with *physical chemistry*, going slightly more in-depth, these correlations will serve you well. ▶ Skoog and ▶ Harris, as already mentioned, also have a lot to say on the subject of "activity and activity coefficients".

Skoog, Appendix B (activity coefficients)

Harris, Section 7.2: Activity coefficients

■ **Potential Differences**

It should be emphasised once again that the potential of a half-cell *cannot be measured directly*, but that a potential difference is always required. The voltage (U) resulting when the two half-cells are conductively connected to each other—via a salt bridge or a semi-permeable membrane (a **diaphragm**)—is then obtained as the difference of the potential of the two half-cells (which is why it is often stated as ΔE). However, some formalisms have to be considered here:

- The standard potentials (which we need here) can be found in corresponding tables (and also in ▶ Harris), usually as *standard **reduction** potentials*, i.e. the table value belongs to the general *reduction reaction* under standard conditions:

Harris, Appendix H (Standard Reduction Potentials)

$$M^{x+} + x\,e^- \rightarrow M$$

Accordingly, the sign of the respective indicated potential also refers to exactly this reduction reaction.

— If one now wants to determine the resulting potential difference for any half-cell, it must be taken into account that in addition to the reduction, an oxidation also takes place. For the associated half-cell, the corresponding partial equation must then be given "in reverse order"; of course, the sign of the associated potential must also be reversed, or the latter must be subtracted from the former (which amounts to the same thing).

In the overall reaction, the electrons then always flow from the partner with the *more negative* potential to the one with the *more positive* potential.

▶ **Example**

For a simple calculation, let us take a well-known example again, in which the less noble zinc is oxidised and the electrons released in the process reduce the copper(II) ions currently present in solution to elemental copper:

$$Zn + Cu^{2+} \rightarrow Zn^{2+} + Cu$$

For the description of this system some customs are to be respected: Separate the **cathode** (where in the galvanic cell the ***reduction*** *of cations* takes place) and the **anode** (where the **oxidation** takes place). For the cathode applies:

Cathode: $Cu^{2+} + 2e^- \rightarrow Cu \quad E^0(Cu^{2+}/Cu) = +0.339$ V

The tables (for example from Harris) also show the value for the anode partner:

$Zn^{2+} + 2e^- \rightarrow Zn \quad E^0(Zn^{2+}/Zn) = -0.762$ V

So, as stated above: The electrons flow from the more negative to the more positive potential (here: from −0.762 V to +0.339 V): Elementary zinc is being oxidised, copper ions get reduced.

However, the numerical value −0.762 V belongs to the *reduction equation*, so in order to determine the resulting total potential, the less noble system (i.e.: zinc—after all, one is looking at the "back reaction to reduction" here!) and thus also the associated sign must be *reversed*:

Anode: $Zn \rightarrow Zn^{2+} + 2e^- \rightarrow Zn \quad -E^0(Zn^{2+}/Zn) = +0.762$ V

This results in:

$$\Delta E^0 = E^0{}_{Cathode} - \left(E^0{}_{Anode}\right) = E^0{}_{Cathode} + E^0\left(\text{Anode with inverted } sign\right)$$
$$= 0.339 \text{ V} + 0.762 \text{ V} = 1.101 \text{ V}$$

This value is of course only valid for standard conditions with corresponding concentrations/activities. For deviations from this (or if the concentrations/activities do not correspond to the standard, but otherwise standard conditions apply, Eq. 4.2), the Nernst equation must be applied again (and, if necessary, the activity coefficient γ must be used for non-ideal solutions). ◀

❗ **Cave**

The positive numerical value may seem surprising at first glance (or even lead to a popular mistake), after all, you have learned in *general chemistry* and *physical chemistry* (keyword: thermodynamics) that for spontaneously occurring reactions the Gibbs energy is *negative*. Please do not confuse the potential difference (ΔE) with the actual Gibbs energy released (or required) (ΔG). The following applies:

$$"G = -z \cdot F \cdot "E \qquad (4.14)$$

Here F is the already known Faraday constant with the numerical value F = 96.495 C/mol, while z is the number of electrons transferred as usual.

Thus, the *larger* the numerical value of the potential difference ($\Delta E > 0$), the *more* energy (change of sign!) is released ($\Delta G < 0$).

The effect of the sign convention is clearly illustrated in Skoog; in addition, Fig. 13.8 from Harris is very helpful.

Skoog, Section 22.3.5: Sign convention for electrode potentials

Harris, Section 13.4: The Nernst Equation

For the description of such galvanic cells, the **cell symbolism** recommended by IUPAC has become accepted:

$$\text{Anode}_{\text{reduced form}} \,|\, \text{Anode}_{\text{oxidized form}} \,\|\, \text{Cathode}_{\text{oxidized form}} \,|\, \text{Cathode}_{\text{reduced form}}$$

(4.15)

Here | stands for the phase boundary between the (dissolved) ions and the (undissolved) solid, while ‖ stands for the separation of the two half-cells, which are separated from each other by a salt bridge or a diaphragm.

The convention requires that
1. first the half-cell in which the oxidation takes place (i.e. the *anode*) is specified,
2. then the half-cell of the reduction (the *cathode*).
3. Within the half-cells, the *reactant* is always named first, then the *product*.

Our example with the total equation $Zn + Cu^{2+} \rightarrow Zn^{2+} + Cu$ would then simply look like this in the cell symbolism:

$$Zn \,|\, Zn^{2+} \,\|\, Cu^{2+} \,|\, Cu$$

❓ Questions

1. Give the oxidation numbers of all the atoms of the following molecules/molecular ions: Sulphite (a), thiosulphate (b), nitrite (c), chlorous acid (d).

    a ⊖O–S(=O)–O⊖   b ⊖O–S(=O)(=O)–S⊖   c ⊖O–N=O   d O=Cl–O–H

2. What are the Galvani voltages for a system in which a rod of elemental iron is immersed in an aqueous solution of ferrous ions whose concentration is (a) 0.1 mol/L; (b) 0.01 mol/L; (c) 0.0023 mol/L? (Let us assume that in all three cases the difference between concentration and activity may be ignored.)
3. What Galvani voltages result for the examples in question 2 if $\gamma = 0.89$ for a), $\gamma = 0.95$ for b), and $\gamma = 0.99$ for c)?
4. State the cell symbolism for a system with the two half-cells cobalt(II)-ions/cobalt and palladium(II)-ions/palladium. Assuming standard conditions, which potential difference would be measured?

Potentiometry

Contents

5.1 Electroactive Analytes – 94
5.1.1 Direct Potentiometry – 97
5.1.2 Potentiometric Titrations – 101

5.2 Ion Selective Electrodes (ISE) – 102
5.2.1 pH Measurement with the Glass Electrode – 102
5.2.2 Other Ion Sensitive Electrodes (ISE) – 105

The basic principles of electrochemistry briefly summarised in ▶ Chap. 4 can also be used in the context of analysis: The potential difference between a half-cell of precisely known composition and one containing the analyte to be investigated allows conclusions to be drawn about its properties, in particular its concentration. Such investigations, which are based on the determination of electrochemical potentials, are summarised under the generic term **potentiometry** (▶ Skoog devotes the whole of Chapter 23 to potentiometry.) It should be obvious that the resulting voltage—which is the consequence of the respective potentials of the half-cells involved—depends on the concentration of the respective dissolved particles, i.e. also on the concentration of the analyte. Accordingly, Nernst's equation (▶ Eqs. 4.1 and 4.2) comes into play here once again.

Skoog, Chapter 23: Potentiometry

The elegance of this type of analysis is not only that the electrodes used are pleasingly small and easy to handle, but also that over the course of time numerous analysis-specific electrodes have been developed in which there are hardly any disturbing influences from other solution components (impurities, etc.).

5.1 Electroactive Analytes

Potentiometric analyses are particularly simple if the analyte under consideration is itself an *electroactive species*, i.e. the analyte can be oxidised or reduced and is thus itself part of a galvanic cell. Accordingly, the solution of unknown concentration can represent one of the two half-cells, which is then connected to an inert electrode (which is therefore not itself oxidised or reduced when it accepts electrons from the analyte or transfers them to it); electrodes made of platinum or similar noble metals are suitable. This is referred to as an **indicator electrode** because it responds directly to the analyte. This is connected to a second half cell of exactly known composition (incl. concentration etc.), which accordingly has an *exactly known* potential; this is referred to as the **reference electrode**.

If electrical conductivity is then established via a salt bridge or a diaphragm, the resulting voltage allows statements to be made about the concentration of the indicator electrode. The schematic structure of such a measuring system is shown in Fig. 23.1 from Skoog.

Skoog, Section 23.1: Potentiometry, fundamentals

- **Indicator Electrodes**

In principle, a distinction is made between two different types of indicator electrodes:
- In the case of the *metal electrode,* one considers the potential that results because a redox reaction takes place on the surface of the metal.
- Ion-selective electrodes, at which *no redox reactions* take place, are discussed in ▶ Sect. 5.2.

The decisive factor for a metal electrode is that it behaves chemically inert towards the analyte, i.e. it is not itself involved in redox reactions of any kind, but "only" accepts electrons (in the case of oxidisable analytes) or transfers electrons from the other half cell to a reducible analyte. As already mentioned, platinum electrodes are particularly common in the laboratory. If an *even more noble* metal is required, often gold is used; silver can also be used potentiometrically: Fig. 14.7 from ▶ Harris, for example, shows a silver electrode that can be used to determine the silver ion content of an analyte solution.

Harris, Section 14.2: Indicator electrodes

5.1 · Electroactive Analytes

It is important to emphasise once again that the measured voltage depends on the reference electrode used in each case: Please note that the tabulated (standard) potentials always refer to the *normal hydrogen electrode* (NHE). However, this is rather unwieldy due to its apparatus design—the half-cell must be constantly supplied with elemental hydrogen, since H_2 is only poorly soluble in water—which is why in the laboratory, usually other reference electrodes are used.

> **The NHE and the Activity**
> If one wants to be even more precise, one does not use the *normal hydrogen electrode* (NHE), but the *standard hydrogen electrode* (SHE). The standard conditions also apply to that one:
> – Pressure $p(H_2) = 1.013$ bar
> – Temperature $T = 298.15$ K (i.e. 25 °C)
>
> Deviating from the NHE, however, here it is not $c(H^+) = 1.00$ mol/L, but $a(H^+) = 1.00$—so here it is not the *concentration* of the hydroxonium ions that is required, but their *activity*, and in such a concentrated solution the activity coefficient $\gamma(H^+)$ will certainly be "< 1".

> **1st and 2nd Order Electrodes**
> In principle, a distinction is made between two different types of electrodes:
> (a) For **1st order electrodes** (in some textbooks also: 1st type) the potential *depends directly* on the concentration of the electrolyte solution surrounding them.
>   – Examples include metal electrodes whose potential depends on the concentration of ions in which the same metal is immersed in elemental form (as a sheet, rod or wire).
> (b) With **2nd order** (or 2nd type) **electrodes,** on the other hand, *the potential is constant*—which is why they are often used as **conductive electrodes.**
>   – Important examples are the silver/silver chloride electrode or the calomel electrode; both of which will be discussed in more detail below.

■ **Reference Electrodes**

Half cells with constant potential are used as reference electrodes. The fact that the potential of this electrode does not change is ensured by working with *a saturated solution of a metal salt* (with associated precipitate), so that the concentration of the counter ions (i.e. the anions) may be assumed to be constant.

In principle, an "ordinary" half-cell can be used as a reference electrode, as you are already familiar with from the general experimental setup for the galvanic cell: Fig. 14.1 from Harris shows, for example, a silver/silver chloride half-cell in which a silver wire (or sheet) is immersed in a saturated potassium chloride solution; the KCl precipitate ensures that $[Cl^-]$ remains constant.

For the sake of simplicity, however, one usually works with *closed electrodes* that are immersed in the analyte solution itself, whereby—precisely because of the closed design of this electrode—contact between the reference material and the analyte solution is minimised. Two such regularly used reference electrodes will be examined here in a little more detail:

Harris, Section 14.1: Reference electrodes

Skoog, Section 23.2.2: The silver/silver chloride electrode
Harris, Section 14.1: Reference electrodes

- **The Silver/Silver Chloride Electrode**

The basis for this electrode is the reaction

$$AgCl_{(s)} + e^- \rightarrow Ag_{(s)} + Cl^-$$

The standard reduction potential of this reaction is $E^0(Ag^+/Ag) = + 0.222$ V; however, this value assumes that the activity (remember?) would actually be $a(Cl^-) = 1$, *which is not true for a saturated solution* of potassium chloride: The actual potential at a temperature of 25 °C is "only" $E(Ag^+/Ag) = + 0.197$ V.

The technical construction of such a closed silver/silver chloride double-chamber electrode is shown in Fig. 23.2b from Skoog and Fig. 14.3 from Harris. Please note that the term "closed" is actually *not quite* correct, since there *must* be some contact with the analyte solution: For this reason, such an electrode is sealed by a semi-permeable (i.e. only permeable to a very limited extent) plug made of a chemically otherwise inert, porous material, which at the same time assumes the role of the salt bridge.

This plug, through which the actual contact with the analyte solution takes place, also represents the "weak point" of such an electrode, because the pores of the plug material can become clogged and for this reason must be cleaned regularly. (Hence, it is also usual to rinse such electrodes thoroughly immediately before use.)

- **The (Saturated) Calomel Electrode (SCE)**

Obviously, this electrode is also based on a redox reaction:

$$\frac{1}{2}Hg_2Cl_{2(s)} + e^- \rightarrow Hg_{(l)} + Cl^-$$

Mercury with oxidation number +1 is reduced to form elemental mercury. As with the silver/silver chloride electrode, the standard reduction potential associated with this reaction is a little misleading, because again the tabulated value ($E^0(Hg^+/Hg) = +0.268$ V) only results when the activity of the chloride ions is $a(Cl^-) = 1$; the *actual* potential in the presence of a saturated KCl solution is a little lower: $E(Hg^+/Hg) = +0.241$ V.

> **Calomel**
> Perhaps you have already encountered this compound in the introduction to *Inorganic Chemistry*, but nevertheless the perhaps unusual (or at least unexpected) name of this electrode should be briefly explained: The compound mercury(I) chloride, a white, crystalline solid, is called "calomel" because it undergoes a disproportionation reaction with ammonia in which, in addition to ammonium chloride, mercury(II) amidochloride is formed on the one hand and elemental mercury on the other:
>
> $$Hg_2Cl_2 + 2\,NH_3 \rightarrow Hg \downarrow + [Hg(NH_2)]Cl + NH_4Cl$$
>
> This finely distributed mercury is deep black—and nothing else means "calomel": καλος = beautiful, μελας = black, thus "beautiful black". Admittedly, this disproportionation reaction has nothing whatsoever to do with the electrochemical processes taking place in a calomel electrode, but it is its name giver and thus shall not remain unmentioned.

If the potential of a calomel electrode is kept constant by potassium chloride, it is also referred to as *saturated calomel electrode*, or—in the technical literature—as a *standard calomel electrode*, **SCE** for short. Schematic representations of the saturated calomel electrode are shown as Fig. 23.2a in ▶ Skoog and Fig. 14.5 in Harris.

Even though it has already been mentioned in Part II of "Analytical Chemistry I", it should be explicitly stated again:

The potential difference of a given redox system (M^{x+}/M) obtained in interaction with these reference electrodes, of course, differs from the respective tabulated value based on the NHE: If one wants to compare such values, the *relative potential* of the reference electrode used in the measurement must be "calculated out" as a constant from the measured value: The measured value thus shifts in each case by the "own potential" of the reference electrode used in comparison to that which would result if the NHE were used. (Fig. 14.6 from the same chapter of ▶ Harris is recommended for illustration.)

Skoog, Section 23.2.1: The calomel electrode
Harris, Section 14.1: Reference electrodes

- **Potentiometric Methods**

Basically, two different measurement methods are available for electroactive analytes:

1. In **direct potentiometry**, the concentration of the analyte to be quantified is simply determined by measuring the voltage resulting in interaction with a reference electrode, i.e. via the *potential difference*.
2. In **potentiometric titration**, on the other hand, the analyte solution is titrated in the classical way, and a titration curve can be drawn up on the basis of the *change in the measurable voltage* (i.e. in the end, again the potential difference).

5.1.1 Direct Potentiometry

In direct potentiometry, the analyte concentration is determined using the Nernst equation (▶ Eqs. 4.1 and 4.2). It is necessary to calibrate the apparatus using standard solutions.

Typical applications of direct potentiometry include the determination of pH values using the glass electrode, but also numerous other measurements using ion-sensitive electrodes (discussed in ▶ Sect. 5.2). The basic principle, however, always remains the same: A galvanic element is constructed, as explained earlier in this section. (Perhaps you would like to look again at Fig. 14.1 from Harris.)

This results in a total cell voltage ΔE_Z, the *open-circuit voltage in the currentless state*. The following applies:

Harris, Section 14.1: Reference electrodes

$$\Delta E_Z = E_{Ind} - E_{Ref} + E_D \tag{5.1}$$

Here E_{Ind} is the electrode potential of the *indicator* electrode, E_{Ref} is the potential of the *reference* electrode and E_D is the **diffusion potential** arising at the boundary between the two half cells. E_{Ind} is again given by Nernst's equation (▶ Eqs. 4.1 and 4.2). Since the electrode material should be as inert as possible, the use of elemental platinum is very common here, but gold or carbon are also occasionally used. A little more attention should be paid to the diffusion potential:

- **The Diffusion Potential**

This potential builds up wherever two different electrolyte solutions come into contact, i.e. in the case of the galvanic cells (and the like) described so far, at the two *ends of the salt bridge*. The resulting voltage lies in the range of a few millivolts, but nevertheless represents an insurmountable limit for the accuracy or correctness of any direct potentiometric measurement.

The reason for this potential difference is to be found in the different mobility of the various ions present in an electrolyte. This mobility, the *migration velocity of the electrolytes* (charge carriers), is commonly abbreviated as u (in some textbooks also µ) and is defined as:

$$\text{Mobility } u = \frac{\text{Velocity}}{\text{Field strength}} \quad (5.2)$$

Here u describes the terminal velocity reached by a charged particle in an electric field of field strength 1 V/m. Thus:

$$u = \frac{v}{F} \quad (5.3)$$

Here v is the velocity, F the field strength.

The unit for the mobility u is then:

$$[u] = \frac{[v]}{[F]} = \frac{m/s}{V/m} = \frac{m^2}{s \cdot V} = m^2 s^{-1} V^{-1}$$

Two factors play an important role in the mobility of different ions:
— the *size*

It should be understandable that smaller ions in principle have a smaller mechanical resistance than larger ones when diffusing through a solution. To give you a rough idea of the size ratios of cations and anions (without their hydrate shell present in aqueous solutions), ◘ Fig. 5.1 shows to scale the atomic and ionic radii of selected main group elements.

— the *charge density*

In any aqueous solution, a hydrate shell forms around each ion. The water molecules orient themselves in such a way that the *negatively* polarised oxygen atoms of the solvent molecules interact with the *cations*, while the *positively* polarised hydrogen atoms of the H_2O molecules make contact with the *anions,* as shown in ◘ Fig. 5.2.

The higher the charge density of the ion, the more water molecules will interact accordingly. This can lead to the (perhaps unexpected) result that an *intrinsically smaller* ion, thanks to its higher charge density, builds up the more voluminous hydrate shell, ultimately becoming so large that it diffuses through the solvent more slowly than an *intrinsically larger* ion with a lower charge density.

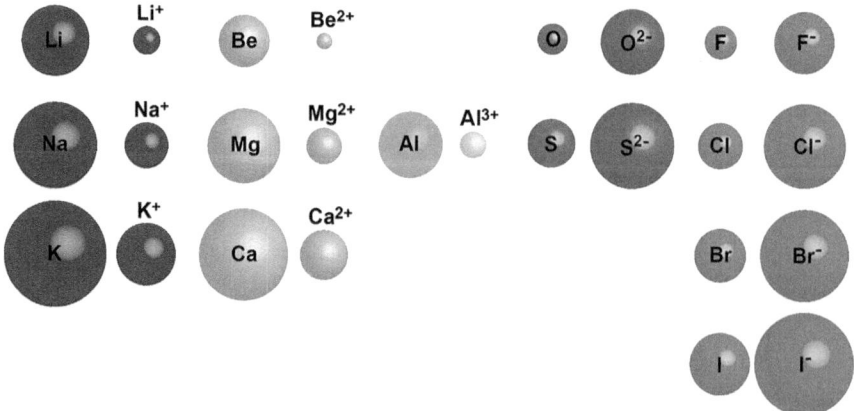

◘ **Fig. 5.1** Size ratios of selected atoms and ions. (Ortanderl S, Ritgen U: Chemie—das Lehrbuch für Dummies, 205. Copyright Wiley-VCH Verlag GmbH & Co. KGaA. Reproduced with permission)

5.1 · Electroactive Analytes

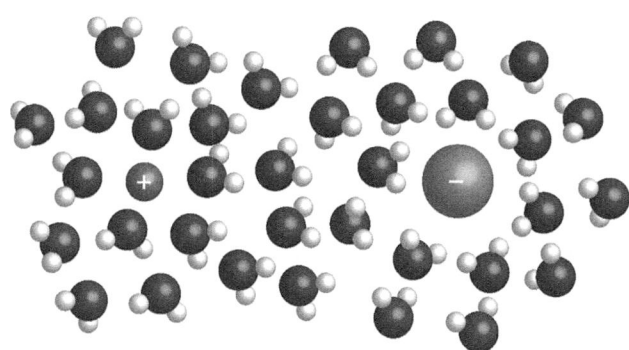

Fig. 5.2 Hydrate shells for anions and cations. (Ortanderl S, Ritgen U: Chemie—das Lehrbuch für Dummies, 325. Copyright Wiley-VCH Verlag GmbH & Co. KGaA. Reproduced with permission)

For example, sodium and chloride ions differ quite significantly in their mobility:
- $u(Na^+) = 5.19 \cdot 10^{-8}$ m^{2}/(s · V)
- $u(Cl^-) = 7.91 \cdot 10^{-8}$ m^{2}/(s · V)

The effect of the different mobility of the ions present in an electrolyte is shown for this ion pair in Fig. 14.9 from Harris.

If we e.g. look at a sodium chloride solution, the hydrate shell of the larger chloride ions is less voluminous than the hydrate shell of the smaller sodium ions: The latter have a higher charge density, so more water molecules attach to it. Accordingly, the mobility of the hydrated Cl^- ions is slightly higher than that of the (also hydrated) Na^+ ions, so that the former diffuse faster. For this reason, there is a slight excess of chloride ions at the diffusion front, so that negative charges accumulate there. A little behind this diffusion front the situation is exactly reversed: A slight accumulation of sodium ions with their associated positive charges is found there. Ultimately, this is precisely what gives rise to the potential difference between the phase of the electrolyte (NaCl solution) and that of pure water.

Due to the different mobility of the ions, a potential builds up. It is important to emphasise that this potential *cannot* be completely eliminated. However, it can at least be minimised by using cations and anions with almost identical mobility, for example in salt bridges. For this reason, potassium chloride is often used here as the electrolyte, because:
- $u(K^+) = 7.62 \cdot 10^{-8}$ m^{2}/(s · V)
- $u(Cl^-) = 7.91 \cdot 10^{-8}$ m^{2}/(s · V)

Harris, Section 14.3: What is a diffusion potential?

> **► Example**
> This is also the reason why potassium chloride is always used in the two reference electrodes from ► Sect. 5.1: The mobility of K^+ ions corresponds almost exactly to that of Cl^- ions, i.e. the diffusion potentials at the salt bridge interfaces are almost negligible here and can therefore really be neglected in most cases. ◄

■ Ionic Strength

The **ionic strength,** which has already been briefly mentioned in ► Chap. 4, is the electric field strength resulting from the ions dissolved in a solvent. The usual formula symbol for ionic strength is I. (However, the formula symbol μ is also found—mainly in older teaching texts—which of course adds to the confusion, after all this Greek letter is also occasionally used for the mobility u (see

Harris, Section 7.1: The influence of ionic strength on the solubility of salts above). We will stick with I.) It should be obvious that the ionic strength of a solution depends on its concentration. The relationship is:

$$I_c = \frac{1}{2} \cdot \sum_i c_i \cdot z_i^2 \tag{5.4}$$

Here c_i is the concentration and z_i is the charge of the ion i in question. Please note that the ion charge enters the equation *exponentially*: Ions with *two* charges thus make *four times* the contribution to the total ionic strength, and so on.

▶ **Example**

Let's look at two examples—sodium chloride and calcium bromide:

$$\begin{aligned}
I(NaCl) &= \tfrac{1}{2} \cdot \left(z^2(Na^+) \cdot c(Na^+) + z^2(Cl^-) \cdot c(Cl^-) \right) \\
&= \tfrac{1}{2} \cdot \left(1^2 \cdot c(Na^+) + (-1)^2 \cdot c(Cl^-) \right) \\
&= \tfrac{1}{2} \cdot \left(1 \cdot c(Na^+) + 1 \cdot c(Cl^-) \right) \\
&= \tfrac{1}{2} \cdot \left(c(NaCl) + c(NaCl) \right) \\
&= c(NaCl)
\end{aligned}$$

$$\begin{aligned}
I(CaBr_2) &= \tfrac{1}{2} \cdot \left(z^2(Ca^{2+}) \cdot c(Ca^{2+}) + z^2(Br^-) \cdot c(Br^-) \right) \\
&= \tfrac{1}{2} \cdot \left(2^2 \cdot c(CaBr_2) + (-1)^2 \cdot 2 \cdot c(CaBr_2) \right) \\
&= \tfrac{1}{2} \cdot \left(4 \cdot c(CaBr_2) + 2 \cdot c(CaBr_2) \right) \\
&= \tfrac{1}{2} \cdot \left(6 \cdot c(CaBr_2) \right) \\
&= 3 \cdot c(CaBr_2)
\end{aligned}$$

You can see that multivalent ions contribute significantly more to the resulting ionic strength than singly charged ions. (You can find more information about this in Harris.) ◀

However, the ionic strength of any ionic analyte is—a fact that is often forgotten or neglected—not only dependent on the concentration of the analyte itself, but also on *any other cations and anions present* in the solution: The interactions of the individual charged particles are, after all, purely electrostatic in nature, so our analyte cations and anions interact just as much with "their own" counter-ions as they do with any other ions present in the solution. For this reason, analyte ions of the same concentration can have completely different ionic strengths (and thus also different activities) depending on the matrix (i.e.: other solution components, here above all those which are themselves ionic).

We will encounter the problem of fluctuating ionic strengths in potentiometric measurements again, for example when covering ion-selective electrodes (▶ Sect. 5.2.2).

However, when we are dealing with a solution of *constant* ionic strength, Eq. (5.1) simplifies to the following general form for calculating the resulting voltage:

$$\Delta E_Z = K + \frac{0.059}{z} \lg c \tag{5.5}$$

5.1 · Electroactive Analytes

In the constant K all constant members from the Nernst equation are summarised, thus:
- E_{Ref} (their potential is constant),
- E^0 of the indicator electrode (i.e. of the analyte),
- E_D, which can also be regarded as constant *and*
- the logarithm of the activity coefficients (when working with activities instead of concentrations).

The cell voltage concentration line then has the slope 0.059 V/z.

5.1.2 Potentiometric Titrations

Here, practically all types of titration we already know from Part II of "Analytical Chemistry I" can be used, i.e.:
1. Acid/base titrations
2. Precipitation titrations
3. Redox titrations
4. Complexometric titrations

During the entire titration process, the resulting potential difference is determined (on a sample basis or continuously), which results from the fact that the concentration of the analyte changes due to the (respective) reaction taking place. Titration curves can then be recorded.

Such potentiometric titrations have several advantages:
- If, for example, the end point of a titration is to be determined potentiometrically, the resulting measurement result is usually much more accurate than when using colour indicators (which, as we remember, do not have a clear *point* of change, but a *range* of change, and in trace analysis in particular, even slight over—or under-titration is often enough to significantly falsify the measurement result).
- Even with solutions of only very low concentration, good results are obtained by potentiometric means.
- While the point of change of a colour indicator is often difficult to detect in tinted analyte solutions or in turbid reaction mixtures, this problem does not arise with potentiometric endpoint determination.
- Only a few accompanying substances (i.e. "non-analytes") prove to be interfering in potentiometry.
- Since the change in potential is *not* measured in *absolute terms,* but rather in *relative* terms, there is no need for calibration.
- Potentiometric measurements can be automated quite easily, which is of considerable importance especially in environmental analysis (e.g. drinking water or waste water samples) as well as in pharmacopoeia analysis.
- With computer-assisted potentiometric titrations it is also possible to quantify several analytes simultaneously. But that would lead too far here.

❓ Questions
5. When does the greater diffusion potential form: When the mobility of the cations and anions of an electrolyte solution differ as *little as possible* or *as much as possible*? Explain your answer.
6. Which salt will have the higher ionic strength in aqueous solution: Sodium sulfate or sodium phosphate? Again, an explanation of your answer is desired.

7. In Part II of "Analytical Chemistry I" you quantified the tin(II) ion content of a hydrochloric acid solution cerimetrically. Was this a potentiometric titration? Again, give reasons for your answer.

5.2 Ion Selective Electrodes (ISE)

Harris, Section 14.4: How do ion-selective electrodes work?

This has already been mentioned in ▶ Sect. 5.1: In addition to the potentiometric investigation of electroactive analytes, there is also the possibility of using *ion-sensitive* electrodes. This analytical method also belongs to the subject area of potentiometry, i.e. information is also obtained here from the concentration dependence of the electrochemical potential. However, and this is the fundamental difference to the method described in the previous section, *no redox processes* take place here. Instead, the method exploits the fact that a *potential difference* builds up at the phase boundary between the electrode material and the electrolyte, which depends on the concentration (more precisely: the activity) of the ions involved:

- **ISE: Mode of Action**

What all ion-sensitive electrodes have in common is that they respond quickly (and ideally also: specifically) to a particular kind of ion.

Some important examples may be mentioned here:
- the glass electrode (which we will discuss immediately after this section)
- the silver/silver chloride electrode (you have already become acquainted with this 2nd order electrode —as a reference electrode—in ▶ Sect. 5.1)
- the hydrogen electrode (already known from part II of "Analytical Chemistry I" and briefly discussed again in ▶ Chap. 4)
- the iodine/platinum electrode (which here shall only be mentioned by name)

The *glass electrode* is of particular importance here because, among other things, it is also used for a real "standard measurement" in the laboratory: for determining the pH value of aqueous solutions.

5.2.1 pH Measurement with the Glass Electrode

Skoog, Section 23.4.3: The glass electrode for measuring the pH value
Harris, Section 14.5: pH measurement with the glass electrode

Ion-sensitive electrodes that respond specifically to H_3O^+ ions are used to determine the pH value. **Single-rod electrodes** are usually used, effectively thin-walled spheres, inside which there is a solution with a known and constant pH value. The contact between the internal solution with a known H_3O^+ ion concentration (which serves as a reference) and the external solution to be analysed, which then leads—as usual—to a potential difference, is established via two conductive electrodes which only "pass on" the potential without themselves being involved in any potential-forming process. The *silver/silver chloride electrode* (from ▶ Sect. 5.1), a second-order electrode, has proved particularly useful as a conductive electrode. Figure 23.3 from Skoog shows how such a glass electrode is constructed. Figure 14.11 from Harris is even more detailed; in the same chapter you will also find a picture of a glass combination electrode (as Fig. 14.12).

Please note: Both the reference electrode (inside the glass electrode) and the analyte solution are concerned with the concentration of H_3O^+ ions. Thus, if the two solutions have different pH values, the resulting potential difference

5.2 Ion Selective Electrodes (ISE)

really depends *only* on the concentration of hydroxonium ions. It is calculated as follows:

$$\Delta E = E^0 - 0.059 V \cdot pH \tag{5.6}$$

If, for example, a silver/silver chloride electrode is used as the conductive electrode, the following cell symbolism results (analogous to the rules described in Sect. 4):

$$Ag_{(s)} \big| AgCl_{(s)} \big| Cl^-_{(aq)} H^+_{(aq)} \text{(analyte solution)} \vdots H^+_{(aq)} \text{(reference solution)},$$
$$Cl^-_{(aq)} \big| AgCl_{(s)} \big| Ag_{(s)}$$

Let's take a closer look at the individual components of this measurement chain:
- $Ag_{(s)} \big| AgCl_{(s)} \big| Cl^-_{(aq)}$ belongs to the *external* reference electrode, which derives the potential of the analyte solution.
- $H^+_{(aq)}$ (analyte solution) refers to the solution to be determined.
  - Together, those two give the "external potential", i.e. the potential of the analyte solution under investigation.
- The *dotted line* represents the *glass membrane*, which ensures that the analyte solution does not come into direct contact with the reference solution.
- $H^+_{(aq)}$ (reference solution) stands for the reference solution with a precisely known concentration of hydroxonium ions (i.e. with a known pH value).
- The three phases of $Cl^-_{(aq)} \big| AgCl_{(s)} \big| Ag_{(s)}$ represent the *internal* reference electrode.

The *resulting total potential is* the sum of the various individual potentials:
- the potential difference of the external reference electrode,
- the diffusion potential at the diaphragm of the reference electrode,
- the potential difference at the phase boundary glass membrane/$H^+_{(aq)}$ (analyte solution),
- the potential difference at the inner phase boundary $H^+_{(aq)}$ (reference solution)/glass membrane *and*
- the potential difference of the inner reference electrode.
- In addition, there is the *potential for asymmetry* resulting from inhomogeneities in the glass membrane used, but that would go too far, here.

Ideally, all these individual potentials *except the potential difference at the phase boundary* glass membrane/$H^+_{(aq)}$(analyte solution) may be regarded as constant; they are thus included in the constant K from Eq. (5.5). Accordingly, for pH measurements where $z = 1$, the following simple relationship is obtained:

$$\Delta E = K + 0.059 \cdot \lg c_{\text{analyte solution}}$$

■ The Role of Glass in the Glass Electrode

There is, of course, a reason why *glass* is used for the glass electrode. In order to understand the role of this material, we should first take a brief look at the structure of glass: In principle, it is—as you certainly remember from *inorganic chemistry*—amorphous silicon dioxide (SiO_2), which forms a three-dimensional network in the form of corner-linked SiO_4 tetrahedra. In addition to this main component, however, glass also contains *additives:* metal ions (especially of the alkali and alkaline earth metals) that interact closely with non-bridging, negatively charged oxygen atoms of the network. (Fig. 23.5 from ▶ Skoog and Fig. 14.13 from ▶ Harris may be helpful here.)

Skoog, Section 23.4.3: The glass electrode for measuring the pH value

Harris, Section 14.5: pH measurement with the glass electrode

Harris, Section 14.5: pH measurement with the glass electrode

Some ions, especially monovalent ones, can, at least to a certain degree, *diffuse into* this silica network; the same applies to water molecules, which then let the surface of a glass wetted by them *swell* to a limited extent: There—only at and just below the surface—this is a partially hydrated gel region in which the diffusion of the monovalent cations succeeds even more easily.

The H_3O^+ ions present in the aqueous medium can also diffuse into this gel zone and displace other monovalent ions there: Thus, an *ion exchange* takes place (just as you may already know from ion exchange chromatography from Part III of "Analytical Chemistry I"). Here, the ion exchange takes place with exactly *those* ions that we consider to be *analyte ions* (the H^+ ions)—and this is exactly what the glass electrode is based on. Again, reference should be made to Harris: Fig. 14.14 clearly shows that in the glass membrane of the glass electrode there are two corresponding gel zones which act as ion exchangers:
— on the inner side of the membrane (there the glass is in contact with the reference solution with precisely known H_3O^+ ion concentration and thus also precisely known activity)
— on the outside, which is wetted by the analyte solution, whose hydroxonium ion content we want to determine with the aid of the glass electrode.

Skoog, Section 23.4.3: The glass electrode for measuring the pH value

Figure 14.15 from the same chapter additionally illustrates the associated ion-exchange equilibrium; the profile of the resulting potentials can be taken from Fig. 23.6 from Skoog.

You will find this basic principle—that the analyte ions interact with the corresponding electrode membrane—in *all* ion-sensitive electrodes, which we will discuss in ▶ Sect. 5.2.2.

- **Sources of Error**

As practical as the glass electrode is, it is unfortunately not completely infallible. In addition to the popular laboratory error of comparing measured values even if they were recorded at different temperatures (although redox potentials are also known to be temperature-dependent), three error factors are to be highlighted in particular:
— At extremely high concentrations of H_3O^+ ions, especially in connection with very strong acids, a slightly *too high* pH value will be measured. The cause of this behaviour, known as **acid error,** has not yet been fully explained, so we will not go into it any further here—but you should know *that* it occurs.
— Depending on the composition of the analyte solution—which will certainly differ from the composition of the reference solution inside the glass electrode—a *diffusion potential* occurs at the diaphragm of the reference electrode; this can obviously falsify the measured value.
— Some foreign ions (i.e. non-analytes) can also falsify the measured value. Usually this happens only within very narrow limits, but *sodium ions* can have a drastic effect: This is generally referred to as **alkali error.**

Skoog, Section 23.4.3: The glass electrode for measuring the pH value

The extent to which acid and alkali errors manifest themselves certainly depends on the glass electrode used in each case: Fig. 23.7 from ▶ Skoog shows the measurement deviations of some commercially available pH electrodes.

- **The Alkali Error**

In the case of alkaline analyte solutions (with a correspondingly very low content of H_3O^+ ions) which also have a high sodium content (such as caustic soda lye, $NaOH_{(aq)}$), the glass electrode also responds to the Na^+ ions; for this reason the resulting voltage then deviates from the theoretically expected value: an

5.2 Ion Selective Electrodes (ISE)

excessively high H_3O^+ concentration is displayed; the pH value determined in this way is therefore *too low.*

Despite the slightly misleading name, this phenomenon is much less pronounced with ions of the other alkali metals. (It would therefore make more sense to speak of the "sodium error", but this term has not yet become established in the technical literature.) However, the fact that the glass electrode responds so specifically to Na^+ can be explained quite easily:

The hydroxonium ion (H_3O^+) has a remarkably high charge density, so it will also have a remarkably massive hydrate shell. In contrast, the charge density of the much larger sodium ion is not as high, so the hydrate shell is much more moderate: The hydroxonium ion with its hydrate shell has almost the same ionic radius and thus the same charge density as the hydrated sodium ion. Thus, it should not be surprising that the glass electrode tends to confuse the two.

❓ Questions

8. The potential difference between a solution with a known pH value (pH = 5.75) and the analyte solution is 0.1062 V. What is the pH value of the analyte solution?
9. A practical laboratory question: Why is it advisable to immerse a glass electrode in a cleaning solution and then dab it dry, rather than wipe it with a cloth (which was soaked in the cleaning solution)?

5.2.2 Other Ion Sensitive Electrodes (ISE)

In the previous section it was already noted that the (glass) membrane of the glass electrode is of considerable importance: It is there—and *only* there—that the measuring electrode comes into contact with the analyte solution. Accordingly, in this section, which deals with other ion-selective electrodes (ISE), we want to look at these membranes in a little more detail. After all, they are at the heart of all ion-selective electrodes, and although they are constructed very differently (more on this in a moment), they all have one thing in common:

The basis for them is always a poorly soluble compound of the ion to be measured.

(This is exactly what the membrane of the glass electrode can be seen as, where metal ions of the glass have been replaced by hydroxonium ions by diffusion and ion exchange processes.)

If the electrode is then immersed in the analyte solution, at least a small part of the membrane substance goes into solution according to the solubility product of the compound used. Consequently, an equilibrium with the dissolved ions is established at the membrane surface. The higher the activity of the ions in the analyte solution, the more analyte ions accumulate on the membrane surface, charging it according to the charge of the analyte ions.

- In the case of cationic analyte ions, a positive charge of the membrane results.
- Anionic analyte ions lead to an accumulation of negative charges on the membrane surface.

But as much as these electrodes may be tailored to the particular analyte being sought, no electrode really responds *exclusively* to a single type of ion.

> The glass electrode from ▶ Sect. 5.2.1, for example, is already quite specific, but the alkali error shows us that at least sodium cations (Na^+) can significantly influence the accuracy of the measurement result.

Accordingly, the *selectivity of* the ISE chosen in each case is of considerable importance.

■ The Selectivity Coefficient K_{Sel}

The extent to which an electrode is "precisely tailored" to the relevant analyte is described by its **selectivity coefficient K_{Sel}**. It always refers to a specific potentially interfering "non-analyte ion" X in comparison to the analyte ion A. Then applies:

$$K_{Sel(A,X)} = \frac{\text{Sensitivity to X}}{\text{Sensitivity to A}} \tag{5.7}$$

The smaller $K_{Sel(A,X)}$, the more specifically the associated ISE responds to analyte A.

■ Various Membranes

Four different types of membranes are commonly used in analysis:
- **Glass membranes** for hydroxonium ions (H^+ or H_3O^+) and selected other monovalent cations
- **Solid state electrodes** (crystal membrane electrodes) based on crystalline inorganic salts
- **Liquid membrane electrodes** comprising a hydrophobic polymer membrane saturated in a hydrophobic liquid ion exchanger
- **Composite electrodes** (secondary electrodes) with a species-selective electrode located behind a membrane that either separates the (analyte) particles of interest from others or produces those particles in a chemical reaction

Here, too, in order to finally generate the galvanic element, which is only able to indicate a potential *difference*, corresponding reference electrodes are required. Once again, the most important ones to be mentioned are:
- the silver/silver chloride electrode
- the calomel electrode (both known from ▶ Sect. 5.1)

Whichever membrane material is used, one basic principle is common to all ISEs:

> *No* redox processes take place here. It is characteristic that only the desired analyte ions bind (more or less) specifically to the respective membrane.

■ Solid State Electrodes (Crystal Membrane Electrodes)

Harris, Section 14.6: Ion selective electrodes (ISE)

Among the solid-state electrodes based on crystalline (inorganic) salts (a general schematic representation can be found in ▶ Harris as Fig. 14.19), the *doped fluoride electrode,* which responds specifically to fluoride ions, is the most important. As an example, let us look at what happens at a lanthanum(III) fluoride electrode doped with europium(II) ions:

If you'd like to take a look at the crystal structure of *undoped* lanthanum(III) fluoride, here's a link: ▶ https://de.wikipedia.org/wiki/Lanthanfluorid

5.2 Ion Selective Electrodes (ISE)

- This doping, in which the Eu^{2+} ions occupy the same positions in the crystal lattice as the La^{3+} ions, leads to vacancies in the lattice, which "in itself" should correspond to the LaF_3 structure, because the divalent cations each bring "one anion too few".
- Although ionic lattices are generally thought to be rather inflexible—it is usually assumed that once an ion has taken its place in the lattice, it will not move from there—this is *not quite true*: The fluoride anions, which are quite small compared to lanthanum cations, can indeed change their position within the lattice—they then migrate from their current position to an adjacent, currently unoccupied lattice site. This re-positioning of negative charge carriers causes some *conductivity within the lattice*, as illustrated in Fig. 14.20 from the same chapter of ▶ Harris. This phenomenon is referred to as **ionic conductivity**.

The construction of a fluoride electrode can be seen in Fig. 14.19 of the same chapter from Harris. A membrane of said doped LaF_3 separates the internal solution of the electrode, a potassium fluoride solution of known concentration (often with $c(KF) = 0.1$ mol/L), from the sample solution. Due to the (limited) diffusion of the fluoride ions, a measurable voltage builds up again, which depends directly on the concentration of the fluoride ions present in the solution. (The Ag/AgCl system again usually serves as the reference electrode.)

> **Tip for the Lab**
> Although the fluoride electrode can be used over a fairly wide range of analyte concentrations (concentrations of 10^{-6}–10^{-1} mol/L are commonly assumed; refinements can be made to increase the sensitivity of the electrode so that the detection limit is lowered by further two to three orders of magnitude), sample solutions with varying *ionic strength* (please recall the relevant considerations from ▶ Sect. 5.1.1) can cause problems to some extent.
>
> For example, the same amount of fluoride ions dissolved once in de-ionised water and once in seawater would lead to different results, because *the activity also depends on the ionic strength*: The matrix "seawater" contains significantly more foreign ions, and the higher ionic strength of this solution will correspondingly reduce the activity of our analyte. (Perhaps you would like to remind yourself of the connection between ionic strength and activity with the aid of Fig. 7.4 from Harris.)
>
> For this reason, small quantities of a special buffer solution known as **TISAB** (*Total Ionic Strength Adjustment Buffer*) are usually added to the sample solution for potentiometric tests. Four components that are present together in aqueous solution are of particular importance here:
> - CDTA (this abbreviation belongs to the historical name 1,2-cyclohexylened initrilotetraacetate; IUPAC prefers the name *trans*-1,2-cyclohexanediamine-N,N,N',N'-tetraacetic acid; see the following page for the structure of this compound).
> - Acetic acid
> - Sodium hydroxide (in small quantity)
> - Sodium chloride

Harris, Section 7.2: Activity coefficients

TISAB fulfils several functions:
- On the one hand, the buffer itself ensures a high ionic strength of the sample solution, so that any fluctuations due to the sample matrix have only a minimal effect (ideally: no longer measurable): it *fixes* the ionic strength.
- On the other hand, the buffer also keeps the pH value more or less constant, since, due to similar charge density, hydroxide ions can influence the measurement if present in too large a concentration (i.e.: too high a pH value).
- In the quantification of fluoride ions (and this section is primarily concerned with them), there is the additional factor that these analyte ions form hardly soluble compounds with trivalent cations (Fe^{3+}, Al^{3+}, etc.); (partial) precipitation of the analyte naturally leads to a measured value for the analyte content that is falsely too low. However, since these trivalent ions are complexed by the components of the buffer, their undesired precipitation does not occur.

In countries where drinking water is fluoridated (such as the USA, Chile, and Brazil), fluoride electrodes are used for routine tests of fluoride content; however, it is also used to determine the fluoride load of biosamples (such as milk or serum).

- **Liquid Membrane Electrodes**

No redox processes take place at these electrodes either: Again, a membrane is used which is able to bind the desired analyte ion preferentially (perhaps even specifically?).

Generally, a hydrophobic polymer membrane impregnated with a viscous organic solution is used. This contains
1. in any case an ion exchanger *and in addition*
2. occasionally an additional ligand that interacts preferentially with the analyte ion (and ideally does not interact at all with other ions originating from the matrix, although this cannot always be completely avoided).

Skoog, Section 23.4.6: Liquid membrane electrodes

The interior of the liquid membrane electrode contains the analyte ion and the corresponding counter ion in a precisely known (and constant) concentration; the schematic structure of such an electrode is shown in Fig. 23.8 from ▶ Skoog.

The potential difference arising at the membrane is again determined by reference electrodes: If the concentration (or the activity) of the analyte ion in the solution under investigation changes, this leads to a change in the resulting voltage.
- If a ligand is required for the mode of action of the liquid membrane electrode, it should—as already mentioned above—ideally have a high affinity only to the analyte ion. However, since other ions can interfere, the selectivity coefficient K_{Sel} comes into play here again.

5.2 Ion Selective Electrodes (ISE)

- After interaction with the ligand, the analyte is bound in a complexed form.
- Hydrophobic anions act as counter ions, which dissolve well in the (also hydrophobic) membrane of the electrode, but are only poorly soluble in water.
- Accordingly, the complexed analyte ion can diffuse through the interface; however, this is not true for the counter ion.
- Once some analyte ions have diffused from the membrane into the aqueous phase, there is a positive excess charge in the aqueous phase, which leads to a potential difference and prevents the diffusion of further analyte ions.

Skoog, Section 23.4.6: Liquid membrane electrodes

An introductory overview of various liquid membrane electrodes, the analytes that can be detected with them, detection limits and, if necessary, interfering influences by other ions is given in Table 23.4 from Skoog.

> ▶ Example
>
> As an example, we use a liquid membrane electrode that specifically responds to potassium ions (K^+). The ligand is Valinomycin, which acts as an **ionophore**, i.e. it assumes the role of an ion transporter. The liquid membrane consists primarily of the extremely nonpolar material polyvinyl chloride (PVC).
>
> Harris, Section 14.4: How do ion-selective electrodes work?
>
> What exactly happens at the liquid membrane is shown in Fig. 14.10 from Harris: Inside the membrane, the analyte cations (generally designated C^+ for *cation* in the figure) are mainly present in the complexed state (as a K—Valinomycin complex, designated LC^+ there). Since the corresponding counterions must also be readily soluble in the membrane, anions that are as nonpolar as possible are used here (in this case: tetraphenylborate, $[B(C_6H_5)_4]^-$; in said figure they are symbolised by R^-).
>
> But even if the equilibrium of ion and ligand (L)
>
> $$K^+ + \text{Valinomycin} \rightarrow [K - \text{Valinomycin}]^+ \quad \text{or general} \quad C^+ + L \rightarrow LC^+$$
>
> lies strongly on the product side, i.e. almost all potassium ions are complexed, *some* of the cations must also be present in their free form (i.e. uncomplexed). These can easily diffuse through the membrane - on *both* sides of the membrane, both into the analyte solution and into the reference solution inside the electrode. This, on the other hand, is not the case for the corresponding anions ($[B(C_6H_5)_4]^-$ or R^-). The result is that an *excess of anions* builds up on both sides of the membrane inside it, while the two solutions in contact with the membrane contain a small excess of *cations*. This (limited) charge separation causes a potential difference to develop, which can then be measured with the help of the conductive electrodes. ◀

> **A Little More About Valinomycin**
>
> You may already be familiar with Valinomycin, whose structural formula (see following figure) can be found in everyday life: It belongs to the class of **macrolide antibiotics** that are used against various bacteria. They have multiple effects:
> - On the one hand, they inhibit the protein biosynthesis of the bacteria (unless those have already developed resistance to Valinomycin in particular or even to macrolide antibiotics in general), so that they have a *bacteriostatic* effect.
> - They also strengthen the patient's immune system, although the mechanism behind this has not yet been fully elucidated.
> - Last but not least, they also have an antiphlogistic effect, i.e. they are anti-inflammatory **drugs.**

Valinomycin

That Valinomycin belongs to the macrolide antibiotics has just been mentioned, but at the same time this molecule is also a **depsipeptide**: It consists of the proteinogenic amino acid L-Valine (Figure a—this is *(S)*-configured at the chirality centre according to the Cahn-Ingold-Prelog nomenclature) and D-Valine (b—this *non-proteinogenic* amino acid accordingly has *(R)*-configuration) as well as of the hydroxy acids L-lactic acid (L-Lactate, Figure c) and D-hydroxyisovaleric acid (D-isovalerate, D-2-hydroxy-3-methylbutyric acid, Figure d). Accordingly, the ring structure is held together by both peptide and ester bonds. (Look again at figure of Valinomycin.)

"Building blocks" of Valinomycin

Immediately adjacent to Fig. 14.10 above, Harris also shows how potassium ions complexed by Valinomycin are "enveloped" by the ionophore molecule:
— A large part of the polar molecular constituents come to lie inside the complex, the "cavity" spanned by the Valinomycin.
— The outer surface of the almost spherical potassium-Valinomycin complex then consists almost exclusively of the nonpolar methyl and isopropyl side chains of the amino and hydroxy acids of which this depsipeptide is composed, so that the complex is readily soluble in nonpolar solvents or even in a nonpolar membrane.
— Understandably, this does not apply to *uncomplexed* potassium ions.

▶ **Example**

As already mentioned, the *selectivity of* an ISE according to Eq. (5.7) must always be taken into account. In the case of our example electrode, for example:

$$K_{Sel}\left(K^+, Na^+\right) = 1 \cdot 10^{-5}$$
$$K_{Sel}\left(K^+, Cs^+\right) = 0.44$$
$$K_{Sel}\left(K^+, Rb^+\right) = 2.8$$

5.2 Ion Selective Electrodes (ISE)

While sodium ions hardly interfere, the influence of any cesium ions is already massive, and the electrode responds even more sensitively to rubidium ions than to potassium ions. The reason for the immense rubidium specificity is that the Rb^+ cation fits *even better* into the Valinomycin cavity than K^+ for steric reasons alone: The interaction between ion and ligand is much more pronounced. ◄

■ Composite Electrodes

In the case of composite electrodes, too, the measuring electrode—which is connected to a reference electrode in the usual way—is surrounded by a membrane which responds to the electrode-specific analyte and isolates it from the matrix components. Frequently, the analyte is first (chemically) converted by interaction with the membrane to the form to which said electrode ultimately responds.

An example is the CO_2 gas electrode shown in Fig. 23.12 of Skoog and Fig. 14.28 of Harris. The actual heart of this apparatus is a pH electrode, as we already know it. This is surrounded by a semipermeable membrane made of
- Polyethylene
- Teflon *or*
- Rubber

Skoog, Section 23.6.1: Gas sensitive probes

Harris, Section 14.6: Ion selective electrodes (ISE)

which the analyte (i.e. CO_2) can easily penetrate. (The reference electrode is—once again—a silver/silver chloride electrode.)

If carbon dioxide now diffuses through the membrane, this lowers the pH value of the filling in the electrode cavity, according to the following reaction equation:

$$CO_2 + H_2O \rightarrow H_2CO_3 \rightarrow H^+ + HCO_3^-$$

This change in pH value then serves as a measure of the CO_2 concentration of the analyte solution.

In the same way other gases reacting acidic can be quantified (SO_2, SO_3, NO_x, etc.).

> **An Outlook**
> There are also composite electrodes whose membrane material contains *enzymes that* act as catalysts for the reactions that the analyte undergoes. However, this already takes us into the field of biosensors, which would once again go beyond the scope of this section. If you are interested, please refer to Section 23.6.2 of Skoog, but we will return to this topic in Part IV.

❓ Questions

10. What could explain the fact that the solid-state electrode of europium(II)-doped lanthanum(III)-fluoride presented in this section measures excessive fluoride concentrations in a strongly basic medium? Is it to be expected that the presence of other halides (in addition to fluoride) in the analyte solution will also have a falsifying effect?
11. Could a composite electrode that specifically responds to acidic gases also be used to quantify ammonia? Justify your decision.

Coulometry

Contents

6.1 The Karl Fischer Titration – 115

6.2 Karl Fischer Titration—Classic – 115

6.3 Karl Fischer Titration—Coulometric – 116

Skoog, Chapter 24: Coulometry
Harris, Section 16.1: Fundamentals of electrolysis

Coulometry, which is dealt with in Chapter 24 of ▶ Skoog, is concerned with the number of electrons that are transferred in the course of a chemical reaction. The basis for this measuring method are **Faraday's laws**:

- If an electric charge passes through an electrolyte, the mass (m) deposited in the process is proportional to the quantity of charge (Q), whereby the quantity of charge can be calculated as the product of current intensity (I, also called "current strength") and time (t, duration of current flow):

$$m \sim Q \quad \text{and} \quad Q = I \cdot t \tag{6.1}$$

The mass (m) is given in grams (g), the current in amperes (A), the time in seconds (s). Correspondingly, the unit A · s (Amperes times seconds) results for the quantity of charge (Q).

- If 96 485 coulombs are passed through an electrolytic cell, the amount of substance of one *equivalent* (with the equivalent mass M/z) is converted at each electrode:

$$m = \frac{M \cdot Q}{z \cdot F} \tag{6.2}$$

Here m is again the mass of the coulometrically deposited substance, M its molar mass, Q the quantity of charge from Eq. (6.1), z the number of electrons required for the deposition per equivalent and F is the Faraday constant—being nothing other than the electric charge of one mole of electrons; it is obtained as the mathematical product of the Avogadro constant (N_A) and the elementary charge e, i.e. the charge of a single electron:

$$F = e \cdot N_A = 1.6022 \cdot 10^{-19} \, C \cdot 6.022 \cdot 10^{23} \, mol^{-1} = 96\,485 \; C/mol. \tag{6.3}$$

Somewhat casually, but descriptively, one could say: In coulometry "one titrates with electrons".

In principle, two different methods can be distinguished in coulometry: It can be operated
- potentiostatically *or*
- galvanostatically.

Skoog, Section 24.3: Potentiostatic coulometry
Harris, Section 16.3: Coulometry

Potentiostatic coulometry (or better: coulometry with *controlled* potential) is much more selective, but also technically more complex: *three* electrodes are required (the working electrode and the reference electrode, plus an auxiliary electrode). The calculation of the total number of electrons transferred in this measurement is somewhat difficult, because although the potential of the working electrode is constant, the current intensity, which depends on the concentration of the analyte, decreases exponentially in the course of the experiment, so that the current intensity must be integrated over time.

Galvanostatic coulometry is much simpler and is also referred to as **coulometric titration**—quite analogous to the above-mentioned casual description. If the current I and the time t required for the coulometric deposition of the analyte are known, the charge quantity (i.e. the number of electrons transferred) can simply be calculated according to Eq. (6.1), which then—via Eq. (6.2)—allows conclusions to be drawn about the amount of the analyte.

6.1 The Karl Fischer Titration

A characteristic application of coulometry is found in Karl Fischer titration for the determination of the water content of non-polar analytes (for example, solvents to be used in analysis, but also in food chemistry the water content often plays an important role). The decisive reagent for determining the water content is sulfur dioxide (SO_2). As you certainly know from *general* or *inorganic chemistry*, this is the anhydride of sulfurous acid (H_2SO_3); accordingly, it is easy to think that the water content of an analyte can be determined simply by letting the sample react with sulfur dioxide and determining the amount of sulfurous acid produced. Unfortunately, it is not quite that simple, because the equilibrium of the reaction

$$SO_{2(aq)} + H_2O_{(l)} \rightleftarrows H_2SO_{3(aq)}$$

lies far on the left-hand side. However, for the quantification of water it can be exploited that sulfites (the anions of sulfurous acid) can be converted to sulfates by suitable oxidising agents—in an aqueous medium as well as in other solvents.

6.2 Karl Fischer Titration—Classic

In the classical variant of the Karl Fischer titration, sulfur dioxide is first brought to reaction with an alcohol (generally: R–OH; methanol is often used, but other alcohols are also employed) to form the simple ester of sulfurous acid:

$$SO_2 + R-OH \rightarrow R-O-S(=O)-O^- + H^+$$

The hydrogen cations (H^+) released in the process are captured by a base; without a base, the conversion does not take place completely.

A suitable oxidising agent then converts the mono-ester of sulfurous acid with water to the mono-ester of sulfuric acid (with the oxidation number of sulfur changing accordingly from +IV to +VI):

$$\text{OX}: \quad R-O-S(=O)-O^- + H_2O \rightarrow R-O-SO_3^- + 2H^+ + 2e^- \tag{6.4}$$

Since hydrogen cations are also released here, the presence of a base is again required.

Elemental iodine acts as an oxidising agent:

$$\text{RED}: \quad I_2 + 2e^- \rightarrow 2I^- \tag{6.5}$$

Overall, using a general base B gives the net equation:

$$R-O-SO_2^- + H_2O + I_2 + 2B \rightarrow R-O-SO_3^- + 2I^- + 2BH^+ \tag{6.6}$$

The consumption of elemental iodine, which is converted to iodide ions here, is accordingly a direct measure of the water content: The molar ratio $n(H_2O):n(I_2)$ is 1:1.

Although it is possible to add the elemental iodine in dissolved form to the analyte via an (automatable) burette, a variant of the Karl Fischer titration is now much more common:

6.3 Karl Fischer Titration—Coulometric

Here, the elemental iodine required for the reaction is generated *in situ* by electrochemical means: The iodine is generated by anodic oxidation of iodide ions in solution according to Eq. (6.7),

$$2I^- \rightarrow I_2 + 2e^- \qquad (6.7)$$

Skoog, Section 24.4.1: Coulometric titrations, instrumentation
Harris, Section 16.6: Karl Fischer titration of water

(which is the errverse of Eq. 6.5), and the elemental iodine thus obtained immediately reacts with the reactant (i.e. the monoester of the sulfurous acid) according to Eq. (6.6). The current flow is maintained only until all the water contained in the reaction mixture has been consumed; the amount of iodine produced coulometrically can then be determined according to Eqs. (6.1) and (6.3). The corresponding experimental set-up is shown in Fig. 24.8 from Skoog and (in a schematic representation) Fig. 16.32 from Harris.

> **Important**
> Please bear in mind once again that coulometry is actually nothing other than **electrolysis**:
> — In the case of Karl Fischer titration, iodide ions are oxidised to elemental iodine at the anode.
> — What happens at the cathode is irrelevant for our purposes here (and also depends on the apparatus used and, of course, the counter ion of the iodide ions). However, this should not detract from the fact that ultimately a simple electrolysis takes place.

> **Questions**
> 12. What is the water content (in mmol/L and as volume concentration $\sigma(H_2O)$ according to DIN 1310) of a sample of V(sample) = 50.00 mL to which 5.00 g potassium iodide (KI) were added in the course of sample preparation if the end point was reached after exactly 13 min and 32 s at a current intensity I = 52.6 mA?

Amperometry

Amperometry is another analytical technique involving *electrolysis*: The current that flows *at constant potential* between the corresponding pair of electrodes is measured, because if one of the substances involved in this electrolysis is the analyte to be quantified, the measurable current is proportional to the analyte concentration (although here, too, the system must of course be calibrated); a material that is as chemically inert as possible, such as platinum, gold or even carbon, again serves as the counter or working electrode.

- **The Clark Electrode**

A common application of amperometry is the Clark electrode, which is used to determine the oxygen content of a solution (and is therefore particularly important in biochemistry and medicine). Here, a composite electrode is used whose silicone coating is highly permeable to molecular oxygen; the oxygen is then reduced to water at the platinum wire underneath (which is usually additionally coated with gold). The following reaction takes place at this electrode:

RED: $O_2 + 4 H^+ + 4 e^- \rightarrow 2 H_2O$

As a working electrode we again use the silver/silver chloride electrode known from ▸ Sect. 5.1, which will not be discussed again here.

The higher the O_2 concentration, the stronger the resulting current flow; the Clark electrode must of course also first be calibrated using solutions with known oxygen concentration.

Harris, Section 16.4: Amperometry

▸ Harris discusses the Clark electrode in Excursus 16.1; the structure of a corresponding microelectrode is also shown schematically there.

- ■ **Biosensors**

The principle of the Clark electrode (or amperometry in general) can also be extended to other analytes. First simple modifications of this electrode were developed, which responded to carbon monoxide (CO) or nitrous gases (NO_x); by now, more complex components also are in use, which respond much more specifically to the respective analytes—often from the field of biochemistry: If, for example, enzymes or DNA come into play, the ATP content of biological tissue can be determined. Since ATP (adenosine triphosphate, ◘ Fig. 7.1) is present in practically all cells—it is the most important energy transporter for almost all biochemical metabolic pathways—the determination of their ATP content allows numerous conclusions to be drawn about the conditions prevailing there, right up to the (often medically not unimportant) question of whether the cell is still behaving as expected. In the same way, electrodes equipped with antibodies allow even more specific investigations.

Blood glucose meters, often indispensable for diabetes patients, are also based on corresponding biosensors. Two carbon working electrodes are used, one of which is coated with the enzyme *glucose oxidase*. In the presence of

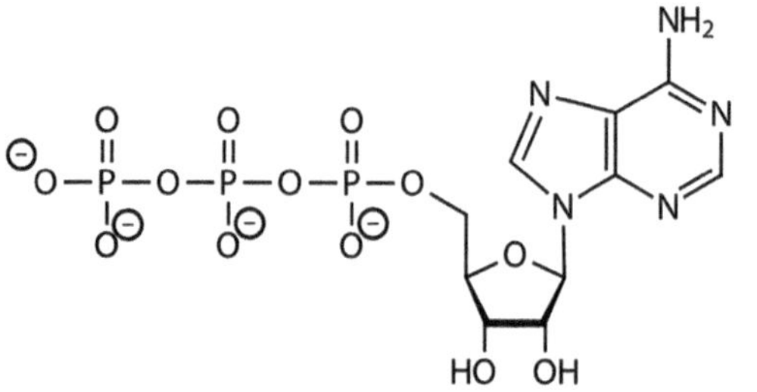

◘ Fig. 7.1 Adenosine triphosphate (ATP)

Chapter 7 · Amperometry

Fig. 7.2 a Glucose, b gluconolactone, c ascorbic acid, d Paracetamol

oxygen, this enzyme catalyses the oxidation of glucose—i.e. blood sugar, Fig. 7.2—to gluconolactone (b), in which the hydroxyl group at the **anomeric centre** has been oxidised to the keto group; hydrogen peroxide (H_2O_2) is formed in the process. The hydrogen peroxide is then oxidised at the working electrode to elemental oxygen and hydrogen cations (H^+); the electrons released in this process provide a current flow. *Thus, it is not actually the glucose itself that is quantified, but the hydrogen peroxide formed during its oxidation.*

Once again, the silver/silver chloride electrode serves as a reference. The schematic structure of the corresponding disposable test strips used in blood glucose meters can be seen in Fig. 16.10b from Harris. (Yes, the corresponding reference electrode is really discarded after use.)

Harris, Section 16.4: Amperometry

Of course, with these test strips the question arises as to why *two* carbon working electrodes are actually used. The reason is that the analyte solution to be tested (i.e. the patient's blood) contains not only glucose but also other substances (= non-analytes) which can also be oxidised, for example vitamin C (ascorbic acid, Fig. 7.2c), or pharmaceuticals such as the **analgesic** Paracetamol® (*para*-acetaminophenol, (d)). Understandably, the electrons released in this process also cause a current to flow, as does the H_2O_2 mentioned above. Now the importance of the enzyme used here becomes apparent, because only under its influence glucose is converted:

— At the coated electrode, the current flow results on the one hand from the conversion of the glucose (or the hydrogen peroxide produced in the process), and on the other hand from the oxidation of the non-analytes mentioned.
— At the non-coated electrode, oxidation of the blood glucose *does not occur*; the current flow measured there therefore results *exclusively* from the non-analytes.
— The *difference* between the two current flows then allows a direct conclusion to be drawn about the concentration of glucose present in the blood.

? Questions
13. What is the basic difference between coulometry and amperometry?
14. Why are coulometric analyses largely independent of ambient temperature, while amperometry shows clear temperature dependence?

Voltammetry

Content

Further Reading – 132

© Springer-Verlag GmbH Germany, part of Springer Nature 2025
U. Ritgen, *Analytical Chemistry II*, https://doi.org/10.1007/978-3-662-68710-9_8

Skoog, Chapter 25: Voltammetry

The term "voltammetry", which is dealt with in Chapter 25 of Skoog, covers several analytical methods, all of which have in common that the relationship between voltage and current during one or other electrochemical process is recorded. In the associated voltammograms, the current density is plotted against the potential of the electrode used; exemplified for the oxidation of hexacyanidoferrate(4-) to hexacyanidoferrate(3-) according to the equation

$$\left[\text{Fe}(\text{CN})_6\right]^{4-} \rightarrow \left[\text{Fe}(\text{CN})_6\right]^{3-} + e^-$$

Harris, Section 16.5: Voltammetry

are the corresponding voltammograms for hexacyanidoferrate(4-) solutions of different concentrations as Fig. 16.14a from Harris. (The reference electrode here is a saturated calomel electrode, which we already know from ▶ Sect. 5.1.) The resulting current is determined by the rate at which the ions in question diffuse to the electrode, thus effectively by their concentration in the solution in question. The linear relationship between current density and concentration is shown in Fig. 16.14b. Therefore, voltammetry can also be considered a special form of electrolysis.

> **For Those Who Want to Be *Very* Precise (Chemically Speaking)**
> In Fig. 16.14a, the concentration of Fe(II) is indicated in the various curves resulting from the different analyte ion concentrations. However, it should be explicitly mentioned again (in *general and inorganic chemistry* as well as in (advanced) courses or textbooks of *inorganic chemistry* it will certainly already have been mentioned) that a solution of hexacyanidoferrate(4-) ions contains "free" iron(II) ions only in negligible amounts, and a $[\text{Fe}(\text{CN})_6]^{4-}$ ion is not a Fe^{2+} ion (even if the iron in it does have the oxidation number +II). A sulfate ion (SO_4^{2-}) is also not an S^{6+} ion (which, unlike Fe^{2+}, is not known in free form), even if the sulfur in both forms has the same oxidation number (+VI).
>
> For hexacyanidoferrate(3-) ions and "free" iron(III) ions, of course, the same applies.

❗ When Typos Distort the Meaning
Please do not confuse voltammetry with volta_m_etry: the latter is a potentiometric titration in which the current flow is specifically influenced. Shockingly, even in some textbooks the "wrong notation" with only one "m" can be found, although voltammetry is really about the voltage-dependent current flow.

■ **Polarography**

The historically most important form of voltammetry is polarography. This method uses elemental mercury, which is why other methods usually are used nowadays, since mercury is highly toxic; but because of its historical significance—and because it allows explaining the principle of voltammetry quite well—it will nevertheless be discussed.

The elemental mercury is part of a **mercury drop electrode,** the schematic structure of which can be seen in Fig. 16.15 in Harris. The voltage resulting when a drop of mercury comes into contact with the analyte solution is measured; a saturated calomel electrode (known from ▶ Sect. 5.1) is used as a reference. The reduction of the analytes (usually metal ions of the general formula M^{x+}) to the elemental metal (M) takes place, and since metals generally dissolve well in mercury and form an **amalgam,** this mercury drop is then "contaminated" and can no longer be used for further measurements. For this reason, the drop is removed (conveniently, thanks to gravity, it comes off the electrode quite quickly all by itself), another drop is formed from a mercury reservoir, and the resulting voltage is measured again, and so on. All in all, a current flow dependent on the concentration of the analyte M^{x+} results, quite analogous to the already mentioned Fig. 16.14a.

There is another good reason for the use of mercury: It is well known that base metals (e.g. alkali metals) are very difficult to reduce to their elemental form (large negative potential!). However, if the reduction takes place in the presence of elemental mercury (i.e. if the result is not the "pure metal" but its amalgam), this significantly reduces the reduction potential. Let us take a look at the alkali metals as an example:

$$E^0\left(Li^+ + e^- / Li\right) = -3.0 \text{ V} \qquad E^0\left(Li^+ + e^- / Li\right)_{amalgamated} = -2.2 \text{ V}$$

$$E^0\left(Na^+ + e^- / Na\right) = -2.7 \text{ V} \qquad E^0\left(Na^+ + e^- / Na\right)_{amalgamated} = -2.0 \text{ V}$$

$$E^0\left(K^+ + e^- / K\right) = -2.9 \text{ V} \qquad E^0\left(K^+ + e^- / K\right)_{amalgamated} = -1.9 \text{ V}$$

$$E^0\left(Rb^+ + e^- / Rb\right) = -2.9 \text{ V} \qquad E^0\left(Rb^+ + e^- / Rb\right)_{amalgamated} = -1.9 \text{ V}$$

$$E^0\left(Cs^+ + e^- / Cs\right) = -3.0 \text{ V} \qquad E^0\left(Cs^+ + e^- / Cs\right)_{amalgamated} = -2.0 \text{ V}$$

However, the mercury drop electrode *cannot* be used to investigate oxidation processes because mercury itself is easily oxidised. For such investigations (or if one would like to avoid the use of mercury if possible) there are now a large number of other working electrodes available, made of platinum, gold or carbon, for example. However, we do not want to overdo it here.

■■ **Tactile Polarography**

In tactile polarography, the principle of polarography is combined with a gradual increase in potential, usually in equidistant steps (a standard value is 0.004 V). A staircase-like voltage profile is then obtained, as shown in Fig. 16.17 of ▶ Harris; commonly, however, the individual measured values are combined to form a "continuous" polarogram for evaluation. As an example of a tactile polarogram, reference is made here to the curve that results from the reduction of a low-concentration solution of cadmium(II) ions ($c(Cd^{2+})$ = 0.005 mol/L) in hydrochloric acid solution (Fig. 16.18a): $E^0(Cd^{2+}/Cd) = -0.402$ V; using Nernst's equation (▶ Eq. 4.2), this therefore gives $E(Cd^{2+} + 2e^-/Cd) = -0.47$ V. However, the reference electrode here is the saturated calomel electrode (SCE, known from ▶ Sect. 5.1) with $E(Hg^+/Hg) = +0.241$ V, so the value of the resulting voltage is shifted accordingly, as already mentioned in the section mentioned above. In the experiment it can then be observed that until the reduction potential is reached ($E(Cd^{2+}/Cd)_{referred\ to\ the\ SCE} = -0.64$ V) *nothing happens at all*, and then, when the reduction takes place, there first is a moderate current flow, which then increases rapidly and finally reaches a maximum, which manifests itself in the polarogram as a *current intensity plateau*. (Anything else that happens after this is called **residual current:** This is not due to the analyte itself, but to impurities in the reaction solution and other factors that are difficult to quantify, but which are not important until the plateau is reached in the curve.)

Of particular importance here is the *inflection point* of the polarographic curve, which lies halfway between the initial measured value and the current intensity plateau: It is called the **half-wave potential ($E_{1/2}$)**.

Harris, Section 16.5: Voltammetry

- Under ideal conditions—where the oxidised and the reduced form of the analyte are present in solution—this value corresponds exactly to the *standard potential* of the redox system under consideration. (Again, of course, the "shift" must be taken into account if something other than the SHE is used as the reference electrode.)
- If, on the other hand, the conditions are *not* ideal (as in the case of our cadmium example; the elementary cadmium is *not* in solution), the result is at least *close to that*.

The match of E^0 and $E_{1/2}$ (under ideal conditions) should not be particularly surprising: As with other analytical methods based on redox processes, polarography is governed by Nernst's equation (▶ Eqs. 4.1 and 4.2), and by the time the half-wave potential is reached, exactly 50% of all analyte ions that were initially present in their oxidised form have been converted to the reduced form (under ideal conditions!), so here: $c(analyte_{oxidised\ form}) = c(analyte_{reduced\ form})$. Thus, the whole "lg term" of Nernst's equation is omitted, and $E = E^0$ holds. (Of course, as mentioned above, this *only* applies to systems where both the oxidised and reduced form are in solution, and thus in equilibrium with each other.)

However, as real ideal conditions practically never exist within real existing systems, some deviation is unavoidable—however, this deviation can also be taken into account (i.e. "allowed for"). So the principle remains the same. (Also the step-heights of the plateaus obtained by these measurements can be evaluated quantitatively, but once again we want to restrict ourselves to the basics here.)

- **Cyclic Voltammetry**

In a variant of voltammetry called cyclic **voltammetry** (explained in detail in Section 25.4 of ▶ Skoog), the voltage is not applied continuously/constantly, but it (i.e. the cathode potential) is increased and decreased within precisely defined time windows (usually a few seconds), as shown in Fig. 16.26 from ▶ Harris. The resulting voltammograms (shown, for instance, in Fig. 16.27a; in this example, elementary oxygen is reduced to superoxide ions: $O_2 + e^- \rightarrow O_2^-$ ... are you able to construct the MO diagram of this paramagnetic ion?) consist of several stages:

(a) Initially, the current increases with increasing voltage.
(b) Before the voltage maximum is reached, there is a *current strength maximum*: The maximum of this part of the voltammogram is called the *cathodic peak*.
(c) Afterwards the resulting current decreases again, although the voltage is increased even more.
  - The reason for this—at first sight probably unexpected—finding is that the analyte in the immediate vicinity of the electrode is almost completely consumed: The charge carriers present in the "residue" of the solution do not diffuse fast enough to its surface, so that a diffusion layer forms close to the electrode surface in which the concentration of the oxidised analyte is *locally* lowered. However, since the current flow is concentration-dependent, this explains the drop in the measured value for the current.
(d) When the *maximum voltage* is finally reached, the potential reversal (now the voltage is lowered again) causes the analytes—present in this diffusion layer in increased concentration—to be oxidised again in their reduced form: The measured value for the current strength drops rapidly.

Skoog, Section 25.4: Cyclic voltammetry
Harris, Section 16.5: Voltammetry

(e) Again, before the potential of the cathode reaches zero again, a *minimum* is observed in the voltammogram: Analogous to the cathodic peak mentioned above, an *anodic peak* is now observed—a voltage value with a *minimum current strength*.
    - This is due to the fact that an excess of oxidised analyte now builds up near the electrode surface.
(f) Due to this diffusion layer of oxidised analytes in close proximity to the electrode, further lowering of the cathode potential leads to an increase in the resulting current.
(g) Finally, after passing through a complete cycle of increased and decreased cathode potential, the initial position is reached again.

Strictly speaking, these steps only apply to rapidly occurring, reversible redox processes; in the case of irreversible and/or slowly occurring processes, it can no longer be assumed that the analytes are still present in their equilibrium concentration in oxidised and reduced form. This complicates the quantitative description considerably and goes beyond the basics to be covered in the context of this book. Let us therefore restrict ourselves to "ideal conditions".

In principle, the cathodic and the anodic peak should occur at exactly the same potential; the concrete value can again be determined on the basis of Nernst's equation (▶ Eqs. 4.1 and 4.2).

Let's consider what exactly actually happens in this process:
1. Initially, the analyte is present only in its reduced form.
2. If voltage is applied, oxidation takes place; the electrons released in the process provide the current flow.
3. At a certain potential, exactly 50 % of the analytes have been oxidised, so—if equilibrium conditions prevail—the following applies: $c(analyte_{oxidised\ form}) = c(analyte_{reduced\ form})$. This is the point at which the potential of the redox system is equal to its standard potential according to Nernst's equation (▶ Eqs. 4.1 and 4.2), i.e. $E = E^0$. This is exactly as in the case of tactile polarography from the previous section: Again, we are dealing with the *half-wave potential $E_{1/2}$*.
    (a) The resulting current strength now increases only slightly with a further increase in potential, then the oxidised form of the analyte predominates in the reaction mixture—especially in the immediate vicinity of the electrode: The diffusion layer mentioned above forms; the current flow becomes weaker.
4. Then the polarisation is reversed: The potential of the reference electrode is lowered.
5. Now the reduction of the analyte, which is now in the oxidised form, takes place; the current flow decreases.
6. If fifty percent of the previously oxidised analytes have been reduced again, we are back at the half-wave potential.

Under ideal conditions, the potential value for the maximum current in the voltammogram ($E_{anodic\ peak\ current}$) should coincide exactly with the potential value for the minimum current ($E_{cathodic\ peak\ current}$). In addition, there is now a diffusion potential, as we already know from ▶ Sect. 5.1. This is calculated as

$$E_{anodic\ peak\ current} - E_{cathodic\ peak\ current} = \frac{0.059\ V}{z} \quad (8.1)$$

Here z is again the number of electrons transferred in the course of this redox process; the similarity with a certain term from Nernst's equation is striking. However, this value for the diffusion potential is only valid for *absolutely ideal* systems in which *completely unhindered charge exchange* takes place. Such ideal conditions, however, do not exist even in the best "real existing" systems. For this reason, a formula corrected by a minimum of *charge exchange inhibition* is usually found in the literature (including ▶ Harris):

$$E_{\text{anodic peak current}} - E_{\text{cathodic peak current}} = \frac{0.057\,\text{V}}{z} \tag{8.2}$$

Since cyclic voltammetry can also be used to study more complex redox reactions in which, for example, various intermediates occur (each of which then has its own redox potential, of course, even if these intermediates occasionally cannot be isolated), this method is also frequently used to elucidate the mechanism of redox processes, for example. This is why cyclic voltammetry is often jokingly called "spectroscopy of the electrochemists". (The associated voltammograms, however, are a little more complicated and go far beyond what can and should be addressed here.)

- **Voltammetry in Trace Analysis—Stripping Analysis**

In **stripping analysis**, once again mercury is used (but also other electrode material, which we do not want to discuss here), and again the analyte (metal) ions are reduced to the elemental metal and become concentrated in the mercury (as amalgam). Only then does the actual *stripping* begin: If the direction of the current is reversed, i.e. the potential is made positive, the metal ions are oxidised again and return to the solution. The current forming during this oxidation process is proportional to the amount of analyte that has accumulated in the mercury. Thus, an *anodic* process is crucial here. Two aspects are particularly interesting:

- On the one hand, because the analyte accumulates in the mercury, even minimal amounts can be detected (this is why stripping analysis is mainly used in trace analysis); by using suitable catalysts, analytes can also be detected in an (original) concentration $<10^{-10}$ mol/L.
- On the other hand, because different metal ions have different redox potentials, several analytes can be quantified in parallel. Figure 16.22 from ▶ Harris shows an example of the stripping voltammogram that resulted from the examination of a honey sample: The heavy metals cadmium, lead and copper were detected, each in the **ppb range.**

Harris, Section 16.5: Voltammetry

Skoog, Section 25.8: Stripping methods

In Skoog you will find further variants of *stripping*. The principle—electrochemical deposition of the analyte on the surface of the working electrode—remains unchanged. This technique owes its name to the fact that the analyte is subsequently *detached from* the electrode—it is "*striped*".

❓ Questions
15. Why is it a requirement for an analyte solution that is to be examined by tactile polarography, that no elemental oxygen is dissolved in it?
16. How could one ensure in the laboratory that the analyte solution really does not contain oxygen?

Chapter 8 · Voltammetry

Summary

• **Potentiometry**

In potentiometry, information about the analyte content of solutions is obtained by measuring the resulting potential difference between a reference half-cell of precisely known concentration and the analyte half-cell under investigation. In the study of electroactive analytes, an indicator electrode, which is made of chemically inert material such as platinum and responds directly to the analyte, is in contact with the analyte solution (it is a 1st order electrode); a reference electrode with a known potential is used to determine the potential difference.

The two most important reference electrodes are the silver/silver chloride and the (saturated) calomel electrode; both are 2nd order electrodes with the potential not only being known but also remaining constant.

The two most important potentiometric methods are direct potentiometry, in which the concentration of the analyte to be quantified is determined by measuring the potential difference, and potentiometric titration, in which the concentration of the analyte is changed in the course of a titration, so that a titration curve can be recorded. Due to the different mobility of different charge carriers, in potentiometric measurements where different electrolyte solutions come into contact with each other, additionally a diffusion potential always forms. The mobility of electrolytes depends primarily on their size and their charge density. The influence of the diffusion potential can be minimised to a large extent by using electrolytes whose cations and anions have similar mobilities.

With ion-selective electrodes (ISE), on the other hand, which are each fitted to a specific analyte, no redox processes take place; these electrodes are based on the principle of ion exchange: Analyte ions diffuse into the electrode material and replace other ions there, which then migrate into the analyte solution. This results in a potential difference at the surface of the electrode. The electrode material can be crystalline or amorphous (glass electrode); liquid membrane electrodes and specially coated composite electrodes are also used. The most important ion-selective electrode is the glass electrode, which responds selectively to hydrogen cations and is used primarily for pH measurements. However, in a strongly basic medium with a high sodium concentration, it indicates too low a value due to the alkali error (similar charge density of $Na^+_{(aq)}$ and $H^+_{(aq)}$!). Comparable errors also occur with other ion-selective electrodes. Their sensitivity to non-analyte ions is described by the respective selectivity coefficient. Numerous biosensors also fall into the field of ISE.

• **Coulometry**

In coulometry, the number of electrons required to chemically deposit the analyte is determined—potentiostatically or galvanostatically—in accordance with Faraday's laws. A particularly important example of coulometry is the Karl Fischer titration for determining the water content of nonpolar analytes. In this method, sulfite ions, formed from sulfur dioxide, are first reacted with an alcohol in the presence of a base, to form the monoester of sulfurous acid and then oxidised with elemental iodine generated coulometrically from iodide ions to form sulfate ions (or sulfuric acid monoesters), whereby the water contained in the analyte is consumed. The amount of electrons transferred in the course of iodide oxidation then allows conclusions to be drawn about the water content.

- Amperometry

Amperometry exploits the fact that the current flowing at constant potential of two pairs of electrodes, with one of the two half-cells based on the analyte, depends on the concentration of the analyte; like coulometry, amperometry involves electrolysis. One of the most important applications of this technique is the Clark electrode for determining the oxygen content of solutions; the oxygen present is reduced to water. Modifications of this technique allow the quantification of other analytes; in combination with appropriate composite electrodes, numerous biomolecules can also be quantified (which brings us back to the field of biosensors).

- Voltammetry

Various analytical methods based on the relationship between voltage and current in electrochemical processes are grouped under the generic term voltammetry; voltammograms are obtained in which the resulting current density is plotted as a function of the concentration-dependent potential of the analyte electrode used. The method of polarography, which is based on the good solubility of most metals in elemental mercury, represents a historically important example of voltammetry, even though its use is avoided wherever possible today for toxicological reasons. (For some analytes, however, it cannot be avoided.) Tactile polarography, in which the potential is gradually increased in the course of the measurement, is to be regarded as a special application of voltammetry. In this case, the half-wave potential ($E_{1/2}$) is of particular importance: It is reached at the inflection point of the measurement curve and corresponds approximately to the standard potential of the redox system under investigation. Cyclic voltammetry is not based on a constant potential but on a cyclically varied potential; the resulting multilevel voltammograms with a cathodic and an anodic peak (at maximum and minimum current flow) are due to diffusion phenomena. The recording of cyclic voltammograms also allows the investigation of more complex, multistage redox processes. In a variant of voltammetry known as stripping analysis, the analyte is selectively enriched in elemental mercury; a further advantage of this technique, which is used primarily in trace analysis, is that several different analytes can also be quantified in parallel.

Answers

1. For sulfite (a) it is still very simple: Since oxygen is more electronegative than sulfur, the oxidation number for all three oxygen atoms is −II, for sulfur thus +IV. In the case of thiosulfate (b), it must be taken into account that *two* resonance structures can (and should) be stated for this molecular ion, in which the oxidation numbers of the sulfur atoms differ: Please have a look at the figure below. However, these differences do not pose a problem when dealing with redox equations, as long as you do not switch from one resonance structure to the other in the middle of working on such a question. With nitrite (c), it is easier again: Both oxygens come to −II, the nitrogen to +III; this ion is also resonance-stabilised, but this does not change any oxidation numbers. Finally, in the case of the chlorous acid (d), the problem does not arise at all: In the case of the free acid, there is no resonance stabilisation. Again, the two oxygens come to −II, the hydrogen to +I and the chlorine thus to +III. (The deprotonated anion, on the other hand, the chlorite ion ClO_2^-—the conjugate base of this acid -, is *indeed* resonance-stabilised again, but just as in the case of nitrite, this does not change the oxidation numbers.)

The mesomeric stabilised thiosulfate ion

2. The most important tool here is the Nernst equation. For simplicity's sake, we assume standard conditions, so we can refer back to ▶ Eq. (4.2). Next, we need the standard potential of the system under consideration. Tabular works (or even Harris) will tell us that for our system the redox potential is $E^0(Fe^{2+}/Fe) = -0.44$ V, wth $z = 2$. Since the activity coefficient may also be ignored, the internal potential difference (i.e., Galvani voltage) for (a) with concentration 0.1 mol/L is $E_a = -0.44$ V $+ (0.059$ V$/2) \times$ lg 0.1 $= -0.44$ V $+ (0.059$ V$/2) \times (-1) = -0.44$ V $- (0.059$ V$/2) = -0.47$ V. (Remember, the solid iron (the reduced form) is *not in solution*, so the denominator of the lg term in ▶ Eq. (4.2) is simply 1.)

For solution (b), which is one power of ten less concentrated, the computer value is then $E_b = -0.44$ V $+ (0.059$ V$/2) \times (-2) = -0.44$ V $- (0.059$ V$) = -0.50$ V.

For solution (c) with $c = 0.0023$ mol/L the calculation is not much more complicated:
$E_c = -0.44$ V $+ (0.059$ V$/2) \times$ lg 0.0023 $= -0.44$ V $+ (0.059$ V$/2) \times (-2.63)$ $= -0.44$ V $- 0.08$ V $= -0.52$ V. (It can be clearly seen that the *lower the* concentration of dissolved ions, the further the potential *decreases*.)

Harris, Appendix H (Standard Reduction Potentials)

3. The activity coefficients must be inserted into the Nernst equation—or more simply: One can simply use the coefficient γ as a "correction factor". The activities of solutions 2a to 2c can then be determined according to ▶ Eq. (4.3): $a_{2a} = 0.89 \times 0.1 = 0.089$; correspondingly, $E_{2a} = -0.44$ V $+ (0.059$ V$/2) \times$ lg (0.089) $= -0.44$ V $+ (0.059$ V$/2) \times (-1.05) = -0.44$ V -0.031 V $= -0.47$ V. From $a_{2b} = 0.95 \times 0.01 = 0.0095$, it follows that $E_{2b} = -0.44$ V $+ (0.059$ V$/2) \times$ lg (0.0095) $= -0.44$ V $+ (0.059$ V$/2) \times (-2.02)$ $= -0.44$ V $- 0.06$ V $= -0.50$ V, and $a_{3b} = 0.99 \times 0.0023 = 0.00227$, it follows that $E_{2c} = -0.44$ V $+ (0.059$ V$/2) \times$ lg 0.00227 $= -0.44$ V $+ (0.059$ V$/2) \times (-2.64) = -0.44$ V $- 0.08$ V $= -0.52$ V. In the context of measurement accuracy (or given the *significant figures* to be considered in these calculations—remember?), the difference between concentrations and activities has no discernible effect.

4. First of all, the standard potentials of the two half-cells have to be found out; here Harris can serve you well: $E^0(Co^{2+}/Co) = -0.282$ V; $E^0(Pd^{2+}/Pd) = +0.915$ V. Accordingly, the less noble cobalt will be the anode, the much nobler palladium the cathode, so that the overall reaction equation is:

$Co + Pd^{2+} \rightarrow Co^{2+} + Pd$

According to ▶ Eq. (4.15), the representation of this galvanic cell in cell symbolism is:

$Co|Co^{2+}\ Pd^{2+}|Pd$

For the resulting potential difference, it must again be taken into account that the *reduction potentials* are listed in the tables, but oxidation takes place at the anode; accordingly, the sign of the value must be *reversed* for the calculation of the difference: $\Delta E^0 = E^0_{Cathode} - \left(E^0_{Anode}\right) = E^0_{Cathode} + E^0_{(Anode\ with\ inverted\ sign)}$. Here the result is: $+0.915$ V $- (-0.282$ V$) = 0.915$ V $+ 0.282$ V $= 1.197$ V

5. To answer this question, one only has to consider how a diffusion potential forms in the first place: This happens whenever particles of opposite charge move/migrate with different velocities, so that local charge separation occurs. The larger the difference in mobility of cations and anions, the stronger the resulting charge separation, whereas if the mobility of the charge carriers is (approximately) identical, the charge separation remains almost imperceptible and thus the diffusion potential can (almost) be neglected. (It should be remembered once again why potassium chloride is so popular in electrodes.)

6. According to ▶ Eq. (5.4), the charge of an ion enters the equation with its value/valence squared. Since the sulfate ion has only two negative charges, whereas the phosphate ion has three of those, sodium phosphate will have a far higher ionic strength. Let's do the math:

$$I(Na_2SO_4) = \frac{1}{2} \cdot \left(z^2 \cdot (Na^+) \cdot c(Na^+) + z^2 \left(SO_4^{2-}\right) \cdot c\left(SO_4^{2-}\right) \right)$$
$$= \frac{1}{2} \cdot \left(1^2 \cdot 2 \cdot c(Na_2SO_4) + (-2)^2 \cdot c(Na_2SO_4) \right)$$
$$= \frac{1}{2} \cdot \left(1^2 \cdot 2 \cdot c(Na_2SO_4) + 4 \cdot c(Na_2SO_4) \right)$$
$$= \frac{1}{2} \cdot \left(6 \cdot c(Na_2SO_4) \right)$$
$$= 3 \cdot c(Na_2SO_4)$$

$$I(Na_3PO_4) = \frac{1}{2} \cdot \left(z^2 (Na^+) \cdot c(Na^+) + z^2 \left(PO_4^{3-}\right) \cdot c\left(PO_4^{3-}\right) \right)$$
$$= \frac{1}{2} \cdot \left(1^2 \cdot 3 \cdot c(Na_3PO_4) + (-3)^2 \cdot c(Na_3PO_4) \right)$$
$$= \frac{1}{2} \cdot \left(1^2 \cdot 3 \cdot c(Na_3PO_4) + 9 \cdot c(Na_3PO_4) \right)$$
$$= \frac{1}{2} \cdot \left(12 \cdot c(Na_3PO_4) \right)$$
$$= 6 \cdot c(Na_3PO_4)$$

Please do not make the mistake of trying to determine the ionic strength of a solution solely based on the *number of particles* going into solution: Their *charge* also plays a major role. (This is different from the osmotic pressure of electrolyte solutions: There it is *only* about the number of particles in solution.)

7. In the experiment, you measured the potential change that resulted from titrating a tin(II) solution against cerium(IV), where the tin(II) was oxidised to tin(IV) and the cerium(IV) was reduced to cerium(III). But what is crucial here is that you tracked the titration process in terms of the redox potential of the solution. Although at that time it may have seemed you were simply using "a meter" there, we now know that you cannot determine a single potential, you only ever measure *potential differences*, so you were using a reference electrode (which was part of the "meter"). In other words, this was clearly a potentiometric titration.

8. Consider again ▶ Eq. (5.6): It states that the voltage of a glass electrode changes by 0.059 V per pH "unit": The higher the pH, the more negative the potential. Thus, the reverse is true: a *more positive* potential compared to the reference solution means that the pH value of the analyte solution must be *lower*. With a potential difference of +0.1062 V this means (calculated by rearranging ▶ Eq. 5.6) that the pH value must be 1.8 pH units lower, i.e. pH = 5.75−1.8 = 3.95 applies to the analyte solution.

9. There is a very simple (and frighteningly un-chemical) reason for this: Frictional electricity can also cause electrostatic charging in glasses, and this would, of course, seriously disrupt at least the first measurement car-

ried out with this electrode. (Once again you see how close chemistry and physics come to each other in this field of analytics!)

10. The interference occurring here due to hydroxide ions in high concentration is analogous to the alkali error of the glass electrode: In its hydrated state, the hydroxide ion has a similar charge density as the hydrated fluoride ion; accordingly, confusion can already occur in the electrode—but only if the concentration of the hydroxide ions is correspondingly high; in this respect, the solid-state electrode described may well be granted a satisfactory selectivity coefficient $K_{Sel}(F^-, OH^-)$. The higher homologues of fluorine in anionic form, on the other hand, do not have sufficient similarity to fluoride either in their un-hydrated or in their hydrated state due to the different charge densities and therefore do not interfere.

11. If one considers that behind such composite electrodes there is actually nothing more than the change in the pH value of the aqueous solution into which the gas in question has diffused, there is nothing to prevent ammonia or other basic-reacting gases (such as volatile amines) from being quantified this way.

12. Let us first eliminate an irrelevant piece of information: The amount of potassium iodide added, from which elemental iodine is coulometrically produced, is completely meaningless here, because the iodine consumed again in the course of the actual redox reaction with the water, which is thereby converted to iodide ions, can (and will) be coulometrically oxidised again to elemental iodine. The first thing to do is to determine the amount of elemental iodine that was produced coulometrically.
   ▶ Equations (6.1) and (6.2) are required here, which are to be converted via the—probably generally known—formula n = m/M:

$$n = \frac{Q}{z \cdot F} = \frac{I \cdot t}{z \cdot F} = \frac{52.6 \cdot 10^{-3} \text{ A} \cdot 812 \text{ s}}{2 \cdot 96{,}485 \text{ A} \cdot \text{s/mol}} = 2.21 \times 10^{-3} \text{ mol.}$$

Since the molar ratio $n(H_2O):n(I_2)$ is 1:1, as mentioned earlier in ▶ Chap. 6, the sample thus contains 2.21×10^{-3} mol H_2O. Given a sample volume of V(sample) = 50 mL = 0.050 L, this gives a water content of 44.2 mmol/L. Assuming a density $\rho(H_2O)$ = 1 g/mL, this amount of substance corresponds to a mass $m(H_2O)$ = 795.6 mg according to the formulae $n_i = m_i/M_i$ and $c_i = n_i/V$ known from Part I of "Analytical Chemistry I" and thus, based on one litre, to a volume $V(H_2O)$ = 0.7956 mL. This leads to the realisation that, according to DIN 1310, $\sigma(H_2O)$ = 0.7956 mL/L = 0.07956 %.

13. Coulometry determines how many electrons must be transferred to bring about a redox process; amperometry measures the current flowing in the course of this redox process. One could therefore say: With coulometry, the number of electrons transferred is used to determine how many analyte particles have been converted; with amperometry, the *speed* at which the redox reaction takes place is measured.

14. In coulometry, the only decisive factor is the amount of charge required for complete conversion of the analytes; since the amount of charge required depends solely on the amount of analyte to be converted, temperature fluctuations or the like have no effect (unless, of course, the temperature rises so high that some of the analytes decompose, but we will not go into that sort of quibble now). However, reaction rates are known to be temperature-dependent (you will certainly remember the rule of thumb: for every 10° increase in temperature, the reaction rate approximately doubles), so temperature fluctuations *do* have an influence on amperometric measurements.

Harris, Section 16.5: Voltammetry

15. Elemental oxygen (O_2) in aqueous solution is easily reduced to hydrogen peroxide (H_2O_2), which is then converted to water (H_2O) in a second step. In both cases, a current will flow under tactile polarographic conditions, so you observe *two* polarographic steps. (You will find the corresponding polarogram as Fig. 16.19 in Harris.)
16. The simplest way is to pass elemental nitrogen through the solution for several minutes. (Usually 10 min are enough.) If you then ensure that nitrogen flows continuously over the analyte solution during the actual measurement in order to prevent atmospheric oxygen from diffusing into the solution under investigation, you are more or less on the safe side. (The fact that the polarographic steps characteristic of oxygen do not occur when the analyte solution is treated with nitrogen can also be seen from the figure mentioned in Question 15.)

Further Reading

Binnewies M, Jäckel M, Willner, H, Rayner-Canham, G (2016) Allgemeine und Anorganische Chemie. Springer, Heidelberg
Cammann K (2010) Instrumentelle Analytische Chemie. Spektrum, Heidelberg
Hage DS, Carr JD (2011) Analytical chemistry and quantitative analysis. Prentice Hall, Boston
Harris DC (2014) Lehrbuch der quantitativen analyse. Springer, Heidelberg
In case of problems with the absolute basics on the chemical processes covered in this part of the book, skimming the Binnewies will definitively not hurt. If you need more examples for the application of the various measuring techniques presented here, you might want to have a look at the copious references in Harris.
Lottspeich F, Engels JW (2012) Bioanalytik. Springer, Heidelberg
Skoog DA, Holler FJ, Crouch SR (2013) Instrumentelle analytik. Springer, Heidelberg

Other Analytical Methods

Requirements

This part, more than any of the previous ones, builds on the knowledge they have gathered—among others—from "Analytical Chemistry I" and the previous parts of this volume. In this respect, one could summarise the prerequisites with "everything that has gone before" … but that might be a bit daunting.

With electrogravimetry, you need:
- the basics of any gravimetric analysis
- redox processes and redox potentials
- Nernst's equation
- the relationship between current and voltage
- what has been explained so far about local potentials and potential changes

For thermal methods, you should be aware of the chemical changes that your analytes will undergo at controlledly raised temperatures, such as
- decarboxylation of carbonates or
- loss of water from hydrates or other compounds that can release water (or other small, volatile compounds) at higher temperatures.

In addition, there is basic information about the structure and cohesion of (semi-) crystalline and amorphous polymers. Phase transformation enthalpies are also important in the calorimetric investigations.

For the chapter dealing with special nuclides, some basic understanding of nuclear chemistry is required:
- isotopes (radioactive and stable ones)
- the different types of radioactive decay
- half-lives

In addition, the (absolute) basics of immunology are also required; however, a (short!) summary of the most relevant aspects can be found at the beginning of the relevant section in this part.

It should not be surprising that you will encounter again everything that has been presented so far on the subject of "fluorescence":
- Fluorescence/phosphorescence
- Inner conversion and intersystem crossing
- Singlet and triplet states
- Rotatory/vibratory excitation
- HOMO and LUMO

You will find a very (very!) brief summary of this in this part as well.

Learning Objectives

In this part you will get to know a selection of different methods of analysis, in which the previous principles are used for the quantification of a variety of different analytes (minerals and biosamples as well as engineering materials of one or the other kind):

The combination of the already known gravimetry with the (also already known) process of electrolysis allows the quantification of electrolytically depositable/metallic analytes.

You will learn how qualitative and quantitative information about the analytes can be obtained from the resulting changes/decomposition processes by selectively increasing the temperature.

Radioactive nuclides can be used to study reaction mechanisms and metabolic pathways; in this context, the information gain is based either on the relatively simple traceability of radioactive isotopes, on the different masses of different isotopes of the same element (so that further information can be obtained by combining them with the—already known—mass spectrometry), or on the radioactive decay of radionuclides already present or of basically stable isotopes that are excited to radioactive decay by neutron activation. Quantification is then based on the amount of detectable radioactivity, also in combination with biochemical methods—such as (radio)immunoassay, where the fields of chemistry and biology coincide with selected aspects of medicine.

In this part, the already familiar concepts and principles of spectrophotometry are extended to include the phenomenon of fluorescence/phosphorescence, which also can be quantified on the basis of Lambert-Beer's law. Finally, you will see how the chemical aspects of fluorescence can be combined with light microscopy, which belongs more to the life sciences, so that it becomes clear once again how much the boundaries of the individual sub-areas of the natural sciences (physics, chemistry, biology) are becoming increasingly blurred.

As in the previous parts on analytical chemistry, the aim this time is also to explain the techniques commonly used in the various analytical procedures with regard to the principles behind them to such an extent that the individual instruments used are not just "black boxes" that deliver a measurement result in one or the other unfathomable way, but that you can understand the processes taking place in them at the microscopic (molecular/atomic) level. At the same time, however, the actual structure of the respective devices used should also become comprehensible; due to the space limitations to which such an introductory book is naturally subject, explanations in this regard are, however, limited to a minimum; if you are interested, you are strongly advised to consult the further literature stated in the appendix to this part.

Contents

Chapter 9   **Gravimetric Analyses – 137**

Chapter 10   **Thermal Processes – 147**

Chapter 11   **Use of Radioactive Nuclides – 153**

Chapter 12   **Fluorescence Methods – 169**

Gravimetric Analyses

Contents

9.1　　Electrogravimetry – 138

9.2　　Thermal Methods: Thermogravimetry (TG) – 143

The principle of gravimetry has already been dealt with in some detail in Part II of "Analytical Chemistry I", and the "special case" of electrolytic deposition of the analyte to be quantified was also briefly mentioned there. However, since further gravimetric methods are to be presented in this part—▶ Sect. 9.2 will deal with *thermogravimetry*—we will again briefly turn to electrogravimetry.

9.1 Electrogravimetry

As already briefly outlined in the section above, electrogravimetry involves electrolysing a solution containing electrolytes so that the analyte ions in question are deposited in elemental form on the surface of a suitable electrode. Two things are a prerequisite for this:
- Inert electrodes must be used, i.e. electrodes that do not themselves interact with the analyte solution or the analyte in any other way.
  Preferably platinum electrodes are used here, and of course—as you will certainly remember from Part II—two electrodes are required:
  - At the **working electrode**—the *cathode*—the separation of the analyte in elemental form takes place: the reduction

    $$M^{x+} + xe^- \rightarrow M$$

    Since this electrode requires as large a surface area as possible, a platinum mesh is preferred. (The use of such shaped working electrodes ensures, among other things, that the time required for the analysis is minimised, and saving time is, as is well known, always desirable.)
- A simple platinum wire, often in the shape of a spiral, is usually used as a *counter* electrode at which *no* (electro-)chemical processes relevant to the analyte take place. The oxidation of the solvent takes place at this *anode*. For aqueous solutions (and these are the preferred solutions for electrogravimetry) this results in:

  $$2H_2O \rightarrow O_{2(g)} + 4H^+ + 4e^- \quad \text{or} \quad 6H_2O \rightarrow O_{2(g)} + 4H_3O^+ + 4e^-$$

  The experimental setup is shown schematically in Fig. 16.5 from ▶ Harris.
- For the quantification of the analyte, the electrolysis must happen quantitatively; accordingly, the progress of the reaction must be checked.
  - This is relatively easy with metals which in their elementary state show a characteristic colour. (You already know the example of copper from Part II of "Analytical Chemistry I"; here at the beginning of the experiment the working electrode is not yet completely immersed in the analyte solution, and after a while it is lowered a little further to see whether further deposition occurs, which would then lead to a discolouration of the silvery surface of the part of the platinum electrode not yet covered with copper.)
  - It is also quite easy to check the reaction progress of analyte ions that cause colouration in aqueous solution (which of course also applies to copper, but also to cobalt(II), manganese(II), chromium(III), or chromate(VI) ions and others). If you do not want to rely on the measuring instrument "human eye", **photometric** observation of the reaction progress is recommended. (You may already know photometry from Part II of "Analytical Chemistry I".)
  - If this procedure is not feasible, one takes a small part of the analyte solution (the volume of which should also be known exactly, otherwise

Harris, Section 16.2: Electrogravimetric analysis

9.1 · Electrogravimetry

determining of the concentration of this solution becomes ... tricky) and checks the presence of the analyte with the help of procedures known from *qualitative* analysis. If this test is positive, there are obviously still analyte ions present, and the electrolysis must be continued.
- However, it is simpler to use a method that you learned about in Part II "Electroanalytical methods": amperometry. You measure the current flow within the analyte solution; detecting the end point **amperometrically** should be rather easy.

Subsequently, as already mentioned in Part II, the resulting mass of the now elemental analyte can be determined by simple comparison of the working electrode mass *before* and *after* electrolysis. Using the usual calculations/formulas (relationship between mass, molar mass, and amount of substance and, if necessary, between amount of substance, volume, and molar concentration of substance), statements can then be made about the mass and/or concentration content of the analyte.

■ **Influencing Factors**

The voltage that has to be applied in each case in order to deposit an analyte elementally depends—understandably—on its electrochemical potential—or, to be more precise: on the electrochemical potential of the corresponding analyte solution. (The relationship between concentrations and the resulting potentials is described by **Nernst's equation**, which you also had to deal with in Part II.) However, the desired electrolysis by no means already occurs from **exactly** the voltage corresponding to the potential difference obtained purely by calculation: There is indeed a connection between the applied voltage and the resulting current flow, but it is obviously influenced by several factors. Figure 16.6 from ▶ Harris shows the discrepancy between what would have to happen if all these factors did *not* occur (labelled "naive expectation" in said figure) and the *actual* resulting current-voltage curve using the example of the electrolysis of a copper(II) sulfate solution (in an acidic medium).

Harris, Section 16.2: Electrogravimetric analysis

(a) If the voltage is too low ($E_{applied} < E_{analyt}$), *nothing at all* can be observed.
(b) Even if the applied voltage corresponds to the expected value of the analyte in question (in its respective concentration), there is still no significant deposition; the resulting current is also very weak.
(c) Only when the applied voltage clearly exceeds the potential of the analyte, i.e. when the **decomposition voltage** is reached, a recognisable current flow and thus also a recognisable analyte deposition (or a recognisable decrease in the concentration of the analyte ions in solution) can be observed.
(d) Once the analyte has been completely consumed, i.e. electrolysis is complete, the resulting current rises sharply because now, instead of the analyte, the *hydrogen* in the solvent is being reduced at the working electrode:

$$4H_2O + 4e^- \rightarrow 2H_2 \uparrow + 4OH^- \quad \text{or (in acidic medium)} \quad 4H_3O^+ + 4e^- \rightarrow 2H_2 \uparrow + 4H_2O$$

Now, of course, the question arises as to *why* the decomposition voltage is so much greater than would be expected from the pure potential value of the analyte. Three factors are of particular importance here:
(a) the overvoltage
(b) the concentration potential
(c) Ohm's potential

Let's look at all three of them in turn.

▪▪ Overvoltage

The term "overvoltage" refers first of all to the activation energy of the redox reaction that is to take place at the electrode in question. The desired reaction rate also has an influence here: The faster the process is to take place, the faster the electron transfer must occur, i.e. the greater the local current density must be, and the greater the local current density (given in A/m^2), the greater the overvoltage will be. As an example, the overvoltages that occur during the reductive release of elemental hydrogen (H_2) at platinum electrodes can be mentioned here:

Current density	10^1 A/m^2	Overvoltage	0.024 V
Current density	10^2 A/m^2	Overvoltage	0.068 V
Current density	10^3 A/m^2	Overvoltage	0.288 V
Current density	10^4 A/m^2	Overvoltage	0.676 V

However, it is not only the current density that plays a role: the electrode material and the type of gas forming at it must also be taken into account. Both can be understood quite easily if you consider what exactly happens during this process.

For this purpose, let us take another look at what happens at the counter electrode during the electrogravimetric analysis of an aqueous metal salt solution (the anions of which do *not themselves* enter into redox processes under the given conditions—just for simplification): Molecules of the solvent are oxidised to produce elemental oxygen. For this to happen, however, the solvent molecules obviously must reach the electrode—but if the surface of the electrode is occupied by oxygen molecules after the oxidation of the first molecules (in technical terms: when the O_2 molecules formed by oxidation are **adsorbed** onto the electrode surface), further H_2O molecules must first diffuse through this layer of gas molecules in order to also enter into this reaction, which naturally poses a certain problem. Now it should also become understandable why the current density has such a striking effect on the overvoltage: The higher the current density is, the faster the oxidation of the first water molecules (which are in direct contact with the electrode material) takes place, and accordingly a comparatively thick layer of adsorbed oxygen quickly builds up there. Also the fact that, depending on the kind of gas produced, identical currents applied lead to to different overvoltages, should be obvious now: Some gases are easier to adsorb onto the surface than others and/or impede the diffusion of other solvent molecules to varying degrees. Let's compare the two gases that most often play a role in the electrolysis of aqueous solutions: At the same current density (e.g. 100 A/m^2), the overvoltage of elemental hydrogen at a platinum electrode is 0.068 V, whereas that of elemental oxygen is 0.85 V, i.e. it is greater there by more than an order of magnitude.

As mentioned above, it is also important which *electrode material* is used: Adsorption is significantly stronger on some materials than on others—which can be attributed to the microscopic nature: While, for example, at the same current density, the overvoltage due to hydrogen generated at the surface of a platinum electrode is 0.068 V, it amounts to 0.762 V for a silver electrode, and even 1.090 V for an electrode made of lead. Compared to all other common electrode materials, particularly low overvoltages occur with electrodes made of **platinised platinum**, i.e. a platinum electrode whose surface has been additionally covered with platinum atoms: Here, the overvoltage (upon formation of elemental hydrogen and the same current density) is only 0.030 V.

9.1 · Electrogravimetry

The influence of current density and electrode material on the overvoltages occurring during the formation of elemental hydrogen and elemental oxygen is summarised in Table 16.1 from ▶ Harris.

Harris, Section 16.1: Fundamentals of electrolysis

Platinised Platinum: A Closer Look

Platinum-plated platinum is elemental platinum on whose (quite smooth) surface elemental platinum produced *in situ is* electrochemically deposited. In this process, the "newly added" platinum atoms do not necessarily follow the natural crystallisation of this metal (pure platinum crystallises in the cubically densest spherical packing, shown two-dimensionally in a), but rather attach themselves primarily at random to its surface (as illustrated in b). This leads to a significantly enlarged surface compared to the "platinum sheet".

The platinum is plated by electrolytic deposition from an aqueous solution of hexachloridoplatinum(IV) acid ($H_2[PtCl_6]$) in the presence of a catalyst (lead(II) acetate, $Pb(O-C(=O)-CH_3)_2$). Because of the high reactivity of the coating, a platinum electrode treated in this way is quite sensitive to air; if it is not store under protective conditions, it rapidly loses its special properties.

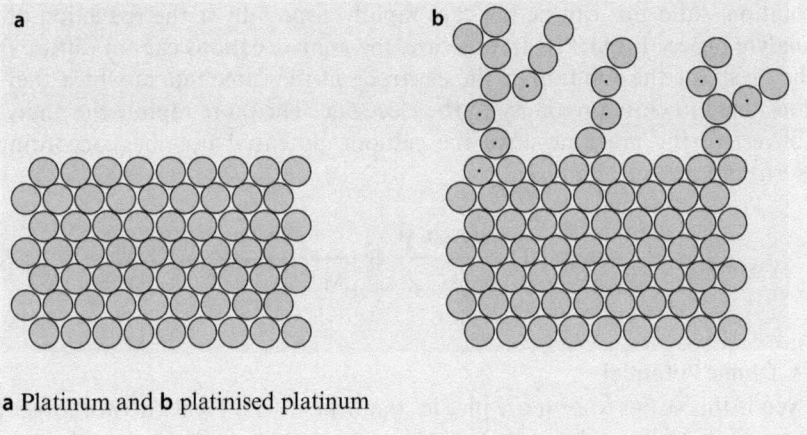

a Platinum and **b** platinised platinum

One could assume that an enlarged surface should offer more possibilities for adsorption of possible gases, so that an *increased* overvoltage should result. This consideration is obvious, but wrong for two reasons:

1. On the one hand, one should not imagine such adsorption processes in such a simplistic way that an H_2 or O_2 molecule simply attaches itself to a single platinum atom (shown in (a) as an example with an H_2 molecule): The interactions between a gas molecule and a single metal atom (van der Waals forces, or more precisely: **London dispersion forces**) are far too weak for this.
2. The illustration marked (b) comes closer to the truth. It shows the **physisorption of** a hydrogen molecule to a surface consisting of several platinum atoms.

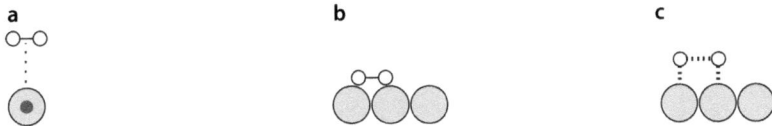

a Hypothetical, **b** real occurring physisorption and **c** chemisorption

3. In fact, however, even this idea is not quite correct: When gases are adsorbed onto a surface, in addition to physisorption, usually **chemisorption** occurs: in this case, the bonds within the individual gas molecules are (measurably) weakened (illustrated in (c) by the dashed bond between the two adsorbed gas atoms), while at the same time real chemical bonds are also formed between the gas particles and the surface of the adsorbing material. (Since these do not have the same strength as "ordinary" chemical bonds, they have also been shown as dashed in the figure)

For this reason, gas molecules adsorbed/chemisorbed onto the surface of a (metal) catalyst, for example, can be made to react more easily than in the unadsorbed state (think of the catalyst of the **Haber-Bosch synthesis**).

▪▪ Concentration Potential

Quite analogously to the polarisation phenomena which we have already dealt with in Part II, a diffusion potential also forms in electrogravimetry as soon as the concentration of the analyte ions in the immediate vicinity of the electrode surface differs from the concentration of the same ions in the "rest" of the solution—and this can be the case rapidly, especially if the reduction of the analyte proceeds quickly: In this case, the analyte cations cannot diffuse from the "rest" of the solution to the electrode at the same rate at which they are consumed in close proximity to the electrode. The more rapidly the analyte is converted, the more negative the cathode potential becomes, according to Nernst's equation (Eq. 9.1).

$$E_{Cathode} = E^0(\text{Analyte M}) - \frac{0.059 \text{ V}}{z} \cdot \lg \frac{1}{\left[M^{x+}\right]_{\text{on the surface}}} \tag{9.1}$$

▪▪ Ohmic Potential

Even if this seems to be more physics than chemistry: Please do not forget that every (half-)cell also has a certain *electrical resistance*. **Ohm's law** also applies in electrolysis systems:

$$U = R \cdot I \quad \text{or} \quad \frac{U}{I} = R \tag{9.2}$$

Here U is the voltage resulting from a potential difference ΔE; thus one can also work with potentials, and then the following applies:

$$\Delta E_{Ohm} = I \cdot R_{Redox-System} \tag{9.3}$$

> ▶ Example
>
> Assuming that the resistance of the test arrangement is 4.2 Ω and the current flow is I = 23 mA, an additional voltage (i.e. an overvoltage) of
>
> 4.2 Ω × 23 mA = 4.2 Ω × 0.023 A = 0.0966 V
>
> must be applied to overcome this resistance. ◀

? Questions

1. What was the concentration of a solution of cobalt(II) chloride with V = 20.00 mL if, in the electrogravimetric test, $m_{\text{(platinum mesh before the start of the experiment)}}$ = 27.30 g and $m_{\text{(platinum mesh at the end)}}$ = 27.68 g?
2. What additional voltage must be applied if, at a current flow of 17 mA, the resistance of a test arrangement is 2.3 Ω?
3. Say something about the concentration potential of the solution from task 1.
4. What is the ohmic potential at a current of 21 mA if the resistance of the experimental setup is 1.86 Ω?

9.2 Thermal Methods: Thermogravimetry (TG)

The collective term "thermal analysis methods" covers various methods in which the temperature dependence of one or other physical property of the sample in question is analysed. Since this section initially covered *gravimetric* analyses, we will first deal with **thermogravimetry**, i.e. the dependence of the mass of an analyte sample on temperature. Further thermal methods will be dealt with in the following chapter; an overview of the various methods can of course also be found in ► Skoog.

Skoog, Chapter 31: Thermal methods

From the introduction to general gravimetry (Part II of "Analytical Chemistry I"), you know that gravimetric quantification of an analyte is only possible (or useful) if the weighing form
- has a known chemical composition *and*
- is stable, i.e. does not change its composition over time
    - neither by reaction with the oxygen (or other gases) in the air
    - nor by hygroscopic behaviour etc.

Unfortunately, by no means all substances (nor all analytes) do us the favour of meeting these conditions. But even in such cases, further information can be obtained: by looking at how the mass of an analyte changes with increasing temperature—in almost all cases, a loss of mass can be observed. For example, Fig. 31.2 from ► Skoog shows the loss of mass of an organic sample that decomposes (as expected) with increased temperature until only non-ashable residues remain. This figure also shows the influence of the atmosphere in the sample chamber:

Skoog, Section 31.1.1: Thermogravimetric analysis, measuring equipment

- When oxygen is excluded by using a *pure* nitrogen atmosphere, the mass loss is only 23.5%.
- If the nitrogen atmosphere is replaced by oxygen, the elemental carbon formed up to this point is oxidised to carbon dioxide, resulting in a significantly greater mass loss: In total, it now amounts to 91.5%.
- This does not change when afterwards the oxygen is replaced by nitrogen again.

In the same way, compounds containing **water of crystallisation** and/or which are converted at sufficiently high temperatures to another, less mass-rich, compound can also be investigated: If they are heated sufficiently, the water of crystallisation and/or the by-products released during said conversion can escape—which understandably leads to a mass loss that can be tracked in the form of a thermogravimetric curve:

Harris, Section 26.2: Precipitation

Figure 26.4 from ▶ Harris shows an example of the corresponding curve for the thermal decomposition of calcium salicylate monohydrate ($Ca(C_6H_4(OH)-COO)_2 \cdot H_2O$), in which several **metastable** intermediates can be detected at the same time with a continuous increase in temperature:

1. Up to about 150 °C the initial substance remains unchanged. Then the first mass loss occurs.
2. A first plateau in the curve occurs at 200 °C when the anhydrous form ($Ca(C_6H_4(OH)-COO)_2$) is present, i.e. the water of crystallisation has been split off, resulting in a mass loss of about 10%.
3. At about 300 °C, an H atom is transferred from the hydroxyl group of one of the two salicylate anions to the other, so that one equivalent of free salicylic acid ($C_6H_4(OH)-COOH$) is split off; the now present negative charge of the now deprotonated hydroxyl group as well as the (also deprotonated) carboxyl group of the remaining salicylic acid form the counterions to the calcium cation.
4. At 500 °C the *first decomposition stage* takes place; calcium carbonate ($CaCO_3$) is obtained.
5. This compound releases carbon dioxide at about 700 °C, leaving anhydrous calcium oxide (CaO, *burnt* or *unslaked* lime, also known as *quicklime*).

Overall, this results in a mass loss of more than 80%.

▪ TG for the Investigation of More Complex Hydrates

A frequently cited example of a compound containing water of crystallisation is copper sulfate, which under standard conditions exists as copper(II) sulfate pentahydrate ($CuSO_4 \cdot 5H_2O$), but in which the five water molecules do not occupy *equivalent* lattice sites: Rather, this compound is the monohydrate of the salt tetraaqua copper(II) sulfate, that is, $[Cu(H_2O)_4]SO_4 \cdot H_2O$. And even the four water molecules that are part of the complex cation cannot be removed simultaneously:

— If the characteristically blue crystals are heated successively, two water molecules are initially split off at about 95 °C, so that the (still blue) trihydrate ($CuSO_4 \cdot 3H_2O$) is formed.
— At 115 °C, two water molecules are split off again, and the (no longer quite so blue) monohydrate ($CuSO_4 \cdot H_2O$) is obtained.
— Only when the temperature is raised above 200 °C does the last water of crystallisation split off; then the (colourless) anhydrous copper(II) sulfate is present.

▶ https://de.wikipedia.org/wiki/Kupfersulfat#/media/File:Cuso4_5h2o.PNG

This gradual "dehydration" of the salt can be traced very well with the help of thermogravimetry; the individual steps and the associated mass losses can be easily understood in the corresponding thermogravimetric curve, which can be found, for example, in the German-language Wikipedia.

▪▪ Micro-thermogravimetry (μ-TG)

A variant of thermogravimetry is also suitable for tiny sample quantities: With **micro-thermogravimetry**, samples with a mass <1 μg can be examined thermogravimetrically; accordingly, a water content of e.g. 0.5% can also be detected (requiring slightly larger sample quantities). This is where *microbalances* come into play, the apparatus design of which is quite complex, since minimal changes in mass are to be registered: The change in sample mass leads to a deflection of the balance beam, which thus comes between a light source and a photodiode. This leads to a weakening of the photodiode current, which is simultaneously compared with the current of a second photodiode *not* covered

9.2 · Thermal Methods: Thermogravimetry (TG)

by the balance beam. The beam deflection is then counteracted by a permanent magnet via the magnetic field resulting from the interaction of the differences of the two photodiode currents with a coil, leading to the *actual* measurement results. Some technical aspects of the experimental setup, which go beyond the basics presented in this part, are well explained in ▶ Skoog.

Skoog, Section 31.1: Thermogravimetric analysis
Skoog, Section 31.4: Microthermal analysis

❓ Questions

5. Why does the splitting off of water of crystallisation in a thermogravimetric investigation occur, in most cases, only at temperatures which are noticeably above the boiling point of water (i.e.: 100 °C)?
6. Why can a (micro-)thermogravimetric analysis of, e.g., an organic sample lead to different results with different atmospheres present in the sample container?

As is so often the case in instrumental analysis, gravimetric analysis (in particular (micro-)thermogravimetry) can also be combined with other measurement methods—even with methods that you are already familiar with: For example, the combination of (μ-)TG with an infrared spectrometer (known, for example, from Part IV of "Analytical Chemistry I"), which is then referred to as TG/IR, allows the identification of the decay products released in the course of the thermal decomposition of the analyte. The use of Fourier transform instruments is particularly popular, in this respect the abbreviation TG/FTIR should not (any longer) be confusing. Alternatively, thermogravimetry and mass spectrometry can be used together: TG/MS is becoming increasingly popular in analytics.

Thermal Processes

Contents

10.1 Differential Thermal Analysis – 148

10.2 Calorimetry – 150

Skoog, Section 31.2: Differential thermal analysis

10.1 Differential Thermal Analysis

In **differential thermal analysis, DTA** for short, a substance sample is continuously heated in presence of a reference material. The respective temperature of both materials is measured, whereby a temperature program calibrated for this purpose causes a linear increase in the temperature of the sample (T_s). The aim of this measurement is to detect any changes that occur in the sample substance with increasing temperature; thus a $\Delta T/T$ curve is recorded. An example of a differential thermogram is shown in Fig. 31.7 from ▶ Skoog.

With such a differential thermogram, exothermic and endothermic processes within the sample substance can be detected:

— For example, it can be seen that a polymer that is amorphous in itself crystallises (partially) after the **glass transition** (which is accompanied by only a slight change in temperature) with a further increase in temperature—an *exo*thermic process.
— At even higher temperatures, the polymer—now (more) crystalline—will *melt*, which is an *endo*thermic process.
— Depending on the atmosphere in the sample chamber, further changes occur when the temperature is increased further (you already know this from ▶ Sect. 9.2):
  – If the sample chamber contains oxygen (from the air, or because an oxygen-rich atmosphere has been deliberately provided), *exothermic* oxidation processes follow.
  – Such processes can also be prevented by purging with nitrogen or argon; in this case, the peak in the differential thermogram that is characteristic of an exothermic process obviously does *not* occur.
— If the temperature of the sample is increased even further, (*endothermic*) decomposition processes can also be detected.

The extent of exothermic or endothermic nature of each of these processes can be determined precisely by comparison with the reference substance, which (ideally) undergoes no changes at all at increased temperature.

The DTA is mainly used for the study of engineering materials, especially in connection with plastics.

> **Some Considerations on Polymers**
> Without going too deeply into the subject of polymers (there are specific introductions to *macromolecular chemistry* for that), this section will at least briefly touch on some fundamental aspects so that the various processes studied in differential thermal analysis become a little more comprehensible:
>
> If one considers the macromolecules obtained in the course of **polymerisation or polycondensation reactions**, one may first of all imagine them to be long (molecular) threads with only Van der Waals forces prevailing between them. (The more complex case of cross-linking between the individual macromolecules will not be discussed further here.) However, as is well known, there is more or less free rotatability around each single bond, and the individual molecules, as already mentioned, are held together only by Van der Waals interactions. Now, Van der Waals forces are not overly strong, so it should come as no surprise that such polymers ultimately present as a kind of three-dimensional tangle of interwoven, largely randomly intertwined molecular chains that might remind of a pile of thoroughly cooked spaghetti. Two aspects are particularly important here:

10.1 · Differential Thermal Analysis

- The individual molecular chains are not connected with each other by covalent bonds (if one disregards the possibility of possible cross-links, as just mentioned), but are *mechanically* intertwined. In addition, Van der Waals forces act between them. The result is a random three-dimensional arrangement that does not bear much resemblance to a regular crystal: the polymer is **amorphous**.
- The cohesion due to the Van der Waals forces is only slightly pronounced, but even if the individual chains *could* at least theoretically change their relative position to the other chains, this does not happen below a certain temperature (which of course depends on which monomers have been polymerised (or polycondensed) into long chains, which functional groups and/or heteroatoms are present, etc.), simply because the available energy is not sufficient to allow such a relocation of individual chains. Thus, there is practically no free mobility of the individual polymer chains, the whole three-dimensional structure of polymer chains is very inflexible and brittle: In many respects it behaves like glass.

If the temperature is then increased, this promotes the mobility of the individual molecular chains: Above the **glass transition temperature (TG),** such a polymer behaves rather rubbery (if the purely mechanical entanglement is not too pronounced, it may even melt), because the chains now can change their relative position (at least to a certain extent: the purely mechanical entanglement of the individual chains still exists). Further energy input then allows the individual polymer molecules to arrange themselves relative to each other in such a way that the intermolecular interactions (Van der Waals forces) are maximised: The substance **crystallises**. (Depending on the type of monomers, this crystallisation can actually occur *completely* or only *partially*: In the latter case, the overall still amorphous polymer exhibits localised crystalline zones; it is then referred to as a **semicrystalline** or *partially crystalline* polymer). Because of the lattice energy released during this crystallisation process, it is an *exothermic* process, which can be clearly seen in the differential thermogram (Fig. 31.7 from ▶ Skoog).

Skoog, Section 31.2: Differential thermal analysis

For each polymer, the associated glass transition temperature can be specified; this also determines the range of use (in the sense of: "At what temperatures can this polymer be used as intended?") of the material concerned. An amorphous (inorganic) polymer of (alumino)silicates (usually referred to as "glass") is extremely brittle. Above the glass temperature (which may well vary depending on the composition, but is in the *order of* 1000–1500 °C for commercially available glasses) it softens; then it shows significantly greater flexibility, but usually no longer fulfils the intended purpose (e.g. as a window pane). Such polymers are therefore *processed above* the glass temperature (shaping of glass, e.g. glass blowing or similar), but *used below* T_G. Other polymers, for example the elastomeric material of which rubber bands are made, are intended for use at $T > T_G$: Only then do they exhibit the desired (thermo)elastic properties.

By the way, you can check this with a simple experiment: Rubber bands used to seal frozen food are elastic when they have come back to room temperature, but if you try to remove them from a frozen food package that has just come out of the freezer ($T \simeq -20$ °C), they will probably crumble.

By the way: The fact that rubber-elastic polymers become brittle below a certain temperature (T_G) was the reason for what is probably the most famous accident in manned space flight, the **Challenger disaster** of 1986: Due to the unexpectedly low temperatures during the launch, a sealing ring (which went down in literature under the name "O-ring") had lost its elasticity and therefore no longer fulfilled its function as a sealing ring, which caused the fuel to flow out uncontrollably and catch fire. The physicist Richard Feynman impressively proved what had happened to the subsequently appointed investigating commission by briefly immersing an identical sealing ring in a glass of ice water and then showing that the now cooled material did not immediately return to its original shape when pressure was applied to it.

10.2 Calorimetry

In calorimetry, heat quantities are measured with the aid of a calorimeter in order to quantify energy conversions and/or to find out which energy quantities are required, for example, to bring about the quantitative phase transformation of a sample of precisely defined mass or amount of substance. Even though this topic is certainly already known from *physical chemistry*, it should be briefly (!) mentioned again, because there is a method of analysis derived from it, which is of particular importance among the thermal methods: **Differential calorimetry** is one of the most important methods for the analysis of engineering materials. It is used for the

(a) determination of the glass transition temperature of amorphous solids, especially polymers (we have just discussed the glass transition temperature in ▶ Sect. 10.1),
(b) determination of heats of melting (a.k.a. heat of fusion, i.e. the energy required for the phase change from solid to liquid aggregate state),
(c) determination of the degree of crystallisation of semi-crystalline substances (this is also known from ▶ Sect. 10.1).

The most common method is *differential scanning calorimetry* (**DSC**), which has the advantage that it can be carried out very quickly and does not present an excessive challenge in terms of equipment. As with DTA, an analyte sample and a reference substance are measured simultaneously; this time, the focus is on how the *heat flow* in the analyte sample differs from that in the reference substance; this method is shown schematically in Fig. 31.11 from ▶ Skoog.

There are two fundamentally different experimental setups, which, however, do not differ in their information content:
(a) One can build up a temperature gradient (with a given heating rate), then one speaks of *temperature scanning*, or
(b) A constant temperature is chosen from the outset (which must of course be higher than the ambient temperature or the temperature of the two substance samples); in this case, the measurement is carried out in *isothermal mode*.

In both cases, the *heat flow* is considered. Here you can already see the fundamental difference to differential thermal analysis (DTA):
1. With the DSC one considers *energy differences*, as such it is a calorimetric method.
2. The DTA measures *temperature differences*.

However, since a heat flux can be measured far more precisely than a temperature, DSC is much better suited for quantitative analysis than DTA is. The exemplary DSC scan of a polymer (which will not be discussed further here) on the basis of a temperature gradient is shown in Fig. 31.13 from ▶ Skoog: As you can see, heat flow differences in the range of only a few milliwatts were recorded. With regard to the statements about exothermic and endothermic processes, the same applies as explained in ▶ Sect. 10.1.

10.2 · Calorimetry

It is important to note that differential calorimetry alone cannot be used to *identify* an analyte: Although the resulting measured values allow statements to be made about the properties of the analyte in question, the values themselves are *not substance-specific*.

❗ Because of the Linguistic Similarity:
Please do not confuse calorimetry, colorimetry and coulometry.
- *Calorimetry* is about measuring quantities of heat (see *physical chemistry*).
- In *colorimetry*, colours are compared with each other (or with a reference scale) (known, for example, from Part II of "Analytical Chemistry I").
- In *coulometry*, one considers the charge quantity required to effect a redox process (you know these from Part II of this book).

Basically, confusing them should be impossible (what it's about in each case can usually be seen in the context!), but it happens much more often than most examiners would like…

❓ Questions
7. What conclusions can be drawn from a positive peak in a differential thermogram, i.e. a temperature that at the time of measurement is higher in the sample than in the reference substance?
8. What do you think of the statement that DTA is about the heat capacity of the sample substance?
9. Why, despite the undeniable similarity of DTA and DSC, only differential scanning calorimetry is considered (or used) as a *quantitative* method?

Use of Radioactive Nuclides

Contents

11.1 Radiochemical Analysis: Neutron Activation Methods – 155

11.2 Radioactive *Tracers* – 157

11.3 Radioactive Age Determination – 159

11.4 Radioimmunoassay (RIA) – 161

You know from *general and inorganic chemistry* that for the majority of elements there are different isotopes: atoms which do not differ in their atomic number (otherwise they would not be atoms of the same type of element!), but in their *mass number*—because they have a different number of neutrons in the nucleus. (You have dealt a lot with this in Part I of this book, especially in connection with mass spectrometry.)

Actually, elemental transformations are *not* allowed in chemistry: When in laboratory jargon it is said that a compound (e.g. HCl upon contact with water) splits off a "proton", this is of course *not* supposed to mean that a proton were released from the atomic nucleus—after all, e.g. chlorine would then become sulfur (change of the nuclear charge and thus atomic number!), and that is … not chemistry. This simplified expression for splitting off a covalently bonded, positively polarised hydrogen atom in the form of a hydrogen cation (H^+) is based on the fact that the nucleus of the by far most common hydrogen isotope (^1H) consists of only one single proton; chemically, however, it makes no difference at all whether the hydrochloric acid solution does not perhaps also contain one or the other deuterium atom: DCl (or ^2HCl) acts as a strong acid as well, so that the pH value of the aqueous solution is also influenced by ($D^+_{(aq)}$) deuterium ions.

However, there is also *nuclear chemistry,* which deals with **radioactive decay** processes, in the course of which something happens to the nuclear "building blocks", so that it really comes to the *transformation of one element into another* (keyword: radioactive decay series). Even if we do not want to go into this topic in too much depth, we need to mention the two most important radioactive decay processes associated with elemental transformation, which can occur in **radionuclides** (i.e. radioactive, unstable isotopes):

- In **α-decay**, an alpha particle is split off. The α-particles are effectively $^4He^{2+}$ ions, so two protons and two neutrons are released as a unit from the atomic nucleus of the radionuclide under consideration. Accordingly, the *mass number* of the nuclide decreases by 4 and the *atomic number* by 2. (Probably the most important example is the decay of the uranium isotope $^{235}_{92}U$, which upon the loss of an α-particle (^4_2He) is converted into the thorium isotope $^{231}_{90}Th$.)
- In the case of **β-decay**, the *atomic number* of the nuclide under consideration (and thus the elemental type) changes without any (detectable) change in mass: a neutron in the nucleus of the nuclide is transformed into a proton; in this case, one (high-energy) electron is released from each particle concerned. (An example, to which we will return in ▶ Sect. 11.3, is the transformation of the radioactive carbon isotope ^{14}C into nitrogen-14.)

> **Very Briefly: The β-Decay, Slightly More Precisely…**
> The fact that in the β-decay of carbon-14, in addition to the β-particle, an (uncharged) *antineutrino* is also released, is indeed correct (and if one wants to really get into *nuclear chemistry,* also really important), but for our purposes here, it is just not relevant. The same applies to the fact that in addition to the "ordinary" β-decay described above, in which an "equally ordinary" electron (e^-, in the technical language of nuclear chemistry: $β^-$) and said antineutrino are released, there is also the variant in which not electrons (e^-) but **positrons** ($β^+$ or e^+) are released; as a by-product, one neutrino is then also produced per positron. *Electron capture,* in which a proton in the nucleus of the nuclide under consideration is converted into a neutron by absorbing an electron (which is why

> the mass number does not change, but the type of atom does—a chromium-48 atom with atomic number 24, for example, becomes an atom of vanadium-48, atomic number 23) and emitting X-rays, will not be discussed further here.

Which decay occurs with a radionuclide is just as isotope-specific as the **half-life $t_{1/2}$** of the nuclide in question.

The basics of radioactive nuclides are briefly summarised again in Skoog.

Of particular importance for our purposes is the fact that many atomic nuclei, once they have undergone the (α- or β-)decay process, are then present in an excited state from which they relax (gradually or spontaneously) into the ground state. The excess energy is then often emitted in the form of γ-radiation, which despite its fancy label, is just extremely high-energy electromagnetic radiation. (Please think back to the electromagnetic spectrum; if needed, you can find it as Fig. 17.2 in ▶ Harris or as Fig. 2.1 in Part IV of "Analytical Chemistry I".) It should be obvious that this secondary process does *not* involve a change in mass or atomic numbers.

Skoog, Section 32.1: Radioactive nuclides

Harris, Section 17.1: Properties of light

11.1 Radiochemical Analysis: Neutron Activation Methods

While, as briefly explained in the previous section, many naturally occurring isotopes are radioactive "by themselves", i.e. subject to one decay or another, radioactive decay can also be brought about in a targeted manner in the case of isotopes that are stable in themselves: by bombarding the isotope in question with neutrons. (There are also variants in which protons, D^+ ions [deuterons, i.e. one proton and one neutron together] or helium-3 cations [two protons, one neutron] are used instead of neutrons. In the context of this part, however, we want to limit ourselves to the activation by neutrons—among other things, because with neutrons, unlike with charged activation particles, no Coulomb repulsion has to be overcome when entering the nucleus due to the nucleus' own charge—any atomic nucleus *always* contains at least one proton.)

The neutron in question will be captured by the target nucleus (i.e. the atomic nucleus of the isotope to be activated), resulting in an excited new atomic nucleus (with a correspondingly increased mass number). This so-called **compound nucleus** will spontaneously release the excess energy in the form of a γ-ray photon. It is important to note that the *wavelength* of this photon (λ_γ) is *characteristic of the target isotope* in question. Thus, one can already see what the analytical method in question is aiming at: Quantification of an isotope based on the number of γ-photons released.

In some, but not all cases, the nucleus created by neutron capture is radioactive, which then undergoes β-decay (almost exclusively, β⁻-decay is observed, i.e. one neutron of the activated nucleus is transformed into a proton [and thus changes the atom's atomic number]). Here, too, if this results in a new excited state, the emission of a γ-photon can occur, the wavelength of which is then also isotopically characteristic. This is referred to as **decay radiation**. However, since this process takes somewhat longer than the photon emission immediately after neutron capture described above, it is referred to as *delayed gamma radiation*.

An already mentioned, gamma radiation is nothing else than "ordinary" electromagnetic radiation, but it is *very* energetic, therefore it cannot be quantified by means of an ordinary photoelectric detector (known from Part IV of "Analytical Chemistry I"). Instead, **photon counters** (or in general: **scintillation counters**) are usually used: Each incident photon generates a (charge) pulse, so

Skoog, Section 32.1.4: Census statistics

that (digital) counts per time unit are registered. The usual unit here is *Counts Per Minute* (**cpm**); the mathematical basis for this, which will not be discussed further here, can be found in Skoog.

- **Neutron Sources**

Of course, the equipment required for this method is not entirely trivial: On the one hand, radioactive nuclides whose decay releases neutrons can be used as neutron sources. (One usually uses the unstable trans-uranium elements; of particular importance here is the californium isotope ^{252}Cf, which leads to a neutron flux of about 10^7 neutrons/cm^2 s).

A significantly larger neutron flux—and with increased neutron flux the detection limits decrease drastically—is obtained by other means, which are rarely found in "routine laboratories":
- reactors *and*
- particle accelerators

With a research reactor, a neutron flux of up to 10^8 neutrons/cm^2 s can be achieved, so that detection limits, depending on the analyte, drop into the micro- to nanogram range. (The detection limits are element-specific and can vary considerably: While dysprosium can already be detected on a *picogram* scale, iron can only be detected when it is present in the high two-digit *microgram* range.)

Skoog, Section 32.3.3: Theory of activation methods

As expected, when the density of the neutron flux increases, a larger number of analyte atoms are activated, as shown in Fig. 32.7 from Skoog.

How long the sample has to be bombarded with neutrons in order to obtain a constant decay rate of the radionuclides produced *in situ* (and thus to obtain meaningful measured values) depends, as this figure also shows, on the *half-life* of the activated analyte: If the irradiation time exceeds the half-life by a factor of 5, there is a noticeable increase in measurement sensitivity (and thus a lowered detection limit). This already shows one of the major disadvantages of this method: In the case of long-lived radionuclides (i.e. nuclides with a long half-life), the measurement time can become unpleasantly long. For this reason, this method can be applied to many elements, but is by no means useful for all of them.

The corresponding measurements can be carried out non-destructively or destructively as required, and a reference value is also essential here, which is why, in addition to the analyte sample, one or more standards are always bombarded with neutrons the same way. Both the analyte and the reference substances can be irradiated in the solid, liquid, or gaseous state, although the analysis of gases via neutron activation is the exception rather than the rule.

- - **Non-destructive Analysis**

Here, the sample—for example, a workpiece that has already been shaped into the desired form—is irradiated in its current state and the photons released are then quantified, which allows conclusions to be drawn about the workpiece's content of the analyte isotope in question.

- - **Destructive Analysis**

For some analytes it is necessary to isolate them first and then activate them, for example in the form of a solution of increased concentration. However, this does not change the actual measuring principle.

11.2 · Radioactive Tracers

> **For Nitpickers**
> Since isotopes activated by neutrons can become radioactive, the term "non-destructive analysis" may be surprising. Strictly speaking, it is even wrong, because the sample substance is indeed (nuclear-)chemically changed (and may even emit radioactive radiation after completion of the analysis). However, the activity induced by the neutrons is usually very limited and therefore poses no problem at all.

■ **Applications**

Especially the non-destructive form of neutron activation is extremely versatile: In addition to the analysis of workpieces already mentioned, archaeological finds and works of art can also be examined this way; in trace analysis, too, neutron activation increasingly represents an alternative to some of the other analytical methods that you have already become acquainted with in the context of this book (and possibly also in "Analytical Chemistry I").

11.2 Radioactive *Tracers*

The fact that radioactive isotopes are chemically (practically) indistinguishable from their stable "counterparts", i.e. they react in the same way, can also be exploited. Thus, by using **radioactively labelled** substances, it could e.g. be shown that a *dynamic equilibrium* exists between a saturated solution and a precipitate present, i.e. that there is a continuous exchange of particles between the undissolved precipitate and the ions in the solution.

> ▶ **Example**
> As an example, consider silver chloride (AgCl), which is quite poorly soluble (as we saw in Part II of "Analytical Chemistry I", for example). The two natural isotopes of silver are ^{107}Ag and ^{109}Ag; both are stable and occur in nature in similar amounts (isotope ratio ^{107}Ag:^{109}Ag = 51.8:48.2). Other silver isotopes are synthetically accessible but have not (yet?) been discovered in nature. The two stable isotopes of chlorine (^{35}Cl and ^{37}Cl) are already known from Part I (isotope ratio ^{35}Cl:^{37}Cl = 75.8:24.2; the radioactive isotope ^{36}Cl occurs in nature only in traces, but can also be obtained synthetically).
>
> Assume to a saturated solution of this salt (with precipitate being present), in which only natural isotopes are present, any amount of solid ^{111}AgCl, i.e. **isotopically labelled** silver(I) chloride, is added. (^{111}Ag is a β-emitter with a half-life of about 7.5 days; that silver isotope in the process converts to cadmium-111.) If one then waits a while and filters the mixture off, it would be expected that any radioactivity should be detected exclusively in the filter residue, after all the "excess" silver chloride—the radioactively labelled one—should actually not go into solution but immediately join the precipitate. In fact, however, it turns out that there must very well have been an exchange of ions between the solution and the precipitate: The radioactivity is also detectable in the filtrate. ◀

In an analogous way, radioactive markers can be used to elucidate metabolic pathways and reaction mechanisms. Because they can be traced quite easily, these markers are also called *tracers*.

You have probably already wondered where some information about the course of complex chemical reaction sequences actually comes from—for example, how the fact was recognised that there is such a thing as keto-enol

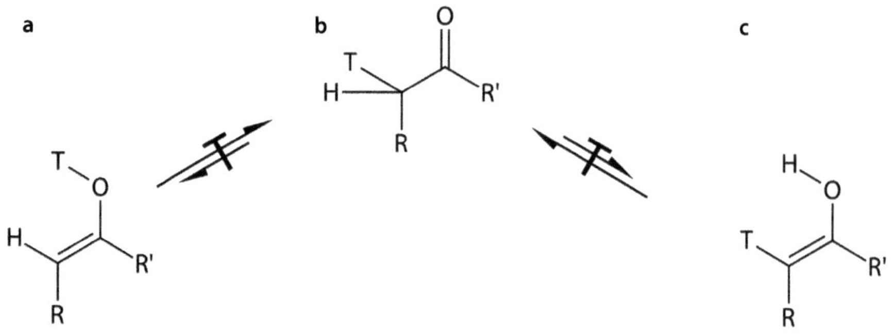

Fig. 11.1 Keto-enol tautomerism on acetone

tautomerism (known from *organic chemistry*). If one e.g. introduces a tritium atom (T, ^{3}H) into an enolisable carbonyl compound by synthetic means (Fig. 11.1a), it can be shown (using, among other things, nuclear magnetic resonance spectroscopy—NMR—from Part I) that this tritium can both become part of the hydroxyl group of the enol form of this compound (Fig. 11.1b) and any other enolisable hydrogen atom (Fig. 11.1c). Synthetically a little more labourious, but equally revealing, is to synthesise specifically the enol form of the carbonyl compound tritiated at the OH group (Fig. 11.1b): Again, after a very short time, ^{1}H-NMR spectroscopy reveals mainly the keto form (a) with a tritium-carbon bond (C-T) being present, but also the enol form, not with an O-T group but an O-H group (Fig. 11.1).

> **Isotope Labelling in the Elucidation of Reaction Mechanisms**
> Even though radioactive isotopes can be traced particularly well, isotope labelling is also possible with *stable* isotopes. (Of course, these isotopes must be easy to detect: Because of the mass difference of isotopes, mass spectrometry is preferably used here—also known from Part I.)
>
> For example, as shown in the figure, it could be demonstrated by labelling with the (stable, but on earth very rarely occurring) oxygen isotope ^{18}O that during base-catalysed ester cleavage the CO bond (the acyl bond) is cleaved., and the reaction is not a simple substitution at the (alkylated) oxygen: 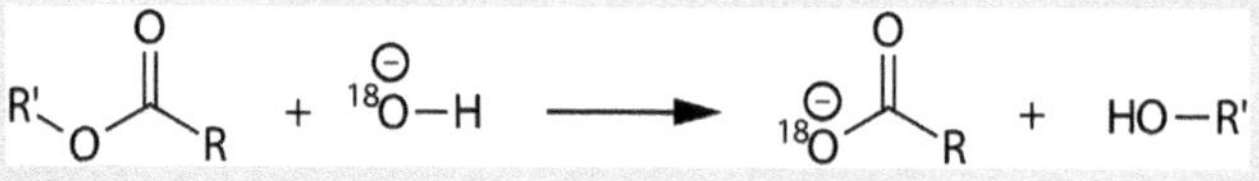
>
> Base-catalysed ester cleavage with isotopic labelling

Analogously, such isotope labelling experiments (such as the specific incorporation of ^{14}C atoms into biomolecules, which, synthetically speaking, is of course anything but trivial) can be used to determine details about the exact course of the individual steps of a biochemical metabolic pathway. (If you are involved in *biochemistry*, you will certainly become familiar with the citrate cycle; some steps of this can be easily traced by using appropriate ^{14}C-labelled substrates.)

Before you start wondering about the technical language: The term *tracer experiment* is commonly only used when the labelling is done by *radioactive* isotopes, simply because these natural emitters are easier to track. So every "tracer experiment" is an "isotope-labelling experiment", but not every "isotope labelling experiment" is a "tracer experiment".

11.3 Radioactive Age Determination

The fact that radioactive nuclides undergo radioactive decay (with half-lives ranging from fractions of a second to several thousand years or more, depending on the isotope in question) can also be exploited to determine the age of some analytes. Of particular importance here is the radioactive carbon isotope ^{14}C.

- **Radiocarbon Dating**

This method of age determination can be used to date samples containing carbon, especially organic materials such as leather, textiles. or paper. There are some simple considerations behind this:
— Compared to the two stable carbon isotopes ^{12}C and ^{13}C (which occur in nature with an abundance of about 98.9% and 1.1%, respectively—you certainly remember the corresponding considerations in the NMR section of Part I), the radioactive carbon-14 (a β-emitter; this carbon isotope decays [half-life $t_{1/2} \simeq 5715$ years] to nitrogen-14 under splitting off an antineutrino) is extremely rare: The natural abundance of ^{14}C is about 10^{-10}%.
— Decay ensures: the older a sample, the lower its carbon-14 content.
— On the other hand, this radioactive isotope is constantly being re-synthesised in the atmosphere. (This is due to a neutron capture reaction in which ^{14}N atoms react to form [excited] ^{14}C atoms, releasing a proton as well as an electron.) These atoms cause the atmosphere to contain a more or less constant amount of $^{14}CO_2$ and other carbon-containing volatile compounds with ^{14}C-atoms in it.
— This way, the ^{14}C content of a living organism remains more or less constant as long as metabolism is still going on (for example, through food intake or—in plants—carbon fixation, etc.).
— The ratio ($^{14}C/^{12}C$) in organism will therefore correspond to the ratio ($^{14}C/^{12}C$) in the atmosphere.
— With the death of the organism, i.e. when all metabolism ceases, the uptake of carbon-14 obviously ceases as well, thus the ^{14}C content continuously decreases exponentially according to the half-life of this isotope. (Please remember the basic information about half-life from the introduction to this chapter.) At any given time t, the ratio of the two isotopes involved can be calculated via:

$$\left(\frac{^{14}C}{^{12}C}\right)_{\text{in the organism}} = \left(\frac{^{14}C}{^{12}C}\right)_{\text{in the atmosphere}} \cdot e^{-\frac{\ln 2}{t_{1/2}} \cdot t} \qquad (11.1)$$

— Accordingly, the quantification of the remaining residual radioactivity (and thus the residual ^{14}C content) of an organic material allows direct conclusions to be drawn about its age.

Since it is relatively difficult to determine the half-life of comparatively long-lived isotopes, this method naturally leads to a certain measurement inaccuracy (the current table value for carbon-14 is $t_{1/2} = 5715 \pm 30$ years), which is why *age periods* are given rather than specific ages. Since, on the other hand, according to the current state of knowledge in physics, the half-life of a material cannot be changed by *any* external influences and radioactive decay can therefore be easily described mathematically, such time periods can be regarded as quite reliable: For example, a sample that is about two thousand years old will not have a measured age of only a few centuries. The general rule of thumb is: measurement result ±200 years. (With older samples, of course, a greater deviation from the measured value is to be expected than with younger ones.) Of course, this method also has its limits: if more than ten half-lives have elapsed, i.e. if the sample is older than about 60,000 years, the remaining ^{14}C quantity is no longer sufficient for a (reasonably) precise age determination.

The actual measurement is carried out either using a scintillation counter, which responds directly to the β-radiation of the sample, or by means of a variant of mass spectrometry, which is, however, quite complex in terms of equipment (among other things, a particle accelerator is required). The latter method is much more sensitive: Milligram-scale samples are sufficient here. (However, this method is also much more expensive. You can't have everything.)

Different Values for $t_{1/2}$ (^{14}C)
In the literature, for the half-life of carbon-14 the value $t_{1/2}$ = 5730 years is often found, given with an accuracy of ±40 years. In 1990, however, IUPAC recommended the use of $t_{1/2}$ (^{14}C) = 5715 ± 30 years. The latter is referred to in the literature as the **Cambridge half-life**; however, since numerous measurement results based on the "old" value had been published prior to this determination, it is still used alternatively as the so-called "**conventional 14C age**" (or, named after the developer of this method, Willard Libby, who received the Nobel Prize for it in 1960, as the "**Libby half-life**"). In the context of this book, however, we work (completely IUPAC-compliant!) with the "new" (Cambridge) value.

However, there are some factors that are detrimental to the measurement accuracy of this inherently well-established method:
(a) Since the concentration of ^{14}C (and thus of $^{14}CO_2$ in the air) depends on the nuclear reaction of neutron capture by the nitrogen isotope ^{14}N that takes place in the Earth's atmosphere, the (energy-rich) cosmic radiation also plays a role, and this radiation fluctuates: The higher the solar activity (keyword: sunspots), the more radiation reaches the Earth's atmosphere; correspondingly, a somewhat greater conversion $^{14}N \rightarrow {}^{14}C$ takes place.
(b) Particularly in the case of younger samples, it should be borne in mind that the amount of carbon dioxide released into the atmosphere has increased markedly since the beginning of industrialisation. This mainly originates from the combustion of fossil fuels, which are significantly older than the 60,000 years mentioned above, so that no detectable carbon-14 can be found in them. Therefore, since that time, the natural ratio ($^{14}C/^{12}C$) in the atmosphere has been distorted by **anthropogenic** "dilution" with ^{14}C-free carbon dioxide. Accordingly, younger samples show a slightly reduced ^{14}C content, which is why the evaluation of the measured values, if this influence is not taken into account, suggests an excessive age.
(c) Another anthropogenic effect that complicates the evaluation of more recent samples is due to the detonation of nuclear weapons: The high-energy radiation released in the process has affected the overall isotopic composition of the atmosphere; in the mid-fifties of the last century, for example, there was a temporary doubling of the ^{14}C content, which even now has not yet returned to its natural/original value. (The extent to which the "dilution" of natural carbon dioxide in the atmosphere by anthropogenic CO_2 compensates for or at least dampens this effect has not yet been conclusively clarified.)

Despite these limitations and also the detection limit mentioned above, the radiocarbon method is an extremely useful tool for approximate age determination of carbon-containing analytes.

■ **Other Isotopes**

Although the ^{14}C method is certainly by far the best known (and most important) age determination technique based on a radioactive isotope, there are several other methods based on the same principle but using different elements:

(a) The decay of uranium to lead, which proceeds via numerous steps with α- and β-decay stages (and also follows different paths depending on the uranium isotope: the isotope ^{238}U, for example, leads via radium-226 to the lead isotope ^{206}Pb [with $t_{1/2} \simeq 690$ million years], while uranium-235 decays via actinium-227 to lead-207; here $t_{1/2} \simeq 4.3$ billion years), is of particular importance for minerals: The first statements about the age of the Earth were based on such investigations; for some rock types, an age of more than 4 billion years was determined in this way.

The accuracy of this method is remarkable: The error is on the order of less than 0.1%. (In *absolute terms*, this still leaves a margin of error of more than a million years for a 2-billion-year-old sample, but in *relative* terms this is truly amazing.) One reason for this accuracy is the fact that in uranium decay, precisely *because* of the two different decay series, two determinations are practically always carried out simultaneously.

(b) The radioactive potassium isotope ^{40}K (which converts to ^{40}Ar by absorbing an electron or splitting off a positron) also provides the basis for age determination; thanks to its long half-life ($t_{1/2}(^{40}K) = 1.28$ billion years), this is also suitable for extremely old sample material.

(c) The same is true for the isotope rubidium-87, which converts to strontium-87 in the course of a β-decay (with $t_{1/2}(^{87}Rb) \simeq 50$ billion years).

❓ Questions

10. Why can sodium be quantified by neutron activation?
11. Could a (radio) labelling experiment be used to determine whether the esterification of a carboxylic acid R-COOH with an alcohol R′-OH splits off the carboxy or the alcohol oxygen as water?
12. What age range must be assumed for a sample of organic origin when the radioactivity originating from the isotope carbon-14 has decayed to one sixteenth of its original value?

11.4 Radioimmunoassay (RIA)

Radioactive isotopes can also be used for biological studies. (Further important analytical methods, especially in the biosciences, will be discussed in ▶ Chap. 12.) A particularly elegant method is the **radioimmunoassay** (**RIA**, for short). However, before we can turn to its measurement principle, we must first address some basic aspects of the functioning of any immune system. (You will certainly learn more details about this in *biochemistry* courses or works.)

The principle of the immune response to a foreign substance (harmful or perceived as harmful), the so-called **antigen**, is that the immune system forms an appropriate **antibody**, which forms a comparatively stable complex with the antigen without interacting noticeably with other (bio)molecules in the same way.

1. The antigen usually is a relatively large biomolecule such as a protein, carbohydrate, or lipid (e.g. from food) or a molecule that is easily biochemically metabolised (e.g. elemental iodine and the like).
2. The antibodies that respond to the antigen are proteins.

The interaction of the antibody with the antigen can lead to several things:
– Ideally, the (harmful) effect of the antigen is contained or completely prevented.
– Alternatively, the complexation of the antigen by the antibody represents a kind of marker for *phagocytes*, which then digest the whole complex and this way also render the antigen harmless.
– The interaction of the antigen-antibody complex with specific cells of the body can motivate certain *leukocytes* (white blood cells) to kill the entire cells (along with the antigen-antibody complex).
– There are other possibilities, but that would lead too far here.

> **A Brief Glimpse on Medicine: Autoimmune Diseases**
> Even though antibodies should actually only react to external material, for example to foreign proteins that the body is unable to metabolise (a much-mentioned example today is gluten, which is found in some types of grain, and which leads to the clinical picture of *celiac disease* in people with a corresponding predisposition), it does happen that the immune system "overreacts", i.e. responds to the body's own biomolecules. This leads to chronic (non-infectious) inflammatory processes, the therapy of which is often very costly (if the disease in question can be treated at all) and which can lead to complete organ destruction. Such autoimmune diseases include, among others (not even nearly claiming completeness!):
> – *Multiple sclerosis*—here the myelin sheaths of the central nervous system are affected
> – *Diabetes mellitus* (type 1)—certain cells of the pancreas are afflicted
> – *Crohn's disease*—affects the entire digestive tract.
> – *Ulcerative colitis*—similar to Crohn's disease, but "only" affects the mucous membrane of the colon.
> – *Rheumatoid arthritis*—destroys the connective tissue, especially the joints, in the long term.
> – *Hashimoto's thyroiditis* (often just called "*Hashimoto's*")—results in chronic inflammation of the thyroid gland, *or*
> – *Psoriasis*—usually affects only the skin, but can develop into a systemic disease, affecting the eyes, heart and joints, along with adjacent soft tissues.

Which endogenous compounds are involved in each case is beyond the scope of this introduction. If you are interested, the work of Heinrich, Müller and Graeve is strongly recommended: The "Löffler/Petrides" (a real classic of specialised literature, which has been in circulation for many years, being revised on a more or less regular basis; it is still named after its original authors, even if in the meantime, after more and more new knowledge has been gained, other scientists are responsible for the actual text) treats in a very vivid and easily comprehensible way which diseases are developed due to which metabolic disturbances or which disruptive compounds—especially those enthusing about organic chemistry will recognise a variety of highly interesting connections here: With this book, you are already right in the middle of the field of "chemical biology".

11.4 · Radioimmunoassay (RIA)

The fact that antibodies react so specifically to the antigens in question can be exploited in analytics, as mentioned above: The radioimmunoassay (for the development of which Rosalyn Yalow was awarded the Nobel Prize in Physiology or Medicine in 1977) is used to quantify an antigen. To do this, antibodies specific to the antigen being sought (usually derived from animals) are first mixed with a well-defined amount of the antigen, which is either radioactive or has been appropriately labelled by deliberate incorporation of one or the other radionuclide. (Do you recognise the parallels with the radioactive *tracers* from ▶ Sect. 11.2?)

Subsequently, this mixture of antigen$_{labelled}$ and antibodies is mixed with the analyte solution containing the non-radioactive/radioactively labelled antigen (antigen$_{natural}$). Since the radioactive labelling does not affect the (bio)chemical behaviour, the labeled and unchanged antigens will compete for interaction with the antibody binding sites during the **incubation period**. Apparently, it should be noted here that the antibodies used are usually fixed to a carrier material by covalent bonds (which, of course, do not alter the part of the protein that is supposed to interact with the antigen!); these are called *immobilised antibodies*. After the incubation period has elapsed, the carrier plate is rinsed off, removing unbound radiolabelled and natural antigens alike. The remaining radioactivity of the labelled antigens bound to the fixed antibodies is then quantified—again using a *scintillation counter* or, if a γ-ray radionuclide was used, a *photon counter*: The greater the number of unlabelled antigens from the analyte solution (i.e., the more concentrated it was), the less radioactivity will be detected. Since the number of possible binding partners is limited by the number of antibodies present on the carrier plate, the ratio of the presence of radioactive and non-radioactive antigens on the antibody carrier allows a direct conclusion on the extent of displacement of the labelled antigens by the unlabelled ones from the sample solution.

In addition to the usually quite large antigen analytes (with molar masses >1000 g/mol; we obviously deal with higher-molecular compounds), smaller analytes can also be quantified in this way. For this purpose, they have to be converted into higher-molecular compounds by coupling to a carrier protein. Such *low-molecular weight* compounds are called **haptens**.

In addition to *radioactively labelled* antigens, *fully radioactive* antigens can also be used (as long as correspondingly specific antibodies exist for them, that is): For example, elemental iodine is also among the antigens used in routine examinations: Both the isotopes ^{125}I and ^{131}I emit γ-radiation, and for both, the corresponding radioactivity can be quantified rather easily, using a photon counter.

One of the great advantages of immunoassay methods in general and also of radio-immunoassays is that they respond to minimal amounts of analyte, i.e. the detection limits are very low—and this can even be improved by orders of magnitude (!) if the previous techniques are combined with fluorescence phenomena; this is precisely what will be discussed in ▶ Chap. 12. (The principle of fluorescence and how it is produced was discussed in detail in Part IV of "Analytical Chemistry I", for example.)

■ ELISA

A variant very similar to the RIA, but without radioactivity, is the *Enzyme Linked Immunosorbent Assay* (**ELISA** for short). This method of analysis is also suitable for low-molecular analytes and is used in routine analysis (which can also be automated) for the qualitative or quantitative detection of e.g. pesticides, toxins, or hormones. The principle corresponds, as mentioned, to that of the RIA:

- The analytes (which act as antigens or haptens) interact with specific antibodies (also referred to as *primary antibodies*).
- A *secondary* antibody (also known as *detection antibody*) is then added, which interacts—again specifically—with the hapten bound to the primary antibody.
- It is important here that this secondary antibody has been labelled in advance with an enzyme that catalyses a colour reaction.
  - This colour reaction is carried out with a dye which is colourless in itself, but which can enzymatically be brought to reaction in such a way that a quantifiable colouration of the solution is obtained.
  - Alternatives to (photometrically determinable) staining are chemiluminescence or, if applicable, fluorescence phenomena, which can also be quantified. (The topic of "fluorescence", which has already been addressed once in Part IV of "Analytical Chemistry I", will be dealt with again in the chapter immediately following.)

Apart from the fact that ELISA does not use radioactive substances, the procedure is similar to RIA:
1. First, the antigen (i.e. the analyte) is bound to an antibody that reacts specifically to the antigen and is usually fixed to the (plastic) matrix of a microtitre plate by covalent bonds.
2. In a washing step, any unbound analytes are removed.
3. Subsequently, a second antibody, which also reacts specifically to the antigen, i.e. the *primary antibody*, is added.

> **For Those Particularly Interested in Biochemistry**
> Please note that our analyte (the hapten) must actually interact with *two different*, specifically reacting antibodies. It is important to note that these two antibodies must interact with different sites of the antigen, because otherwise they would mutually interfere with each other's activity. (One can imagine that this may lead to difficulties, particularly with small haptens.)

- The detection antibody, which in advance has been *labelled with an enzyme*, now interacts with this primary antibody. A "sandwich" complex is formed: Fixation antibody—antigen—primary antibody—detection antibody.
- If a substrate is finally added, which thanks to the enzymatic activity of the labelling enzyme is converted, the (quantifiable) staining results.

Even though the antibodies must be analyte-specific (i.e.: hapten-specific) in each case, only a rather limited number of enzymes are used for labelling so far; those are called **reporter enzymes**. Particularly popular are:
- *β-Galactosidase*: This enzyme hydrolyses the glycosidic bond of β-galactopyranosides (β-Gal-R, ◘ Fig. 11.2a), so that galactose is cleaved off and the free alcohol R-OH is formed (b). The substrate for this enzyme in ELISA is usually the dye X-Gal (◘ Fig. 11.2c), in which galactose is first cleaved appropriately by enzymatic action; the resulting intermediate (d) dimerises under the influence of atmospheric oxygen to the deep blue dye 5,5'-dibromo-4,4'-dichloroindigo (e). However, X-gal is light-sensitive and therefore predominantly suitable for qualitative studies. For true *quantitative* measurements (and this is what we are primarily concerned with here), the dye *o*-nitrophenyl-β-D-galactopyranoside is preferred (◘ Fig. 11.2f), in which *o*-nitrophenol (g) is formed by the action of the same enzyme, lead-

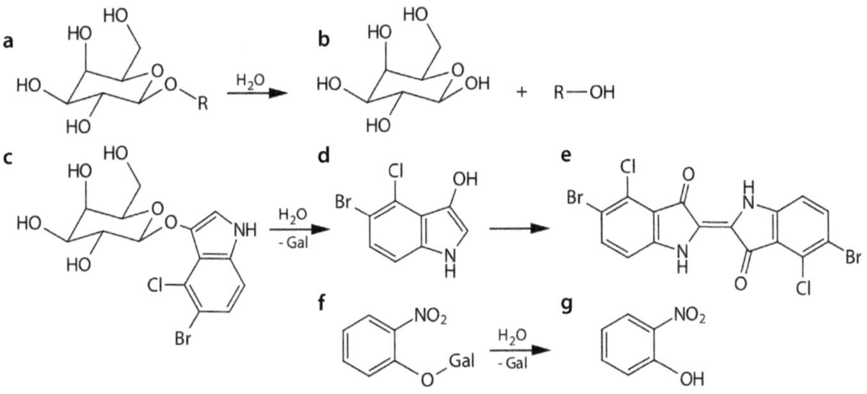

11.4 · Radioimmunoassay (RIA)

Fig. 11.2 Mode of action of β-galactosidase

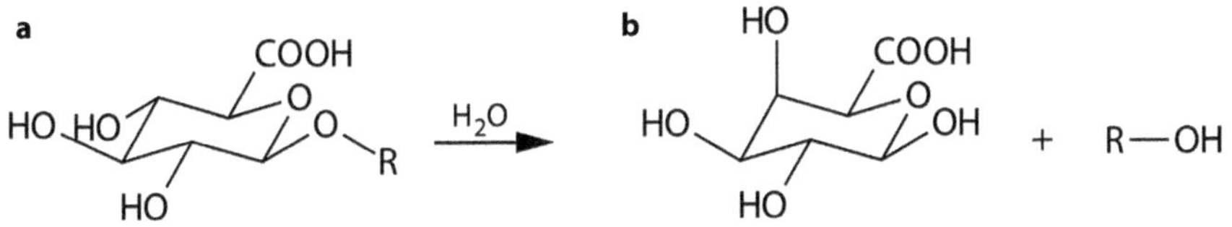

Fig. 11.3 Substrates of β-glucuronidase

ing to a characteristic (and readily quantifiable) yellow colouration of the reaction solution.
- *β-Glucuronidase*: Analogous to β-galactosidase, this enzyme also hydrolyses a glycosidic bond, but it is specialised for glucuronides (Fig. 11.3a), i.e. compounds with glucuronic acid. (This acid, usually abbreviated GlcA, is simply a glucose molecule whose C^6 [i.e., the exocyclic carbon] is oxidised to a carboxylic acid.) In the case of the dye X-Gluc, which is commonly used in ELISA experiments, the sugar is linked to the same **aglycone** (i.e., the "non-sugar part" of the molecule) as in the case of X-Gal; accordingly, the enzyme first catalyses the cleavage of glucuronic acid (Fig. 11.3b), and the aglycone (R-OH) dimerises to the blue dye 5,5′-dibromo-4,4′-dichloroindigo, as described above (Fig. 11.2e). If the β-glycoside of glucuronic acid and *o*-nitrophenol is used as the dye (analogous to Fig. 11.2f), the yellow colouration caused by the enzyme activity can again be readily used for quantification by the *o*-nitrophenol that is released (Fig. 11.2g).
- *Alkaline phosphatase*: This enzyme cleaves phosphate groups; 5-bromo-4-chloro-3-indoxyl phosphate (Fig. 11.4a) is used here for staining; after dephosphorylation (cleavage of phosphoric acid) and subsequent oxidative dimerisation, we again obtain the dye 5,5′-dibromo-4,4′-dichloroindigo from Fig. 11.2e.
- *Horseradish peroxidase* (HRP): This enzyme catalyses the chemiluminescence reaction of luminol (Fig. 11.4b); thus, no dye whose *colouration* would be quantified is produced here, but rather a luminescence reaction is induced, in which the number of photons produced/emitted is important. (You will learn more about the quantification of photons in the next chapter; the luminescence mechanism of luminol will not be discussed any further here.)
- Of course there are other reporter enzymes, but the principle should have become clear by now, and once again we don't want to overdo it.

☐ **Fig. 11.4** **a** 5-Bromo-4-chloro-3-indoxyl phosphate and **b** Luminol

> ▶ Example

To show you how such an ELISA looks "in nature", here is the quantification of the anti-GPI antibody (GPI = glucose-6-phosphate isomerase) in a certain mouse model (K/BxN arthritis mice—this in itself leads too far, because it is quite "biological", but for those interested it should at least be mentioned) via ELISA.

A microtitre plate (in the following colour illustration with 96 wells) is used, onto which two different references are applied in addition to the analyte solutions themselves (serum from a K/BxN arthritis mouse) in different concentrations. One after the other:

- In rows A and B, the mGPI antibody standard ("m" simply stands for "murine", meaning "mouse") is found in decreasing concentrations: The highest concentration is 100 ng/mL, the lowest is just under 0.2 ng/mL. (Such a double determination is useful to exclude possible measurement artefacts, if possible.) These two lines thus represent the *positive blank sample*. (In the context of life sciences, this is also referred to as **positive controls**.)
- The wells in rows C and D contain, also for duplicate determination, the mouse serum *to be examined,* i.e. the actual analyte. Here, a serial 1:2 dilution series was prepared, with the highest concentration (on the microtitre plate on the left) being 1:120,000. (Such a strong pre-dilution is necessary because otherwise the concentration of the antibody to be examined is too high to allow any measurement at all.)
- Series E and F (again a duplicate determination) contain—also as dilution series—the comparable serum of a wild-type mouse, which does *not* contain (or at least *should not* contain) the antibody in question. These two series therefore represent, as it were, *blank samples* (in the context of the life sciences, these are referred to as **negative controls**), which are to be as similar as possible to the actual analyte solutions (i.e. the K/BxN sera in different dilutions), mixed with all the necessary reagents, *without* containing the analyte to be quantified.

The final photometric quantification is the colour intensity resulting from the reaction of horseradish peroxidase (HRP, we know it from above) with the colourless substrate 3,3′,5,5′-tetramethylbenzidine (Fig. a; usually abbreviated TMB) when the analyte under investigation has the anti-GPI antibody. The peroxidase initially converts TMB to a blue dye (VIS absorption maximum: 650 nm) (Fig. b); after addition of sulfuric acid (to stop the enzymatic reaction), the formerly blue solution then appears yellow (VIS absorption maximum: 450 nm).

3,3′,5,5′-Tetramethylbenzidine (TMB) in **a** reduced and **b** oxidised forms

11.4 · Radioimmunoassay (RIA)

As expected, the deep yellow colouration is directly dependent on the antibody concentration; accordingly, blank rows E and F remain colourless, as can be seen in the image below.

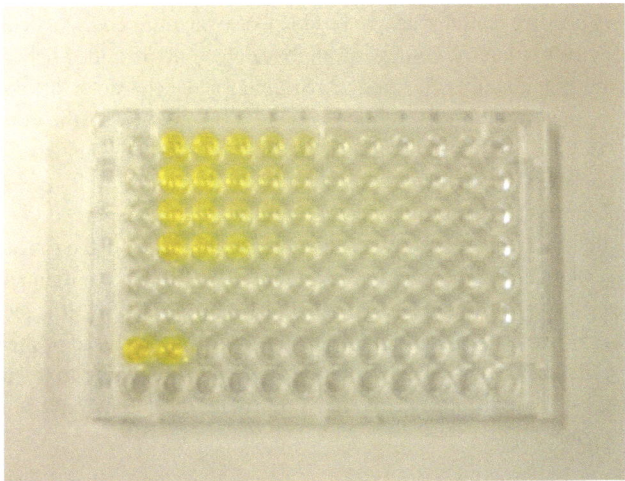

ELISA (with HRP and TMB). (Kindly provided by D. Kockler, Bonn-Rhein-Sieg University of Applied Sciences)

The individual steps in this experiment were:

- Coating the microtitre plate with the antigen GPI-GST (GST stands for glutathione S-transferase; this protein—consisting of 211 amino acids—is often used for the labelling of proteins; this is actually going too far here, but for those who are interested, please refer to the adjacent link [which is not to be understood as "product placement"!]; this labelling protein has no effect on the ELISA and therefore does not need to be removed for the actual detection).
- First wash-out step
- Reaction of the antigen with a blocking solution (overnight at 4 °C or alternatively 1 h at 37 °C); this solution saturates any previously unoccupied binding sites, preventing false-positive results and minimising "background noise".
- Application of the samples; incubation for 1 h at room temperature: Here the binding of the primary antibody to be quantified to the antigen takes place. Second wash-out step
- Addition of the detection antibody (here: HRP; this peroxidase is coupled to an antibody that also recognises the species, in this case we are talking about the *peroxidase-conjugated and affinity purified Goat anti-Mouse IgG (H + L)*—for those particularly interested in biochemistry: the "H + L" stands for the fact that the *whole* antibody is really present, with the heavy [*heavy* = H] *and* the light [*light* = L] chain); incubation 1 h at room temperature.
- Third wash-out step
- Addition of the substrate (TMB) in buffered solution (we do not want to go further into this, now)
- Stopping the enzymatic reaction by addition of sulphuric acid (one-molar)
- Photometric measurement at λ = 450 nm.

Obviously, performing an ELISA is not without effort—but it is also often *extremely* informative. ◄

▶ https://www.thermofisher.com/de/de/home/life-science/protein-biology/protein-biology-learning-center/protein-biology-resource-library/pierce-protein-methods/gst-tagged-proteins-production-purification.html

Tip for the Lab

For technical reasons, the first and the last column of a microtitre plate are usually *not* used for the analysis of samples: In order to clearly ensure that there are really no (false) positive readings and to be able to see the difference particularly clearly, all reagents used in the context of analyte detection are added here, with the mixture containing neither the analyte required for the characteristic colour reaction nor its negative-control counterpart. This is therefore "all the more" a *blank sample,* and this is referred to in the life sciences really as the **blank sample**.

To be on the safe side, highly concentrated positive blank samples (i.e. positive controls) have been added to the first two wells of the row marked G—this should make the difference between positive blank samples and the (negative) blank samples, which are to be found in all other wells of the first column, even more obvious. (If you want to check the reliability of your ELISA, it is recommended to use the same stock solution of the antibody that was used to prepare the dilution series mentioned above. It should be understandable that such a highly concentrated solution results in a particularly intensively stained blank/positive control.)

For Crime Fiction Fans

The chemiluminescence of luminol is also used in forensic science: In the presence of hydrogen peroxide (and also other oxidising agents, but H_2O_2 in aqueous solution has proven effective), even minimal amounts of blood sprayed with an alkaline luminol solution show intense chemiluminescence and can thus easily be detected. The reason for this is that the reaction of luminol with oxidants leading to luminescence proceeds particularly rapidly (and therefore with particularly intense light appearance) in the presence of some *catalysts*, and the complex-bonded iron from haemoglobin (both with oxidation number +II and + III) provides an extremely efficient catalyst. (However, it should not be concealed that copper(II) ions also have a catalytic effect, so corresponding tests in the presence of copper ions may lead to *false positive* results.)

Fluorescence Methods

Contents

12.1 Basics of Fluorescence: A Brief Review and Some Additional Information – 170

12.2 Fluorescence Spectrometry – 176

12.3 Fluorescence Microscopy – 181

12.4 Summary – 183

Further Reading – 188

© Springer-Verlag GmbH Germany, part of Springer Nature 2025
U. Ritgen, *Analytical Chemistry II*, https://doi.org/10.1007/978-3-662-68710-9_12

Although the basics of **fluorescence** have been covered quite extensively in Part IV of "Analytical Chemistry I", some aspects will be touched upon again—especially, of course, those of importance for the quantification of the analytes in question. First of all, it should be clarified/repeated exactly what the term "fluorescence" actually means.

12.1 Basics of Fluorescence: A Brief Review and Some Additional Information

A system (molecule/molecular ion) is brought into an excited state by radiation of wavelength λ_1, i.e. an electron is moved from its current orbital to an energetically higher orbital; in the process, a spin reversal *can* occur, i.e. the electron in question *might* change its spin state.

After system-internal electron transitions, which—as mentioned—*can be* accompanied by a change of the **spin state** of the excited electron, but this is not necessarily the case, and in which also *radiationless vibrational relaxation processes* play a role, the electron finally returns to its ground state (or a *vibrationally/rotationally* (vib./rot.) *excited* ground state) and emits the excess energy in the form of a photon of wavelength λ_2, where always holds:

$$\gamma_2 > \gamma_1$$

It could not be otherwise, after all, energy is also dissipated during the radiationless transitions. The phenomenon that such fluorescence processes always emit lower-energy, longer-wavelength radiation than was previously used to excite the system is known as the **Stokes shift**. (You might already be familiar with the Stokes effect from the explanations on Raman spectroscopy in Part IV of "Analytical Chemistry I".)

For the aforementioned *radiationless transitions*, there are first of all *two principally different* possibilities:

- *Internal conversion* (**IC**)—In this case, a vib./rot.-excited state changes to a less energetic, but still vib./rot.-excited state *without changing the spin state of the excited electron*. (The excess energy is transferred to the environment—usually: the solvent—without radiation.) This radiationless transition, which can be traced back to **vibrational relaxation**, is **symmetry-allowed**.
- *Intersystem Crossing* (**ISC**)—Here a vib./rot. Excited system *changes its spin state* into a still vib./rot.-excited state. This transition is **forbidden by symmetry** and therefore occurs much less frequently. However, if it occurs nevertheless, this usually leads to **phosphorescence**.

Let's take another look at the different spin states that result—presented in a way you may already know from Part IV of "Analytical Chemistry I". Furthermore, we assume that the analyte under consideration is *not* a radical:

- In the ground state, the system is in the singlet state S_0: $_{(\uparrow\downarrow)}{}^{()}$.
- *Inner conversion* causes the transition of one of the two ground-state spin-paired electrons in the HOMO to orbital which in the ground state was the (obviously vacant) LUMO: $_{(\uparrow\downarrow)}{}^{()} \rightarrow {}_{(\uparrow)}{}^{(\downarrow)}$. This results in an *excited* singlet state S_1^* results.
- If subsequently *intersystem crossing* occurs, starting from the state S_1^*, the spin state of the electron in the energetically less favourable orbital (the former LUMO) changes, resulting in an excited *triplet* T_1^* state: $_{(\uparrow)}{}^{(\downarrow)} \rightarrow {}_{(\uparrow)}{}^{(\uparrow)}$.

12.1 · Basics of Fluorescence: A Brief Review and Some Additional Information

- Since it is well-known that the ↑ state is energetically more favourable than ↓, it is not surprising that some amount of energy dissipated (radiationless) in the process.

It follows that the relative energy content of these three states can be summarised as follows:

$$S_0 < T_1^* < S_1^* \tag{12.1}$$

In addition, there is also *external conversion* (**EC**), in which the conversion is caused by direct interaction with a solvent molecule in the first place (usually by direct collision of the particles involved). If the molecules of the solvent used are easily brought to excitation, this radiationless phenomenon occurs very frequently; accordingly, the solvent used has a strong influence on the intensity of the observable fluorescence: The more frequently external conversion occurs, the weaker the fluorescence.

Another factor that reduces the quantifiable fluorescence is the phenomenon of **predissociation** (**PD**): In this case, the transition of an electron from an electronically and possibly additionally vib./rot. Excited state to an *electronically less excited* state occurs; however, the vib./rot. Excitation of the system is so strong that in the course of this internal conversion a bond is broken, which (of course) does not cause any fluorescence. The more labile (unstable) the analyte under consideration, the more significant predissociation becomes.

Of course, the excitation radiation itself can also be energetic enough to directly cause the breaking of a bond, which (understandably) also does not lead to fluorescence. This case, called **direct dissociation** (**DD**), should of course be avoided, if possible. In general, the higher the energy of the excitation radiation used, the greater the probability that it will be sufficient to cause one or the other breaking of a bond within the analyte.

As you can see, there are still numerous radiationless alternatives to the (desired) fluorescence phenomenon. Therefore, it is advisable to first consider what maximum fluorescence is theoretically possible. The quotient of *factually achieved fluorescence* and *theoretically conceivable fluorescence* leads us to the *quantum yield*.

▪ Quantum Yield

The **quantum yield Q** (also called quantum **efficiency**) describes the ratio of molecules excited to show fluorescence to their total number:

$$Q = \frac{\text{Number of fluorescent analytes}}{\text{Number of the theoretically excitable analytes for fluorescence}} \tag{12.2}$$

The easier it is to excite a system to fluorescence, the greater Q becomes; in the case of "fluorescence-optimised" analytes, such as fluorescein (◘ Fig. 12.1a; the unstable spiro-lactone form b also exists, but will not be considered any further here) or quinine (◘ Fig. 12.1c), Q can almost assume the value 1.

However, even a (hypothetical) quantum yield of 1 does not mean that every excitation really leads to a fluorescence photon, because in addition to the actual (and in the context of fluorescence analysis: desired) fluorescence (return from the excited state to the ground state with emission of a photon of wavelength λ_2), the above-mentioned *radiationless* transitions must also be taken into account. For this reason, it makes more sense to focus on the **fluorescence quantum yield Φ**. This describes the ratio of the number of fluorescence phenomena to the number of *possible* corresponding electron transitions.

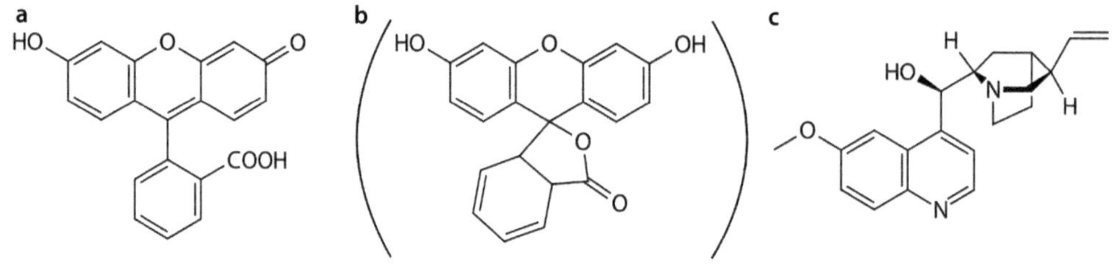

Fig. 12.1 Fluorescein **a**, **b** and quinine **c**

In other words: How large is the fraction of the irradiated excitation photons that actually cause fluorescence? In general

$$\Phi = \frac{I_{Fluorescence}}{I_0} \qquad (12.3)$$

Here I_0 is the intensity ($\triangleq$ number of photons irradiated), $I_{Fluorescence}$ is the intensity of the fluorescence. It should come as no surprise that this ratio, and hence the value for Φ, is <1 in nearly all cases. However, one can describe this even more precisely, since various transitions are possible (radiationless or not) that have deactivating effects with respect to fluorescence. By treating these **deactivation processes** and also the desired fluorescence as independent chemical reactions in each case, a rate constant k for each of the processes can be stated, fully in the sense of kinetics and thermodynamics. We must therefore consider:

k_F = Speed of the (desired) fluorescence phenomenon

k_{ISC} = Speed of *Intersystem Crossing*

k_{IC} = Speed of the Inner Conversion

Add to that:

k_{EC} = Speed of the External Conversion

k_{PD} = Speed of Predissociation

and, where applicable

k_{DD} = Speed of direct dissociation (with high-energy excitation radiation)

Thus, we need to introduce a correction factor ϕ_{corr} that also accounts for the deactivating processes. This results in:

$$\phi_{corr} = \frac{k_F}{k_F + k_{ISC} + k_{IC} + k_{EC} + k_{DD} + k_P} \qquad (12.4)$$

Applying Eq. (12.4) to Eq. (12.3), it follows:

$$\Phi = \frac{\phi_{corr} \cdot I_{Fluorescence}}{I_0} \qquad (12.5)$$

From Eqs. (12.4) and (12.5), it follows that the faster the desired fluorescence occurs, the greater the fluorescence quantum yield, and the more the other, radiationless, processes recede into the background: In the ideal case (solely fluorescence, no other processes occurring), the computational result is $\phi_{corr} = 1$. When considering this quantum yield, one has to take into account that fluorescence can only occur after a first interior conversion (otherwise, it would be a simple return to the ground state, for which applies: wavelength $\lambda_2 = \lambda_1$). However, such considerations would lead too far here; there is a reason why there are whole textbooks on the subject of "fluorescence".

■ **Lifetime of Excited States**

It has already been explained that $\sigma \rightarrow \sigma^*$ transitions occur much less frequently than $\pi \rightarrow \pi^*$ and $n \rightarrow \pi^*$ transitions: the *smaller* the energy difference between ground state and excited state, the *more likely the* associated electron transition. Moreover, the comparatively high-energy $\sigma \rightarrow \sigma^*$ transitions pose the problem of predissociation or direct dissociation, because due to the large energy difference between the two orbitals, these transitions require relatively high-energy excitation radiation, which may well directly cause (DD) or "prepare" (PD) breaking of a bond. Thus, these transitions are responsible for fluorescence phenomena only in rare cases: decisive are $\pi \rightarrow \pi^*$ and $n \rightarrow \pi^*$ transitions; here, the transition with the smaller energy difference between ground state and excited state then occurs preferentially. (It should be emphasised again that $n \rightarrow \pi^*$ transitions can only occur for those analytes that also have *at least one free electron pair*.)

Experiments have shown that fluorescence is particularly common (and particularly fluorescence-quantum efficient) in systems where the $\pi \rightarrow \pi^*$ transition is lowest in energy. This is clearly seen in the **molar extinction coefficient**: This is a factor of 10^2 to 10^3 higher than for the $n \rightarrow \pi^*$ transitions. (The molar extinction coefficient provides a direct measure of the probability of a transition.)

At the same time, the lifetime of the excited π^*-state is significantly shorter if the excited electron then returns to a π-orbital (where it will still be vib./rot.-excited) than if it relaxes to an n-orbital:
— Lifetime π^*/π 10^{-9}–10^{-7} s
— Lifetime π^*/n 10^{-7}–10^{-5} s

The shorter the lifetime of an excited state, the less likely it is that the undesirable "side reactions" such as ISC will occur. For this reason, $\pi \rightarrow \pi^*$ transitions preferentially lead to fluorescence ($S_1^* \rightarrow S_0^*$), while the much longer-lived excited states that have emerged from an $n \rightarrow \pi^*$ transition tend to favour *phosphorescence* (the ISC causes $S_1^* \rightarrow T_1^*$). The ISC is particularly favoured for analytes containing heavy atoms (mainly Br and/or I for organic molecules, and heavy cations such as Sr, Cs, or Ba for inorganic samples); this is called the **heavy-atom effect**. (This effect can be explained by spin-orbit couplings, which are, however, far beyond the scope of this introduction. If necessary, the Hesse-Meier-Zeeh and the more detailed literature will help.) Also, **paramagnetic** substances that can interact with the analyte solution favour ISC and other radiationless transitions, thus reducing the number of fluorescence processes (they reduce the *fluorescence yield*). The phenomenon that the fluorescence of whatever system is attenuated by interaction with another atom or molecule (ion) is called **quenching**. It can even be exploited for analytical purposes itself. (You will encounter this again in Part IV of this book.)

The quantification of fluorescence phenomena plays an increasing role in modern analytics, especially in the biosciences: biomolecules tagged with fluorescent markers allow more precise elucidation of intracellular processes, such as the interaction of proteins or individual steps of redox chains, etc. (A little more about this can be found in ▶ Sect. 12.3.) But fluorescence is also important in "chemical" analytics: for example, corresponding analytes can also be quantified by their fluorescence. This is the subject of the next section.

> **A Practical Hint**
> Even the "ordinary" oxygen in the air shows paramagnetic behaviour. (O_2 is a diradical—remember its MO diagram? When in doubt, the Binnewies will help.) For this reason, in any analytical method based on fluorescence, it is essential to work in the absence of air whenever possible; after all, any interaction of the potentially fluorescent analytes with air would lead to massive attenuation of the phenomenon to be quantified.

- **Unusual Only at First Glance: The *Upconversion Fluorescence***

In Part IV of "Analytical Chemistry I" the phenomenon of fluorescence was dealt with, you learned the general rule that the wavelength of fluorescent radiation is always longer than that of the excitation radiation:

$$\lambda_{\text{Fluorescence}} > \lambda_{\text{Excitation}}$$

After all, the radiationless transitions required for fluorescence inevitably dissipate a certain amount of energy elsewhere (generally in the form of (hardly measurable) heat to the environment or by direct two-particle impact "targeted" at another molecule—which is why the solvent may have such a drastic effect on the quantum yield). However, there is a fluorescence phenomenon that often causes confusion because it seems to contradict this very rule, at least at first glance:

If molecules (or ions) excited to fluorescence undergo **upconversion**, the corresponding system emits fluorescence radiation with a *shorter* wavelength than that of the light used for excitation: *Less* energetic radiation causes the emission of *more energetic* radiation. (An example is shown in Colour Plate 19 from Harris: Here, excitation with green light causes blue fluorescence.)

Let us look a little more closely at a system that exhibits this unusual phenomenon: The system used here is, on the one hand, a cationic ruthenium(II) complex (4,4′-dimethyl-2,2′-bipyridinyl)ruthenium(II), ◘ Fig. 12.2a (the spatial structure of this octahedral complex is stylised in ◘ Fig. 12.2b; the counter ions are not taken into account in either representation), and, on the other hand, 9,10-diphenylanthracene, which is also excitable to fluorescence (◘ Fig. 12.2c).

The fact that ultimately higher-energy photons are emitted than those used to excite the system is based on a multi-stage process in which the—by now—usual orbital representations are again to be used:

1. First, the cationic ruthenium complex (abbreviated here simply as [Ru]) is transferred from its singlet ground state to an electronically excited singlet state: $^1[\text{Ru}] \rightarrow {}^1[\text{Ru}]^*$ or $_{(\uparrow\downarrow)}{}^{()} \rightarrow {}_{(\uparrow)}{}^{(\downarrow)}$

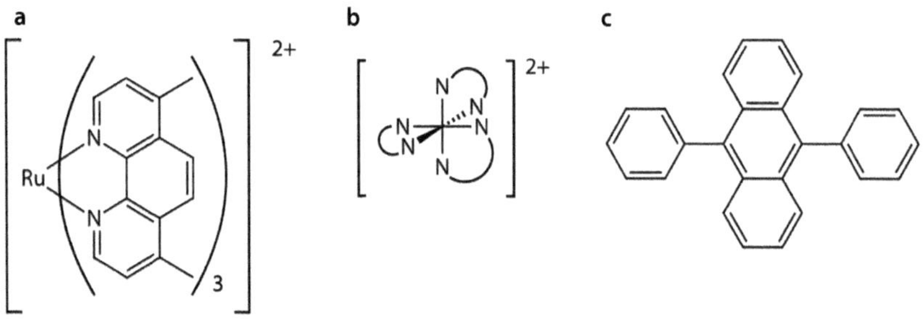

◘ Fig. 12.2 4,4′-Dimethyl-2,2′-bipyridinyl)ruthenium(II) **a**, **b** and 9,10-diphenylanthracene **c**

12.1 · Basics of Fluorescence: A Brief Review and Some Additional Information

2. Since such excited singlet states are even more energetic than the corresponding excited triplet states (Eq. 12.1), this will lead to *Intersystem Crossing*:

$$^1[Ru]^* \rightarrow {}^3[Ru]^* \quad \text{or} \quad {}^{(\downarrow)}_{(\uparrow)} \rightarrow {}^{(\uparrow)}_{(\uparrow)}$$

3. If a ruthenium complex cation that has been electronically excited this way collides with a currently non-excited 9,10-diphenylanthracene (DPA)—in this case, a singlet state is present that can be represented by $_{(\uparrow\downarrow)}^{(\,)}$—external conversion occurs, which we are already familiar with from ▶ Sect. 12.1: The "excess" energy of the excited ruthenium complex is transferred to the anthracene, which thereby enters an electronically excited triplet state: $^3[Ru]^* + {}^1DPA \rightarrow {}^1[Ru] + {}^3DPA^*$.
   (a) For the cationic ruthenium complex, the orbital representation is: $_{(\uparrow)}^{(\uparrow)} \rightarrow {}_{(\uparrow\downarrow)}^{(\,)}$.
   (b) What is effected in the case of 9,10-diphenylanthracene can be represented by $_{(\uparrow\downarrow)}^{(\,)} \rightarrow {}_{(\uparrow)}^{(\uparrow)}$.

4. The lifetime of the excited triplet state $^3DPA^*$ is comparatively long, so that collisions between *two electronically excited* 9,10-diphenylanthracene molecules can also occur. This *again* leads to external conversion: one of the two molecules returns to the (singlet) ground state, while the other is *further* excited to an excited singlet state (see Eq. 12.1): $^3DPA^* + {}^3DPA^* \rightarrow {}^1DPA + {}^1DPA^*$.
   (a) For the molecule returning to the ground state, this means: $_{(\uparrow)}^{(\uparrow)} \rightarrow {}_{(\uparrow\downarrow)}^{(\,)}$
   (b) For the other, however: $_{(\uparrow)}^{(\uparrow)} \rightarrow {}_{(\uparrow)}^{(\downarrow)}$

5. If this 9,10-diphenylanthracene, which is now even more excited by the collision, returns to the ground state—which can be represented as $_{(\uparrow)}^{(\downarrow)} \rightarrow {}_{(\uparrow\downarrow)}^{(\,)}$—*more* energy is released than was required for the excitation from the singlet ground state to the excited triplet state: The emitted photon is correspondingly more energetic, and according to the now certainly sufficiently familiar formula

$$E = h \cdot \nu$$

this results in a shorter wavelength than was used to excite the original system.

> ❗ Please do not think that the *upconversion* violates the law of conservation of energy! The fact that here urexpectedly *more* energetic light is emitted than was used for excitation is due to the fact that here *two* electronically excited particles (the $^3DPA^*$ present after the External Conversion) interact with each other in such a way that *one of* the two was, so to speak, doubly excited after a second External Conversion. (Harris explains the phenomenon of *upconversion* in Excursus 18.2 on the basis of this very example, describes it very vividly as "it takes *two green* photons to produce *one blue* photon".)

Harris, Section 18.6: Sensors based on fluorescence quenching

❓ Questions

13. Wwhat does a fluorescence quantum yield of 4.2% mean?
14. Why is the quantum yield of a fluorescence phenomenon based on upconversion significantly lower than that of an "ordinary" fluorescence process? What is the maximum quantum yield that can be achieved by an upconversion process?

12.2 Fluorescence Spectrometry

In a way, fluorescence spectrometry is "just" a variant of spectrophotometry (which you might already know from Part IV of "Analytical Chemistry I"). The only difference is that instead of determining how much of the originally irradiated light passes through the sample unobstructed (i.e., you determine the **absorbance** or **transmission**), you measure the *intensity of the light emitted* by the analytes excited to fluorescence. Again, the light intensity is directly correlated with the analyte concentration, and in the end, **Lambert-Beer's law** is behind it again, which you already know from Part II. Let us first look at the extinction; after all, in order for fluorescence to occur, at least part of the excitation radiation of intensity I_0 must first be absorbed, so that the intensity of the radiation passing through the sample vessel containing the excitable analyte ($I_{passing\ through}$) applies:

$$E = \lg \frac{I_0}{I_{passing\ through}} \tag{12.6}$$

If the sample vessel contains the analyte in sufficient dilution so that Lambert-Beer's law is applicable (remember?), and the analyte has the (wavelength-dependent) extinction coefficient ε_λ, then for a sample layer thickness d the already known relation holds:

$$E = \varepsilon_\lambda \times c \times d = \lg \frac{I_0}{I_{passing\ through}} \tag{12.7}$$

Alternatively, you can write:

$$\frac{I_0}{I_{passing\ through}} = 10^{\varepsilon_\lambda \cdot c \cdot d} \text{ or } \frac{I_{passing\ through}}{I_0} = 10^{-\varepsilon_\lambda \cdot c \cdot d} \tag{12.8}$$

The following then applies to the intensity of the absorbed radiation:

$$I_{absorbed} = I_0 - I_{passing\ through} \quad \text{or} \quad I_{absorbs} + I_{passing\ through} = I_0 \tag{12.9}$$

Thus:

$$\frac{I_{absorbed}}{I_0} = \frac{I_0 - I_{passing\ through}}{I_0} = 1 - \frac{I_{passing\ through}}{I_0} = 1 - 10^{-\varepsilon_\lambda \cdot c \cdot d} \tag{12.10}$$

> **A Small Insertion for Those Who Have a Look at the Recommended Further Literature**
>
> Especially on the subject of fluorescence, the Cammann is very helpful (and offers plenty of opportunities for further study, both in terms of the apparative set-up and with regard to the various possible applications). However, a small but confusing error has crept into this book: In formula 5.15 of that book, which refers to $I_{absorbed}$ and should have the same information content as "our" Eq. (12.10), instead of the term
>
> $$\frac{I_{absorbed}}{I_0} = \ldots \text{ the expression } \frac{I}{I_a} = \ldots$$
>
> I is used. In the Cammann, I stands for $I_{passing\ through}$ and I_a for $I_{absorbed}$.
>
> With this ratio of the two intensities of radiation (which makes no sense at all in terms of content …) one gets nowhere with the reformulation, of course. This is only mentioned, so that you do not get confused.

12.2 · Fluorescence Spectrometry

Now it must be determined how many of the absorbed photons actually cause fluorescence, i.e. how great the fluorescence intensity I_F is. This is where the factor Φ from Eq. (12.3) comes into play. With it, taking into account Eq. (12.10), the following applies:

$$\frac{I_{Fluorescence}}{I_0} = \Phi \cdot \frac{I_{absorbed}}{I_0} = \Phi \cdot \left(1 - 10^{-\varepsilon_\lambda \cdot c \cdot d}\right) \quad (12.11)$$

If the concentration of the analyte responsible for the fluorescence is sufficiently small, the general rule is:

$$1 - 10^{\varepsilon_\lambda \cdot c \cdot d} \simeq 2303 \cdot \varepsilon_\lambda \cdot c \cdot d \quad (12.12)$$

Thus, the following applies to corresponding analytics solutions:

$$I_F \simeq 2303 \cdot \phi \cdot I_0 \cdot \varepsilon_\lambda \cdot c \cdot d \quad (12.13)$$

The intensity of the fluorescence radiation therefore depends on
- the analyte concentration of the sample c
- the (wavelength-dependent) extinction coefficient ε_λ *and*
- of the quantum yield achieved.

However, fluorescence spectrometry is rarely limited to a simple intensity measurement. Instead, one resorts to a *calibration* and performs a relative measurement with respect to a standard of known concentration of the same fluorescence system. Given the same measurement conditions (Φ, I_0, ε_λ), the following then simply applies:

$$\frac{I_F(\text{sample})}{I_F(\text{standard})} = \frac{c(\text{sample})}{c(\text{standard})} \quad (12.14)$$

The experimental setup for fluorescence spectrometric analysis largely corresponds to that of an "ordinary" photometric measurement, except that the fluorescence radiation with its characteristic wavelength λ_F must be quantified at right angles to the excitation radiation, so that the radiation used for excitation is not unintentionally detected as well. Otherwise, the same devices are required that you are already familiar with from Part IV of "Analytical Chemistry I", i.e. monochromators, beam splitters, and photomultipliers. ◘ Figure 12.3 shows the schematic diagram of a spectrofluorophotometer.

■ Fluorescence Spectra

At first glance, emission spectra based on fluorescence bear striking resemblance to the excitation spectra (i.e. the "normal" spectra obtained by photometry) of the same compounds. However, if the two spectra are superimposed, as shown in ◘ Fig. 12.4 for the compound 7-(*N*-methylamino)-4-nitro-2,1,3-benzooxadiazole (MNBDA), which has an extremely strong tendency to fluorescence and whose structure you can see in ◘ Fig. 12.5a, you can see the already mentioned Stokes shift: For the wavelength of the (fluorescence) radiation emitted by the sample λ_F and the wavelength used for excitation λ_e, the following always applies: $\lambda_F > \lambda_e$, accordingly, the resulting fluorescence spectrum is shifted to longer wavelengths.

MNBDA (◘ Fig. 12.5a) enjoys extreme popularity as a basic structure for fluorescent labels: This compound fluoresces very strongly and thus leads to a very satisfactory fluorescence quantum yield Φ. The same applies to the other two **fluorophores** shown in this figure: both dansyl chloride (b, the systematic name of this compound is 5-(dimethylamino)naphthalene-1-sulfonyl chloride) and fluorescein isothiocyanate (c, usually abbreviated as FITC) are shown here in the form in which they can be readily brought to reaction with biomolecules:

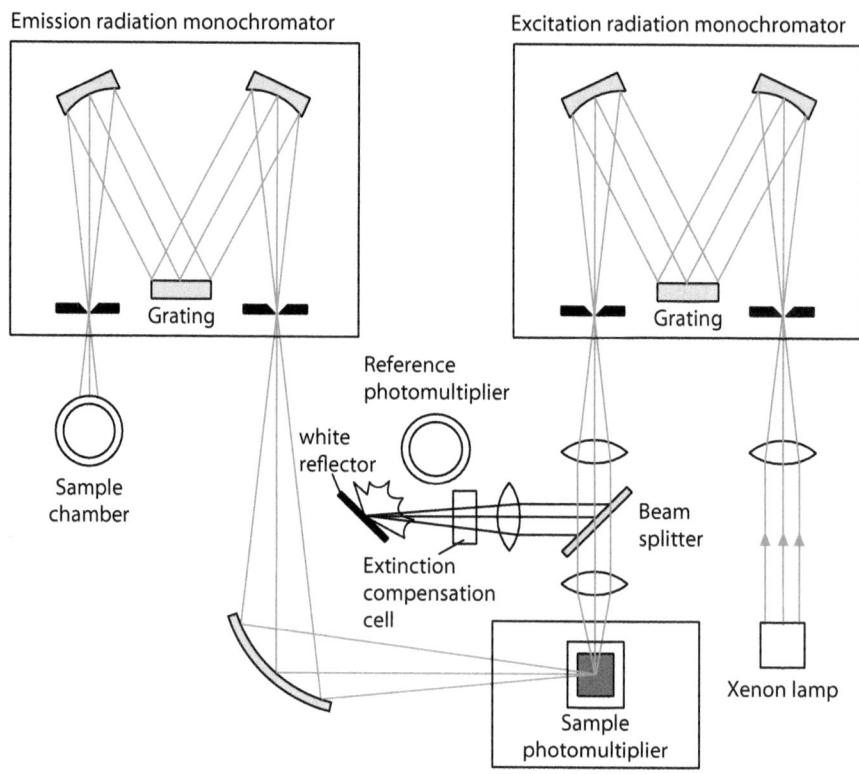

Fig. 12.3 Schematic diagram of a spectrofluorophotometer (K. Cammann (Ed.): Instrumentelle Analytische Chemie, pp. 5–16, Fig. 5.15, 2010 Copyright Spektrum Akademischer Verlag Heidelberg. With permission of Springer)

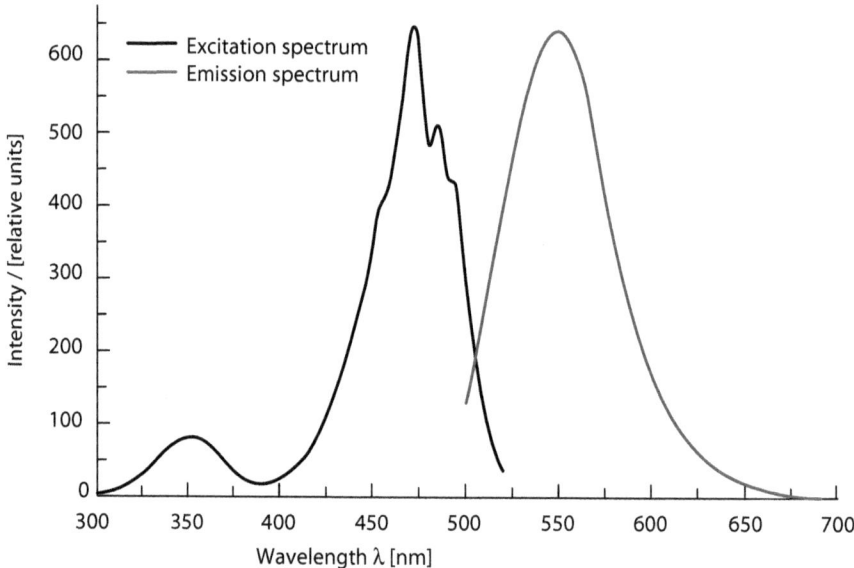

Fig. 12.4 Fluorescence spectrum of MNBDA (K. Cammann (Ed.): Instrumentelle Analytische Chemie, pp. 5–17, Fig. 5.17, 2010 Copyright Spektrum Akademischer Verlag Heidelberg. With permission of Springer)

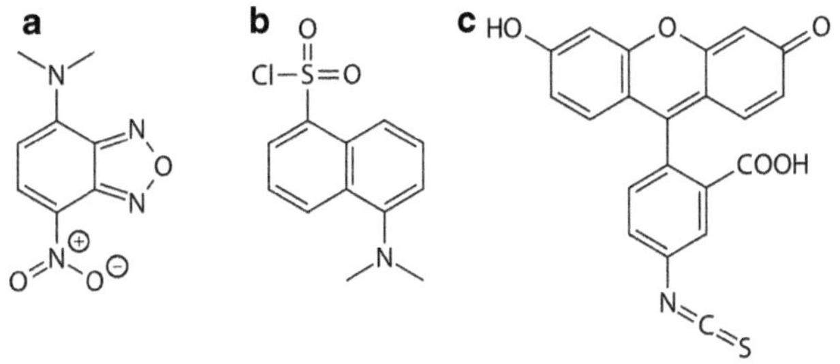

12.2 · Fluorescence Spectrometry

Fig. 12.5 Selected fluorophores

- Dansyl chloride (Fig. 12.5b) reacts primarily with the terminal amino group of peptides; the corresponding **adduct**, a sulfonamide, can be readily quantified because of its fluorescence properties.
- Fluorescein isothiocyanate (Fig. 12.5c) also reacts well with amino groups, but also with other nucleophilic functional groups. Therefore, it is extremely versatile in analytics.

■ **Time-Resolved Fluorescence Spectrometry**

This variant of fluorescence spectrometry is particularly suitable for the type of sample material in which matrix components make the measurement difficult—which is anything but rare, especially with biological samples. In most cases, no (organic) molecules that can be excited to fluorescence are used here, but instead ions of rare earth metals such as europium (Eu) and terbium (Tb), but also dysprosium (Dy) and samarium (Sm); they are usually used in the form of chelate complexes.

The radiation emitted by these (chelated) ions after photochemical excitation is actually (strictly speaking) not to be regarded as fluorescence, but represents a (often remarkably long-lived) *phosphorescence*; for historical reasons, however, such experiments are also attributed to *fluorescence spectrometry*. The excitation and emission mechanism corresponds in principle to that already repeated on this subject in the introduction to this chapter; however, *Intersystem Crossing* is necessary for the emergence of the desired luminescent phenomena of these compounds, because only from the (first) electronically (and also vibrationally/rotationally) excited triplet state (T_1^*), which is present after ISC, can energy be efficiently transferred to the lanthanide ion. For this reason, it is necessary to increase the probability of an ISC—exactly opposite to "ordinary" fluorescence spectrometry—which is why additional heavy ions (such as caesium) are added. (For some systems, strontium is already massive enough to cause the heavy-atom effect described in this chapter.) After the *Intersystem Crossing* mentioned above, the excess energy of the system is partly dissipated by radiationless relaxation processes, but mainly by the emission of phosphorescence radiation. Behind this lies an *internal conversion* in which one of the f-electrons of the lanthanides involved is transferred to an orbital that is unoccupied in the ground state and from there returns to its ground state. However, because such transitions are *symmetry-forbidden*, the lifetime of the corresponding excited states is much longer than in the examples considered so far. (The phosphorescence of such lanthanoid systems can actually last for several hours.)

Two things in particular are remarkable about lanthanide phosphorescence:

- On the one hand, because the number of possible electron transitions is kept within narrow limits, it leads to comparatively narrow bands, which of course simplify the evaluation of the resulting spectra immensely.
- On the other hand, the strong Stokes shift of such systems should be emphasised: The wavelength difference between excitation radiation and phosphorescence radiation can be more than 200 nm.

At least at first glance, it is a technical problem that such systems, in addition to the phosphorescence desired in this case, also show "real" fluorescence, as expected, which, however—the difference between fluorescence and phosphorescence has already been emphasised several times, by now—lasts much less long (as a reminder: real fluorescence usually ends only fractions of a second after the end of excitation). For this reason, in *time-resolved fluorescence spectrometry*, the emission spectrum is not measured immediately after excitation, but after a short waiting time (25–100 µs, depending on the measurement system). Since the "true" fluorescence (which is often referred to as *background fluorescence*) has already decayed after a period of time that is at most in the two-digit *nano*second range, it then no longer affects the measurement. The same applies to **scattered radiation** (again, reference is made to Part IV of "Analytical Chemistry I", in which this phenomenon was discussed in connection with Raman spectroscopy), whose lifetime is *even shorter*.

After the waiting time, the lanthanoid phosphorescence radiation is detected over a period of several hundred to a thousand *micro*seconds. In time-resolved fluorescence spectrometry, the decay of the phosphorescence is then also detectable.

The influence of the various luminescence phenomena,
- scattered radiation,
- background fluorescence, *and*
- lanthanide phosphorescence.

is shown in ◘ Fig. 12.6 by means of an exemplary fluorescence spectrum (it is irrelevant which analyte was measured here; after all, it is "only" a matter of principle).

Time-resolved fluorescence spectrometry is increasingly used in medical diagnostics because, among other things, it can be used to quantify metabolic products that occur only in selected diseases. In most cases, such analyses are only carried out after quite complex sample preparation and in combination with substance separation techniques that you are already familiar with, such as chromatography or (capillary) electrophoresis.

❓ Questions

15. What is the origin of the Stokes shift observed in all fluorescence- and phosphorescence-based analytical methods?
16. Why does time-resolved fluorescence spectrometry, which does not use ordinary (organic) fluorophores but chelate complexes of lanthanides, require a wait of several microseconds before usable measured values can be recorded?

The principle of the immunoassay (from ▶ Sect. 11.4) can also be combined with fluorescence spectrometry: Such fluorimetric immunoassays (FIA, also known as *time-resolved fluoroimmunoassay*, **TRFIA**) have become so much of a standard in the life sciences that commercially available kits are available for a variety of analytes.

Now that we have ventured quite far into the analysis of biological samples, especially thanks to the immunoassay and the variants derived from it, we

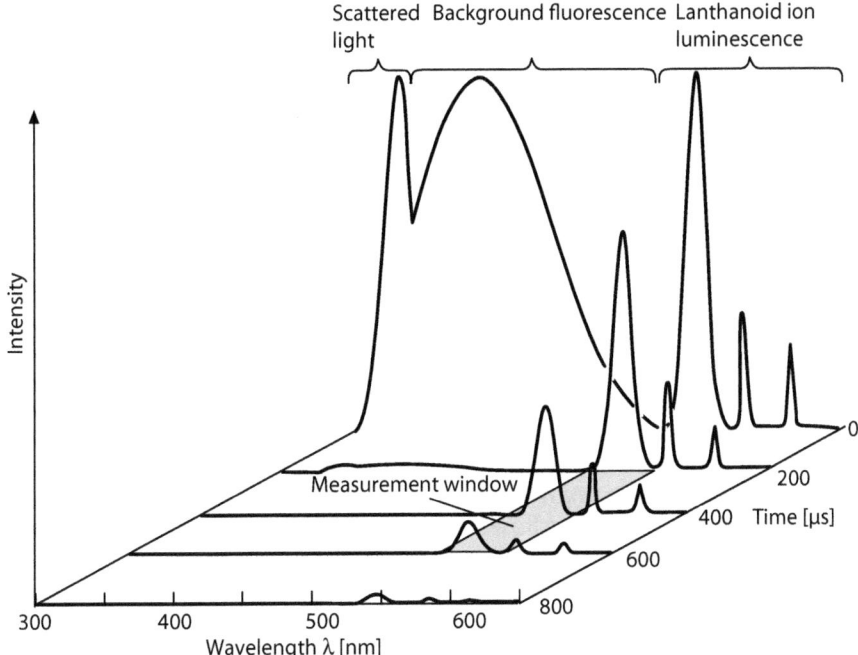

Fig. 12.6 Time-resolved fluorescence spectrometry (K. Cammann (Ed.): Instrumentelle Analytische Chemie, pp. 5–21, Fig. 5.21, 2010 Copyright Spektrum Akademischer Verlag Heidelberg. With permission of Springer)

would like to conclude by explaining another principle based on fluorescence, which—at least at first glance—does not seem to have much to do with "chemistry":

12.3 Fluorescence Microscopy

In this variant of classical light microscopy, the sample to be examined (usually cells or organelles) is first stained using fluorescent dyes (some biomolecules are even *auto-fluorescent,* i.e. contain their own fluorophores), which are then motivated to fluoresce in the usual way with the aid of excitation radiation. Just as in "ordinary" fluorescence spectrometry, the instrumental setup must be taken into account: The source of the excitation radiation must be at right angles to the light path of the fluorescence photons so that the excitation radiation does not obscure the fluorescence radiation or otherwise complicate the detection of the fluorescence photons. For this reason, fluorescence microscopes are almost exclusively designed as **reflected light microscopes**: The excitation radiation comes from the objective; a beam splitter then reflects the radiation specifically onto the object under examination. (Any excitation radiation that does reach the object directly can also be excluded by a blocking filter, but that would be going too far here.)

Cell components can be identified by the fluorescence of the appropriately labelled molecules; proteins are labelled particularly frequently. Various techniques are used for this purpose:

1. One can use fluorescence-labelled ligands that specifically (or at least preferentially) attach to a particular protein species.
2. Covalent bonds can also be formed: The C- and especially the N-terminus of the protein to be investigated/traced can fuse with a fluorescence-labelled peptide with cleavage of water and the formation of a new peptide bond.

3. Some (bio)molecules that act as antigens can be detected by fluorescence-labelled specific antibodies.
4. The tracking of unlabelled, autofluorescent biomolecules has also been the basis for the elucidation of numerous cell-internal or even cross-cell processes.

In plants, however, the autofluorescence of certain biomolecules can also become a problem: Chlorophylls, for example (and also some other plant pigments) fluoresce quite strongly and can therefore mask other fluorescence signals.

With biomolecules labelled this way (mostly proteins), it is then possible to find out exactly where in the cell these proteins occur. (Typical questions: Are the molecules under consideration part of the cytoplasm? Are they part of the cell membrane? Can they even migrate from one cell to another?) The way in which different proteins interact could also be elucidated in this way.

And this is only the beginning: By now, in the biosciences a large number of different markers are used whose special properties are not limited to fluorescence:

- Fluorescent pH indicators can be used to detect pH gradients within a cell.
- There are markers which themselves enter into redox processes, so that the redox potentials of oxidisable (or reducible) cell components can be determined.
- Similarly, information can be obtained with dyes that change colour when an electrical voltage is applied. For some dyes, this property is not particularly pronounced: The (average) human eye does not yet perceive any change, but a shift of the absorption maximum is still detectable in the fluorescence emission spectrum. An example is represented by the compound designated di-4-ANEPPS in ◘ Fig. 12.7a. (Incidentally, the IUPAC name of this compound is (1-(3-sulfonatopropyl)-4-[β-[2-(di-*n*-butylamino)-6-naphthyl]vinyl]pyridiniumbetaine. No, you don't have to memorise that one.)
  - In a subspecies of voltage-sensitive dyes, application of a voltage does not cause a shift in the emission spectrum, but rather an increase in fluorescence. The best-known example is arguably merocyanin 540 (◘ Fig. 12.7b).
  - With this marker, the membrane potential (remember Part II and the electroanalytical methods?) of a bio-membrane could be visualised for the first time.
  - Why the application of a voltage leads to a shift in the fluorescence wavelength or to increased fluorescence has not yet been fully explained, but the phenomenon can still be exploited.
  - Incidentally, the designation "Merocyanin 540" has nothing to do with the wavelength at which an absorption or fluorescence maximum occurs; this is a derivative of the compound merocyanin in which an oxygen atom has been replaced by a sulphur. (Why this chemical

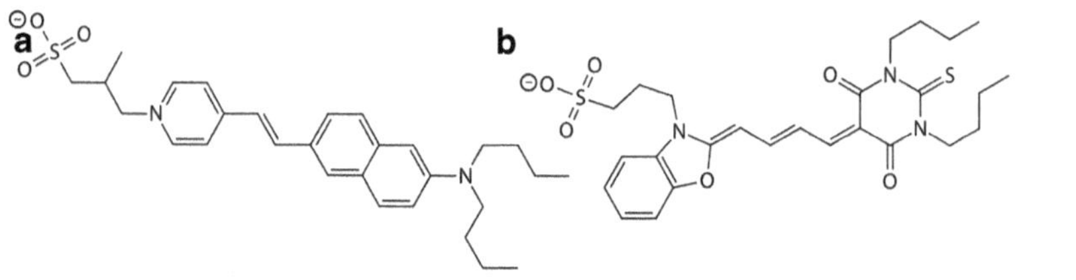

◘ **Fig. 12.7** Fluorescent marker

modification deserves the suffix "540" cannot be inferred from the relevant technical literature.)
— Still other fluorescence markers respond to the presence of calcium(II) ions and allow the transport (influx and efflux) of this important neurotransmitter to be quantitatively traced.

The further and new development of (analyte-specific) fluorescence markers is progressing steadily, and the technology of high-resolution fluorescence microscopy is also being developed further and further. It is to be expected that in the long term this method will be able to provide a great deal of information about inter- and intracellular processes that have not yet been (fully) elucidated.

❓ Questions
17. As important as it is to me that you get to know the possibilities of the application of fluorescence phenomena not only on a small scale (i.e. on the molecular level), but also to see how closely the various natural sciences are (by now) linked, the procedure of fluorescence microscopy is nevertheless "un-chemical" in itself. For this reason, there is no exercise question here. Consider it a free period.

12.4 Summary

■ Gravimetry

In electrogravimetry, an electrolysable analyte is deposited in its elemental form on an inert electrode and the analyte content is determined on the basis of the mass difference of the inert electrode before and after electrolysis. The decomposition voltage required for electrolysis is greater than the voltage that would be expected according to the theoretically calculated potential difference. The reasons for this are the overvoltage, the concentration potential building up near the cathode and the ohmic potential resulting from the resistance of the half-cells involved.

Thermogravimetry (TG) is used to determine the mass loss resulting from thermally induced transformation of the analyte and the release of highly volatile cleavage products (water, carbon dioxide, etc.); the measured values are usually plotted in the form of a thermogravimetric curve. Micro thermogravimetry is a variant of TG that is suitable very small sample quantities.

(Micro-)thermogravimetry is often combined with other analytical methods, such as IR spectroscopy (TG/IR) or mass spectrometry (TG/MS).

■ Thermal Processes

Differential thermal analysis (DTA) can be used to detect and investigate processes induced by temperature rise, such as (partial) crystallisation or the melting process of amorphous solids. For this reason, DTA is one of the standard methods when it comes to investigating non-metallic/amorphous materials (such as glasses or polymers). Depending on the atmosphere prevailing in the examination chamber, oxidative decomposition processes can also be quantified.

A calorimeter can be used to quantify the amounts of heat that are released or have to be spent during energy conversion. In differential calorimetry, which also plays an important role in materials analysis, phase transformation temperatures can be detected accordingly, such as heat of melting or glass temperatures. In this way, the degree of crystallisation of semi-crystalline solids can also be investigated. Dynamic differential scanning calorimetry (DSC) is particularly common. It involves direct comparison with a reference substance, either in the presence of a temperature gradient or at constant temperature. Compared to DTA, DSC is significantly more precise.

- **Radioactive Nuclides in Analytics**

Radionuclides are subject to decay processes in which alpha particles (α-decay) or beta particles (β-decay) are emitted, resulting in elemental transformation. The rate at which this occurs is isotope-specific and directly correlated with the half-life of the isotope under consideration. The released energetic (α- or β-) particles can either be quantified directly with a scintillation counter, or the quantification is carried out via the γ-photons also released in the course of this process. Both allow conclusions to be drawn about the content of the radionuclide in question.

In neutron activation, atoms are brought to collision with neutrons, leading to a high-energy compound nucleus; this newly-formed nucleus emits its excess energy in the form of (quantifiable) γ-radiation with isotope-characteristic wavelength; further decay processes can follow, which then lead to decay radiation that is also isotope-characteristic.

In addition, it can be exploited that radioactive isotopes are practically indistinguishable in their chemical behaviour from their non-radioactive counterparts, i.e. they undergo the same chemical reactions, but can be easily traced—precisely because of their radioactivity: This is the basis of the principle of radiolabelling; corresponding tracers are used, among other things, to elucidate metabolic pathways or other complex reaction mechanisms.

Radioactive isotopes are also used for age determination: The radiocarbon method takes advantage of the fact that in still-living organisms there is an equilibrium between the radioactive carbon isotope ^{14}C, which is present in the atmosphere in traces, and the two stable isotopes carbon-12 and carbon-13, while after the death of the organism in question there is no exchange of substances; the extent of the residual radioactivity still present due to the radioactive isotope carbon-14 then allows the sample to be dated quite precisely. In a similar way, other radionuclides (^{238}U, ^{235}U, ^{40}K, ^{87}Rb) can be used for age determination.

Radioimmunoassay (RIA), mainly applied to biospecimens, combines radioisotopes with antibody-antigen interaction: The antigens to be quantified are brought to reaction with antibodies fixed on a carrier material in the presence of a known amount of substance of radioactive or radiolabelled antigens; then, using a scintillation counter, the level of radioactivity present on the carrier material is quantified: The competition of labelled and unlabelled antigens for the binding sites of the antibodies allows quantitative statements about the content of the analyte (i.e. the unlabelled antigen).

- **Fluorescence Method**

In any analytical method based on fluorescence, it must be taken into account that, in addition to the desired fluorescence phenomenon, deactivation processes also play a role; it is therefore necessary to determine the quantum yield in such studies.

Fluorescence spectrometry may be regarded as a variant of spectrophotometry; Lambert-Beer's law still applies. However, it is not the number of absorbed photons of the excitation radiation that is quantified here, but the amount of emitted fluorescence radiation; therefore, the fluorescence quantum yield must be determined. If the analyte itself cannot be excited to fluores-

12.4 · Summary

cence, it is possible to covalently couple it to a suitable fluorophore via bonds to be newly formed; generally, this does not change the other chemical behaviour of the analyte.

In the time-resolved variant of fluorescence spectrometry, which is primarily used for analytes within interfering matrices, no organic fluorophores are used, but rather lanthanide ions, usually used in the form of chelate complexes, which can be excited not only to fluorescence, but also to phosphorescence. Since phosphorescence phenomena are much more long-lived, the intensity of this light phenomenon is measured after a short waiting period: after any fluorescence of the system has decayed. This analytical method is increasingly used in medical diagnostics; since the analytes are usually biomolecules (possibly labelled by a fluorophore) that can act as antigens, this form of spectrometry is often combined with immunoassays.

✅ Answers

1. The mass difference after/before is $\Delta m = 27.68 - 27.30 = 0.38$ g. It is important to note here that of course *elemental* cobalt is electro-deposited, so $m(Co) = 0.38$ g. With $M(Co) = 58.933$ g/mol, this corresponds (according to $n = m/M$) to $n(Co) = 0.00645$ mol = 6.45 mmol. However, since the elemental cobalt was (desirably) deposited quantitatively: $n(Co) = n(CoCl_2)$, and according to $c = n/V$, for $V = 20.00$ mL, we get: $c(CoCl_2) = 6.45$ mmol/20.00 mL = 0.323 mol/L = 323 mmol/L.

2. According to ▶ Eq. (9.2), $U = 2.3\ \Omega - 0.017\ A = 0.0391$ V.

3. From task 1 we know: $c(CoCl_2) = 323$ mmol/L. Since $c(CoCl_2) = c(Co^{2+})$ and $E^0(Co^{2+}/Co) = -0.282$ V (you can take this from Harris), according to Nernst's equation (analogous to ▶ Eq. (9.1), taking into account $\lg(1/x) = -\lg x$ and thus $-\lg 1/x = +\lg x$) then:

    Harris, Appendix H (Standard Reduction Potentials)

$$\begin{aligned}
E_{Cathode} &= -0.282V + (0.059/2)\cdot \lg\left[Co^{2+}\right] \\
&= -0.282V + (0.059/2)\cdot \lg 0.323 \\
&= -0.282V + (0.059/2V)\cdot(-0.49) \\
&= -0.282V - (0.0295V)\cdot(0.49) \\
&= -0.282V - 0.014V = -0.296V.
\end{aligned}$$

   This is at least the cathode voltage resulting from the *initial* concentration of the solution. Since in the immediate vicinity of the cathode the concentration of the solution is lower than the initial concentration due to the local depletion of cobalt(II) ions, the actual value of this voltage will be *higher* after the deposition has started. However, more precise statements cannot be made so easily, because the actual local concentration of the ions cannot be calculated directly, and the respective concentration potential—similar to the overvoltage—also depends on the applied current flow.

4. Here we need ▶ Eq. (9.3), and we have to consider (*dimensional analysis!*) that $1\ \Omega = 1$ V/A. Then with $I = 0.021$ A and $R = 1.86\ \Omega$ we finally get

$$E_{Ohm} = 0.021A \cdot 1.86\Omega = 0.021A \cdot 1.86V/A = 0.039V.$$

5. Please bear in mind once again that water of crystallisation is an integral part of the crystal lattice and by no means only "moisture trapped in it": If a (mono-, di-, tri- etc.) hydrate is "dehydrated", this requires the complete reconstruction of the entire ion lattice. For this reason, removal of the water of crystallisation often (but not always!) requires temperatures well above the boiling point before "ordinary water".

Skoog, Section 31.1: Thermogravimetric analysis

6. If the (micro-)thermogravimetric study of a sample containing (amongst other elements) carbon, is carried out under oxygen exclusion, the carbon will remain (as soot/ash). If, on the other hand, oxygen is present (perhaps even in excess, for example in a pure oxygen atmosphere), it will react with the carbon to form carbon dioxide and therefore disappear into the gaseous phase. For this reason, depending on the atmosphere, different mass losses result. This can be seen well, for example, in Fig. 31.4 from Skoog, where the polymer polyethylene [$(-CH_2-CH_2-)_n$] was analysed first under nitrogen atmosphere and then under oxygen atmosphere: While the carbon remains as soot in the nitrogen atmosphere (the mass loss amounts to 75% in total) and a mass plateau results even with further increased temperature, the further heating under oxygen atmosphere finally leads to the complete volatilisation of the sample substance: At 750 °C, neither polyethylene nor any residues were detectable any more.

Skoog, Section 31.2: Differential thermal analysis

7. If the temperature of the sample substance rises higher than that of the reference substance at the same heating rate of the sample and reference substance, the sample substance must undergo an *exothermic* process: In this process, energy (= heat) is released, which is then correspondingly dissipated to the environment and thus ensures a higher temperature. The opposite is then true for a lower temperature compared to the reference substance: In an *endothermic* process, the thermal energy added is absorbed by the sample substance and causes a change (such as melting); the energy required for this phase transformation is then, of course, no longer available for heating the substance. (Perhaps you would like to have another look at Fig. 31.7 from Skoog?)

8. In fact, you should think quite a lot of it, because in the end that is what differential thermal analysis is all about: The heat capacity is nothing more than a measure of how much the temperature of a substance changes when heat is added. If this substance undergoes phase transformations (for example during the melting process, or—in the case of polymers—when the glass transition temperature T_G is reached), this transformation is not associated with a change in the enthalpy of the substance, but the heat capacity of the molten sample differs from the heat capacity of the sample in the (semi-)crystalline state. In this respect, it is not at all wrong to regard the DTA as a calorimetric method, even if this is actually only dealt with in ▶ Sect. 10.2.

Skoog, Section 31.2: Differential thermal analysis
Skoog, Section 31.3: Differential scanning calorimetry (DSC)

9. For the simple reason that energy amounts can be quantified incomparably better/clearer than temperature differences. Please compare Figs. 31.7 and 31.13 from Skoog (in the former only up to the temperature just above the negative peak marked "melting"): In principle, both curves show the same thing, but while the differential thermogram merely indicates the temperature ranges at which exothermic or endothermic processes occur, the DSC scan of the same sample shows the heat flux associated with the changes, and while it is commonly quite difficult by apparatus to detect temperature changes in the range <0.1 °C (or K, respectively) as such at all, changes in an energy flux (indicated in W or mW, possibly in even smaller "portions") are much easier to quantify.

10. Sodium belongs to the pure elements: Only the isotope sodium-23 occurs naturally on Earth. Accordingly, any atom ^{23}Na that comes into collision-contact with a neutron is converted to the radioactive ^{24}Na. The energy released in this process is emitted practically spontaneously in the form of a γ-photon:

$$^{23}Na + {}^1n \rightarrow {}^{24}Na + \gamma$$

12.4 · Summary

These energetic photons can certainly be quantified, but what is much more interesting is that the sodium isotope produced is unstable: $t_{1/2}(^{24}Na) \approx 15$ h. In the process, β-decay occurs:

$$^{24}Na \rightarrow {}^{24}Mg + e^-$$

This decay is even easier to quantify. However, it must be taken into account that relatively long irradiation times are required due to the comparatively long half-life.

11. This is one of the purposes of such labelling experiments: If, for example, ^{18}O-labeled alcohol (R′–^{18}OH) is used, it can be shown that only the product R–C(=O)^{18}O–R′ is formed; the water released in the process does *not* contain *any* oxygen-18.

12. Within a half-life, the radioactivity is halved (it should be stressed again: *independent of the amount of radioactive material present*), i.e. after two half-lives it has decayed to a quarter, after three half-lives to an eighth, and after four half-lives then to a sixteenth. If we take the Cambridge half-life of carbon-14 as a basis (of $t_{1/2}(^{14}C) = 5715 \pm 30$ years), we get a statistical age of 22,860 ± 120 years, i.e. the sample may as well be "already" 22,980 or "only" 22,740 years old. As far as the number of significant figures here is concerned (you still remember Part I of "Analytical Chemistry"?), however, you had better be careful, because the problem of the accuracy of this method (keywords: sunspots, anthropogenic dilution of the natural carbon-14 content, nuclear weapons) has already been addressed.

13. Regardless of the factors that are all taken into account in order to make statements about the expected quantum yield (discussed in Eqs. 12.4 and 12.5), there is a simple calculation behind their purely quantitative description: What is the ratio of the number of fluorescence photons obtained to the number of photons used for excitation? (Eq. 12.3 states exactly that.) In the (unfortunately only theoretically achievable) case that really every excitation photon also produces a fluorescence photon, the ratio is then 1:1, given in percent would be here Φ = 100%. The value Φ = 4.2% then means that Φ = 4.2/100, i.e. 1000 excitation photons lead to the emission of 42 fluorescence photons.

14. Since *upconversion* requires *two* excitation photons at a time to cause the emission of *one* fluorescence photon, the fluorescence quantum yield naturally drops drastically: values of Φ > 50% are not even theoretically achievable. In practice, one commonly obtains much lower values. (In the example from Harris, whose colour plates provide an illustration of this phenomenon, the fluorescence quantum yield is a staggering 3.3%: Each 1000 excitation photons lead to the emission of 33 fluorescence photons.)

15. Please recall again the principle of both fluorescence and phosphorescence: A system is put into an excited state and then initially releases/dissipates part of this excess energy to the environment radiationless (whether this energy release is associated with *intersystem crossing* is not important here for the time being). This means that some of the excitation energy is "lost", so when fluorescence/phosphorescence occurs, the system cannot *possibly* emit light of exactly the same wavelength with which it was previously excited. Since the energy content of the fluorescence or phosphorescence photons must therefore be (slightly or significantly) lower, the wavelength λ associated with these photons must be greater than the excitation wavelength, according to the formula (by now certainly well known) $E = h \cdot \nu$. In phosphorescence, this effect is usually even more pronounced, which is why, for example, time-resolved fluorescence spectrometry, which is actually based on *phosphorescence,* results in a significantly larger Stokes shift.

16. While fluorescence phenomena cease practically at the moment when the excitation of the system is stopped and last at most in the two- to three-digit microsecond range, phosphorescence phenomena, especially when lanthanides are involved, are much more long-lived. Although fluorescence phenomena do occur during excitation in time-resolved fluorescence spectrometry (which, of course, reduces the quantum yield with respect to phosphorescence, but that's just by the way), these fade away very quickly, while phosphorescence lasts much longer. If one waits a few microseconds, the (now decayed) fluorescence will no longer hinder the quantification of the phosphorescence radiation.

Further Reading

Ashby MF, Jones DRH (2006, Nachdruck 2013) Werkstoffe 1: Eigenschaften, Mechanismen und Anwendungen. Springer, Heidelberg

Ashby MF, Jones DRH (2007, Nachdruck 2013) Werkstoffe 2: Metalle, Keramiken und Gläser, Kunststoffe und Verbundverkstoffe. Springer, Heidelberg

Bienz S, Bigler L, Fox T, Meier H (2016) Hesse–Meier–Zeeh: Spektroskopische Methoden in der organischen Chemie. Thieme, Stuttgart

Binnewies M, Jäckel M, Willner, H, Rayner-Canham, G (2015) Allgemeine und Anorganische Chemie. Springer, Heidelberg

Brückner R (2004) Reaktionsmechanismen. Spektrum Akademischer Verlag, Heidelberg

Cammann K (2010) Instrumentelle Analytische Chemie. Spektrum, Heidelberg

Christen HR, Vögtle F (1988) Organische Chemie—Von den Grundlagen zur Forschung. Band I. Salle, Frankfurt

Christen HR, Vögtle F (1990) Organische Chemie—Von den Grundlagen zur Forschung. Band II. Salle, Frankfurt

Christen HR, Vögtle F (1994) Organische Chemie—Von den Grundlagen zur Forschung. Band III. Salle, Frankfurt

Harris DC (2014) Lehrbuch der Quantitativen Analyse. Springer, Heidelberg

Heinrich PC, Müller M, Graeve L (2014), Löffler/Petrides Biochemie und Pathobiochemie. Springer, Heidelberg

Koltzenburg S, Maskos M, Nuyken O (2014), Polymere. Springer, Heidelberg

Lottspeich F, Engels JW (2012) Bioanalytik. Springer, Heidelberg.

Skoog DA, Holler FJ, Crouch SR (2013) Instrumentelle Analytik. Springer, Heidelberg

In this text, the explanation of the various analytical methods concentrated on the corresponding (electro-)chemical processes. The technical aspects of the instruments used are quite complex and thus were in most cases just generally outlined. If you are interested in this part of instrumental analysis as well, I suggest to have a look at the book by Skoog, Holler, and Crouch.

Sensors and Automation Technologies

Requirements

Behind the analytical methods covered in this part are mainly principles that you already know. Thus, more than any of the previous parts of this book, this part builds on the knowledge you have gathered—among other things—from the first book, "Analytical Chemistry I", and Parts I–III of this book. In this respect, you could summarise the prerequisites as "everything that has gone before" … but that might read a bit daunting.

In the field of electrochemical sensors you need, above all:
- the principle of quantification of electrochemical processes (in particular amperometry/voltammetry), primarily with the aid of Nernst's equation, and using reference electrodes.
- We will turn to the Clark electrode again, so its basic principle should be understood.
- You will also encounter some important biomolecules again.

Optical sensors are concerned with processes that are accompanied by light phenomena or in which dyes are produced. Accordingly, the following criteria are required:
- basic knowledge of fluorescence phenomena and fluorescence quenching (especially by paramagnetic compounds; of course, this also assumes that you recognise them),
- the principle of photometry, and
- the different types of interaction between analyte and specific sensor constituents (which here may also be from the field of biochemistry).

For flow injection analysis, you should once again visualise what happens inside thin capillaries and what flow conditions prevail therein.

In addition, the analytes are chemically modified both in the topic of "optical sensors" and in flow injection analysis, so it can do no harm here either if you are familiar with the basics of organic chemistry (such as azo coupling).

Learning Objectives

In this part you will learn the basics of sensor technology and experience basic principles and selected areas of application of electrochemical and optical sensors. Technically significant examples such as the (already well-known) Clark electrode for quantifying oxygen in biosystems (also in clinical analytics) and the lambda probe used in the automotive industry for optimising the fuel/oxygen ratio are discussed in more detail.

You will learn about different forms of optical sensors based on fluorometry (including fluorescence quenching) and photometry; exemplary different fields of application from "classical" analytics will be discussed as well as the use of corresponding probes in current research in the field of biochemistry and biology.

With flow injection analysis, you are familiarised with a method that can be easily automated and has by now become indispensable in routine analysis.

The various methods presented in this part are based almost without exception on procedures that you have already become familiar with in Parts I to IV of "Analytical Chemistry I" and/or Parts I to III of the present book. In this respect, this part IV is primarily intended to demonstrate to you the combinability of different methods and to show you the detection limits up to which some analytes can be detected by now. Last but not least, this part should also give you an outlook on the direction in which further progress in analytics can be expected.

Contents

Chapter 13   General Information About Sensors – 191

Chapter 14   Electrochemical Sensors – 193

Chapter 15   Optical Sensors (Optodes) – 203

Chapter 16   Flow Injection Analysis (FIA) – 209

General Information About Sensors

Summary

In the previous parts you have become familiar with numerous methods for the qualitative and/or quantitative detection of a wide variety of analytes, whether they were inorganic compounds (or individual cations or anions thereof), low-molecular- or even high-molecular-mass organic compounds (from simple hydrocarbons to biopolymers such as polysaccharides and proteins to macromolecular substances such as DNA and RNA). Many of the analytical methods described so far are used in the context of routine investigations where several factors are desirable:
- Speed
- Reproducibility
- Automation capability

In sensor technology, one also likes to speak of the "3 S" in this context:
- Sensitivity
- Selectivity
- Stability (especially in the sense of reproducibility)

Ideally, a measuring system of any kind should provide reliable and easily comparable measurement values within the shortest time possible. If the corresponding analysis can additionally be automated, so that the analysts themselves only have to do a minimum of work, this facilitates the workflow immensely.

There are special analytical methods for numerous analytes. You have already become familiar with some of these, such as the ion-selective electrodes (from Part II). In many, but not all cases, the analyses are based on amperometric or voltammetric measurements (which you also know from the aforementioned part of this book). Devices are used which are often referred to as "electrodes" or "detectors"—although, strictly speaking, they represent *complete potentiometric or voltammetric systems*, since a reference electrode always belongs to the measuring electrode in question. In this respect, it would make sense to consistently use the term "sensors" instead.

You have already learned about some sensors of this type in Part II:
- The Clark electrode is used to quantify the oxygen content of aqueous solutions. (In ▶ Chap. 14 we turn to this sensor again.)
- As examples for biosensors, we already looked at ways of determining the ATP content of biological systems and the glucose content of blood samples (important in diabetes diagnostics). (We will deal with further examples of biosensors in ▶ Chap. 15.)

However, other measurement principles also form the basis for selected sensors: ▶ Chap. 15 will deal with optical sensors, while ▶ Chap. 16 will discuss flow injection analysis, which has now become the "everyday standard" for numerous analytes.

However, the basic principle of sensors always remains the same: During a process (which *can be* a chemical reaction but *does not have to be*), changes occur in the physico-chemical properties of the system under consideration, which do not necessarily have to be visible to the "unaided" human eye. These changes in properties are then converted by a *transducer* into a measurable (often quantifiable) signal, such as the conductivity of the system, its colour (keyword: Photometry) and the like.

Electrochemical Sensors

Contents

14.1 Classical Inorganic Sensors – 194

14.2 Amperometric and Voltammetric Biosensors – 197

Electrochemical sensors are mainly based on potentiometry, amperometry, or voltammetry, i.e. they use
- the electric current *and/or*
- the electrical voltage

which occur during electrochemical processes. The corresponding processes can be attributed to "classical inorganics"—we have already dealt with such systems in great detail in Parts I–III (we need only recall the silver/silver chloride electrode and the calomel electrode, or indeed the Clark electrode, to which we shall return in ▶ Sect. 14.1)—but biochemical or even "purely biological" processes can also be investigated amperometrically or voltammetrically, provided that an appropriate sensor system is available. We will turn to this subject in ▶ Sect. 14.2.

14.1 Classical Inorganic Sensors

Let's first return to a system that you already know from Part II, as mentioned earlier:

▪ The Clark Electrode

The Clark electrode represents an oxygen-specific sensor; the measured values obtained with it are based on amperometric analysis. A membrane made of **silicone rubber**, an organosiloxane polymer (silicon units linked via oxygen atoms, which also carry one or more organic substituents –R) is used here; this material is highly permeable to elemental oxygen. Behind this, in a basic-buffered potassium chloride solution, there is a platinum cathode coated with gold, at which any oxygen diffused into the electrode is reduced to form water:

$$\text{RED}: \quad O_2 + 4H^+ + 4e^- \rightarrow 2H_2O$$

This measuring electrode is combined in the usual way with a silver/silver chloride electrode, which serves as a reference and at which the oxidation takes place:

$$\text{OX}: \quad Ag + Cl^- \rightarrow AgCl + e^-$$

(The fact that and why the potential of the reference electrode does not change despite the consumption of chloride was discussed at length in Part II.)

The total current flow I is directly proportional to the partial pressure of the dissolved oxygen $p(O_2)$:

$$I \sim p\left(O_{2(aq)}\right)$$

These oxygen sensors are used in bioreactors as well as in medicine (for example in blood gas analysers to check whether a patient is sufficiently supplied with oxygen or requires artificial respiration), but also to examine the oxygen content in eutrophic (i.e. over-fertilised) waters or in aquariums. With appropriate modifications, this system can also be used to quantify carbon monoxide (CO) or nitrogen oxides (NO, NO_2, or generally: NO_x) and other water-soluble gases.

Harris, Section 16.4: Amperometry
Skoog, Section 25.3.4: Applications of hydrodynamic voltammetry

As has already been pointed out in Part II: If you would like to know a little more about this electrode, take a look at Excursus 16.1 in Harris; it not only shows the construction of a corresponding microelectrode (schematically), it also goes into selected details of the apparatus, for example how to prevent atmospheric oxygen, which does *not* come from the analysis solution, from falsifying the measurement result. And ▶ Skoog also has something to report on this subject.

The Lambda Sensor

The lambda sensor (a.k.a. lambda probe) also functions as an oxygen detector: It is a component of the catalytic converters of internal combustion engines (particularly in use in vehicles) and measures the residual oxygen content of the exhaust gas in comparison to the oxygen content of the ambient air; in this way, the optimum mass ratio of fuel and air can be determined for the respective combustion process under consideration. (This is why it is also called a *lambda* probe: In combustion theory, λ is the (dimensionless) key figure for this ratio). We already know the measuring principle of this probe: It is a *solid-state electrode*, whose mode of operation is already familiar from the fluoride electrode (from Part II). The membrane used here consists of zirconium dioxide (ZrO_2). This crystallises
- monoclinic at room temperature (α-ZrO_2),
- above 1100 °C tetragonal (β-ZrO_2), *and*
- above 2300 °C cubic (γ-ZrO_2). Its crystal structure corresponds to that of calcium fluoride. (γ-ZrO_2 thus crystallises in the *fluorite type:* a cubic close packing of zirconium(IV) cations is present, in which all tetrahedral cavities are occupied by oxide ions.)

Above a temperature of about 600 °C, zirconium(IV) oxide becomes conductive due to the (limited) free mobility of the oxide ions, which are small compared to the zirconium(IV) cations. This behaviour is particularly evident in the high-temperature modifications mentioned above, which is why, in principle, a fairly high operating temperature is required. However, the high-temperature modifications can also be stabilised at lower temperatures by doping with yttrium(III) oxide (Y_2O_3) or, alternatively, with calcium or magnesium oxide; this way, the operating temperature of this probe can be lowered to about 300 °C.

> **Note**
> Yttrium-stabilised zirconia also plays an important role in the field of high-performance ceramics; if you are more involved in materials science, especially non-metallic materials, you will undoubtedly encounter this zirconia variety again sooner or later.

If elementary oxygen diffuses into the membrane material, this leads in the usual way to a *diffusion voltage*, the extent of which is proportional to the partial pressure of the oxygen in the atmosphere in direct contact with the membrane. If the other side of the membrane is in contact with a reference material (i.e. a sample with a known oxygen content), the resulting potential difference (i.e.: voltage) can be used to determine the difference in oxygen content. Let's take a look at what happens at the cathode and the anode, respectively:

$$\text{Cathode}: \quad O_2 + 4e^- \rightarrow 2O^{2-}$$
$$\text{Anode}: \quad 2O^{2-} \rightarrow O_2 + 4e^-$$

Ultimately, the lambda sensor represents just a potentiometric concentration chain: On the side of the (oxygen-rich) outside air, more oxygen atoms can diffuse into the membrane in the form of their ions and thus increase the conductivity there, while on the (oxygen-poorer) exhaust gas side the conductivity is lower—this is precisely what leads to the potential difference. Platinum applied by vapour deposition then serves as a discharge electrode for the resulting voltage, so that currentless measurement is possible. If stoichiometric oxygen conversion occurs during combustion (i.e. if the oxygen content of the exhaust gas

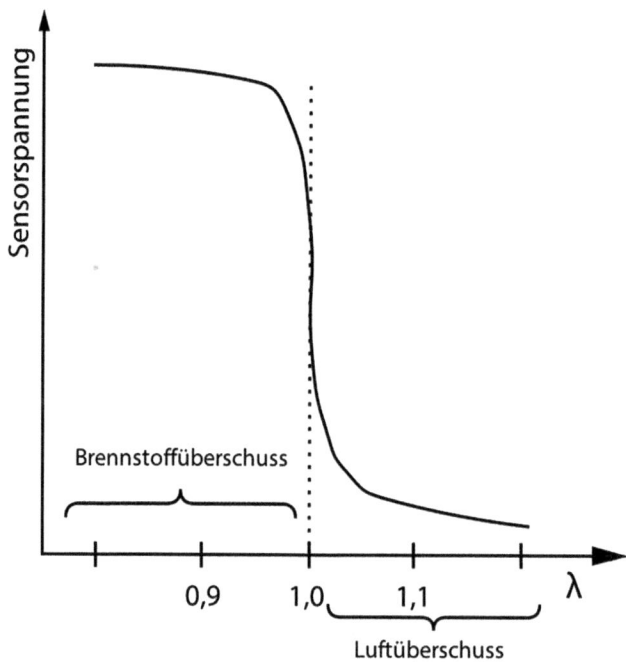

Fig. 14.1 Change in potential difference with decreasing oxygen partial pressure

drops to practically zero), this leads to a drastic potential-jump, as shown in ◘ Fig. 14.1.

On the basis of the respective oxygen partial pressures, the resulting voltage (or the drop in this voltage with decreasing oxygen content in the exhaust gas) could also be calculated according to the—by now well known—**Nernst equation**, but unfortunately this (and thus also the resulting voltage) is temperature-dependent. (It is easy to forget this!) For this reason, the sensor is preferably operated *amperometrically:* The oxygen to be quantified (in the exhaust gas) is reduced (electrochemically) at a platinum electrode; the resulting ions then migrate through the zirconia lattice and cause the current to flow.

> **For the Technically Interested**
> It shall not be concealed that there is also a variant of the lambda probe which is not based on a solid-state electrode, but on the *temperature dependence of the electrical resistance* of semiconducting ceramics, in particular appropriately doped titanium dioxide (TiO_2). Here, too, the conductivity is ultimately decisive, but in contrast to the measuring principle described above, the conductivity in this case is based on the presence of *defects in the lattice* in the form of vacant sites; these act practically like positive charge carriers. In contact with an oxygen-rich atmosphere, said vacancies are occupied by oxide ions originating from the oxygen, which thus *reduce* the conductivity: If the material is in contact with an atmosphere in which there is a high oxygen partial pressure, this increases its electrical resistance. It should be obvious that this also results in usable measurement values, but its use as a *resistance probe is* much rarer and is only mentioned here "for the sake of completeness".
>
> Such conductometric sensors (in this case for gaseous analytes) also include the so-called MOX sensors, in which the sensor is coated with a metal oxide (hence generally MOX or MO_x—yes, *actually the "x" should be subscripted*, but practically nobody does this; tin(IV) oxide (SnO_2) is often used). In this case, the corresponding oxides are usually somewhat oxygen-poorer, i.e. they are slightly

sub-stoichiometric in composition. If a (gaseous) analyte mixture contains reducing or oxidising substances, this affects the conductivity of the sensor through interaction with the sensor surface. An example of this type of sensor is the Taguchi sensor, which responds to the presence of fire-promoting gases (such as methane) and has thus probably already prevented numerous house fires or explosions, primarily in Japan, where natural gas (= methane) is used extensively.

Shown in the figure: A sensor based on a gas-sensitive layer (as mentioned above: with the coating material tin(IV) oxide) in the typical layer structure with an inert ceramic body on which a heating element (in this case: platinum-based) has been applied. (The insulation layer and the interdigital structure (for reading out the resistance—and this is the *actual* measured value) will not be further discussed here.) Thanks to the scale indication built into the photo, the extremely moderate space requirement of such a sensor element should become clear.

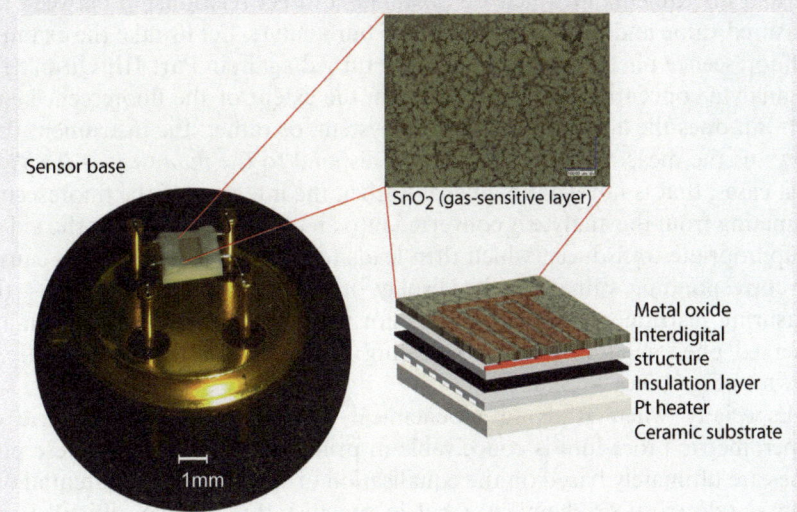

Design of a metal oxide semiconductor gas sensor (MOX sensor). (Kindly provided by J. Warmer, Bonn-Rhein-Sieg University of Applied Sciences)

? Questions
1. It was stated that the measurable current flow I is proportional to the oxygen content of the mixture under investigation or the difference between the oxygen content of the sample and the reference substance. Which additional factors must be taken into account in corresponding quantitative calculations?

14.2 Amperometric and Voltammetric Biosensors

Part II already gave you the first insights into the field of biosensor technology: Enzyme-coated electrodes can also be used to quantify biomolecules. The examples you already know were related to the ATP content of biological systems and the glucose content of blood samples used in diabetes diagnostics. (You can find more on the latter topic in Harris.)

Harris, Section 16.4: Amperometry

In these examples, electrochemical processes were again exploited "directly", so to speak, i.e. the chemical reactions that take place here lead directly to an evaluable measurement result: the concentration-dependent potential difference within a system leads directly to a quantifiable electrical voltage; there is a direct relationship between concentration and quantified value. This applies to many electrical/electrochemical processes.

However, this does *not necessarily* apply to *every* response of a sensor to an analyte: In the case of more complex interactions of the sensor with its analyte, a **transducer** or *converter* is often required to convert the "actual" measured value—a physical quantity that in itself is basically arbitrary—into another quantity, which is then used to quantify the analyte.

This step of instrumental analysis, and thus the importance of such signal transducers, is often overlooked, and we do not want to pay excessive attention to them here either, because behind it lies mostly "purely physical" principles, but it is worthwhile to think at least briefly about what actually happens in the context of instrumental analysis:

In Parts I–III, you have repeatedly seen (or read about) measurement set-ups and instruments in which we postulate a direct relationship between the measured value and the concentration of our analyte. Let us take the example of fluorescence phenomena, to which we turned again in Part III: Ultimately, the analyte concentration is inferred from the extent of the fluorescence elicited. But does the actual quantification system, or rather: the instrument that shows us the measurement result, really respond to the *fluorescence itself*? In most cases, that is not what happens: Rather, the intensity of the fluorescence emanating from the analyte is converted into an *electric current* with the aid of an appropriate transducer, which then leads to a *voltage*, and *this* then causes the corresponding values on the display of the measuring instrument—the measuring instrument, therefore, does not respond directly to the "actually" observed phenomenon. (A very revealing illustration of this relationship is shown as Fig. 1.3 of ▶ Skoog.)

Skoog, Section 1.3.2: Non-electrical data areas

Especially when studying biochemical processes, a potentiometric or amperometric procedure is conceivable in principle, since most of these processes are ultimately based on the equalisation or building-up of potential differences (electrical or chemical)—but in practice, this is quite difficult (even impracticable). If the experimental setup is changed, a different measured value will probably be quantified in the end, and then (at least) one transducer is needed again.

> **Illustration**
> Basically, you are already familiar with such a signal conversion: Please remember the radioimmunoassay from part III. There, it is a specific antigen that is *actually* to be quantified. Instead of directly counting the particles present (which is theoretically conceivable, but difficult to realise), the interaction of the antigen with a suitable immobilised antibody is exploited, but the actual measurement signal is introduced via a third step—in the form of the radioactively- (or fluorescence markers-)identified competitor antigens. Only the quantifiable signal resulting from that (the measurable radioactivity or intensity of the quantifiable fluorescence) allows conclusions to be drawn about the concentration of the analyte antigen; in this respect you have *converted* the "desired signal" (number of analyte antigens) into the "quantifiable signal" (radioactivity/fluorescence). And since this is again not an *electrical* signal, at least one more transducer now comes into play: In the case of radioactivity, you only get the "actually quantified signal" by means of a scintillation counter, and in the case of fluorescence phenomena, we come back again to Fig. 1.3 from Skoog.

14.2 · Amperometric and Voltammetric Biosensors

As long as the biosensors described here are "purely physically"-functioning transducers, they will not be discussed any further; however, if (bio-)chemical principles are behind them, we will take a closer look.

■ Areas of Application

The two systems already described (for the quantification of glucose and ATP, respectively) fall into the field of clinical analytics; numerous other biosensors are also used there. However, they are also used in process engineering (e.g. in fermenters) and in environmental analysis. One of the reasons why they are becoming increasingly popular is that they are
- extremely versatile,
- very specific,
- comparatively cheap, *and*
- in many cases easily transportable.

There is no need for complex equipment (please think back to the blood glucose test strip from Part II); the analyses can be carried out on site within a short time (usually: a few minutes).

In principle, biosensors consist of
- a selective biological/biochemical component, *and*
- a transducer.

The biological/biochemical component is usually fixed on the transducer, which is why a very compact design can be achieved. In principle, there are no limits to the variety of bio-components that can be used (as long as they can be obtained or produced purely in sufficiently large quantities to bring the corresponding sensors onto the market).

In principle, they are all based on the (ideally specific) interaction with the analyte, which is optionally covalently bound or complexed, resulting in a change of one or the other (physico-)chemical property of the analyte system. Typical biological components are
- antigens and antibodies (we already know this from Part II),
- proteins/enzymes (also in combination with **saccharides**, so that *glycoproteins* are present); of particular importance in natural product analysis are **lectins**, which interact specifically with saccharides,
- nucleic acids,
- DNA or RNA sequences (also referred to as **aptamers**), *but also*
- selected organelles, *and even*
- complete cells (the selective interaction with the analyte takes place mainly via the glycoproteins of the cell membrane), *and*
- living microorganisms.

Of course, these bio-components are not infinitely insensitive; their use under extreme conditions (especially with respect to temperature) is therefore only possible to a (very) limited extent. In the meantime, however, synthetic analogues have been developed, at least for some systems, whose robustness exceeds that of bio-components by far.

With such a variety of bio-components, it should come as no surprise that there are also correspondingly different *transducer systems* that respond to the respective consequences of the interaction between sensor and analyte. This chapter, of course, focusses on amperometric and voltammetric methods, but there are also electrochemical sensors based on potentiometry or conductometry. In ▶ Chap. 15 we will deal a little with systems using *optical* transducers, but in the meantime acoustic, calorimetric and piezoelectric transducers are also in use. (However, these will not be discussed further in this introductory section.)

Of particular importance are enzyme-based sensors, where the enzymes used are commonly re-immobilised and can often be reused many times (which of course significantly reduces the cost). There are various immobilisation methods:
- Covalent binding of the enzyme in question to an inert surface (glass, selected polymers)
  - A particularly elegant variant of covalent binding is the co-polymerisation of the enzyme with a suitable monomer (which, of course, must first be found; this is not trivial, as the physicochemical properties of the enzyme in question must be taken into account here)
- Purely physical adsorption to an (usually inorganic) inert substance with a sufficiently large surface area (porous aluminium oxide has proved successful)
- Confinement in a gel matrix

The enzymes then catalyse reactions in which at least one easily detectable and quantifiable product (such as H^+ ions, CO, CO_2 or NH_3) is formed. These then cause a concentration-dependent change in the (physico-)chemical properties of the system under investigation; this is where the membrane electrodes or gas-sensitive probes already known from Part II are used.

A major advantage of these enzymatic sensor systems is that they rapidly detect/quantify even complex analytes, with the reactions taking place under mild pH and temperature conditions. Due to their specificity, a very low detection limit results: Even minimal analyte amounts (down to the femtomole range!) are detected.

Skoog, Section 23.6.2: Biosensors

A brief overview of the combination of (immobilised) enzymes with ion-sensitive electrodes is given in Table 23.6 from Skoog; however, new systems are constantly being developed, so please do not expect "completeness" here.

Skoog, Section 23.6.2: Biosensors

> ▶ Example
>
> As an example, consider the quantification of urea in human blood with the aid of an enzyme electrode (the schematic structure of such an electrode is shown in Fig. 23.13 from Skoog). The enzyme *urease* is used here, which catalyses the hydrolysis of urea to ammonium and hydrogen carbonate ions in a weakly acidic medium:
>
> $H_2N - C(=O) - NH_2 + H_2O + H_3O^+ \rightarrow 2NH_4^+ + HCO_3^-$
>
> In principle, there are two ways of quantifying the ammonium ions produced in this process:
> - with the help of a glass electrode (we already know that one from part II; there it was also mentioned that this kind of electrode does not exclusively respond to hydroxonium ions in aqueous solution, but also to NH_4^+-ions because of the similar charge density)
> - with a gas electrode (not to be confused with the g_l_ass electrode!) which responds to the molecular ammonia which in the usual way is in equilibrium with the ammonium ions in aqueous solution. (Inert metal electrodes are used and the gas—in this case ammonia—is bubbled over their surface. The electrodes serve exclusively to supply or accept electrons; they themselves are not involved in any reaction.)

14.2 · Amperometric and Voltammetric Biosensors

However, difficulties arise in both cases:
- If you remember the selectivity coefficient K_{Sel} (also from Part II), you will immediately see the problem with the glass electrode: It responds to other monovalent ions (besides hydroxonium ions), and both sodium and potassium ions are practically ubiquitous in organic samples.
- A pH problem also arises with the gas electrode:
  - The catalytic effect of the enzyme is strongest at a pH value just below 7.
  - However, the sensor is most sensitive at pH = 8–9; in the weakly basic medium, the ammonium ions resulting from the hydrolysis of urea are almost quantitatively deprotonated to ammonia.
  - The solution to this problem is once again the use of an immobilised enzyme: first—at pH = 7—the sample is brought into contact with it; then the pH is raised so that the quantification of the molecular ammonia now present can be carried out. ◄

Precisely because such sensor systems can be miniaturised to a considerable extent, they are increasingly being combined with one another, so that multifunctional sensors are now also available for a wide variety of analytes. In these, various sensors are combined on chips (which may even be interchangeable) so that several assays can be carried out simultaneously within a very short time. The i-STAT analyser shown in Fig. 23.14 of ▶ Skoog, with which numerous clinical parameters can be checked simultaneously on the basis of a small blood sample (for example, the partial pressure of oxygen and carbon dioxide in the blood, plus the urea, potassium, and glucose content, and much more) does not quite correspond to the medical tricorder from Star Trek, but it is clearly already heading in this direction. Of course, even with such a multifunctional device, each individual sensor must first be calibrated, but this is done with a buffered standard solution, which is located in a reservoir inside the device, being refilled as needed.

Skoog, Section 23.6.2: Biosensors

The combination of potentiometric sensors with other techniques has recently led to a rapid increase in new sensor systems; in particular, the combination with semiconductor technology or microsystem technology offers completely new possibilities. The technical aspects of such sophisticated methods would once again exceed the scope of this merely introductory section; therefore, as a first introduction, the interested reader is only recommended to read the section "Light-Responsive Potentiometric Sensors" from the aforementioned ▶ Skoog chapter; the fundamentals of semiconductor technology required there can, if necessary, be looked up in Sect. 2.3 of the same book. For those interested in the more biological aspects of biosensors, I suggest reading ▶ Chap. 18 of the Lottspeich.

? Questions
2. Why is it essential to buffer the analyte solution when quantifying ammonium ions/ammonia using a glass electrode?

Optical Sensors (Optodes)

Contents

15.1 An Inorganic Example – 204

15.2 A Bio-Organic Example – 206

15.3 An Inorganic Example in Living Cells – 206

The fact that electrical signals are suitable for carrying information (e.g. to a measuring device) was discussed (again) in the previous chapter. However, processes that are *not* accompanied by an electrical voltage and/or current, but by *light phenomena*, can also provide quantifiable signals. In such a case, optical sensors are required, also known as **optodes** (a linguistic shorthand for "**opt**ical electr**odes**"). These are based on optical fibres (analogous to fibre optic cables), the (physical) principle of which will not be further discussed here; if needed, Harris provides a brief summary of the theory required for this.

Harris, Section 19.4: Optical sensors

Of particular interest to us are the chemical processes involved in the use of such optics—where, of course, the fluorescence processes presented in Part III are of particular importance.

The structural design of optodes is quite similar to that of electrochemical sensors: Here, too, reagents that react specifically with the analyte of interest—inorganic complexes or biological/biochemical components such as antibodies, aptamers, proteins/enzymes, as well as complete cells or microorganisms—are immobilised, and then they interact with the analyte to produce quantifiable light phenomena. The resulting light is transmitted through the optics to a measuring device, which then displays the corresponding measured value (after the number of photons received has been converted back into a voltage via a photodiode or photomultiplier) and allows the analyte under consideration to be quantified—again on the basis of calibration.

15.1 An Inorganic Example

When the complex cation tris-(1,10-phenanthroline)ruthenium(II) (◘ Fig. 15.1a, b shows, in stylised form, the spatial structure of the complex) is fixed in a polymer matrix (usually polyacrylamide; the corresponding structural formula is shown in ◘ Fig. 15.1c) and excited with relatively low-energy light; this complex emits orange-red fluorescence radiation. (See Colour Plate 23 of ▶ Harris for what this looks like.) This fluorescent radiation can, of course, be quantified: If the matrix-immobilised fluorescence system is coupled to fibre optics (which in this case serves primarily to spatially separate the *transducer* from the signal *converter*), the fluorescence intensity is relayed to an appropriate measuring device.

Harris, Colour plates

In Part III we have seen that paramagnetic substances, such as the common dioxygen O_2, favour the non-radiative *intersystem crossing*, so that the fluorescence is noticeably attenuated. This phenomenon, called **quenching** (also known from part III), is the reason why the ruthenium-based fluorescence system described above is suitable as an *oxygen sensor*: If the immobilised fluorescence

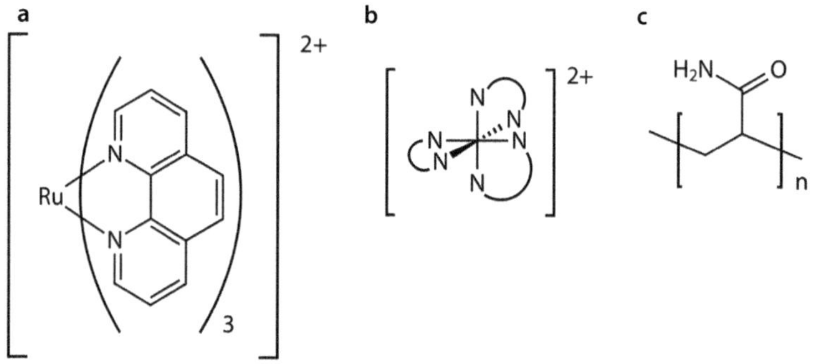

◘ Fig. 15.1 Tris-(1,10-phenanthroline)ruthenium(II) **a, b** and polyacrylamide **c**

15.1 · An Inorganic Example

emitter is brought into contact with a solution containing dissolved oxygen, the *extent of the quenching* allows conclusions to be drawn about the oxygen content of the analyte solution. The higher the partial pressure (pO_2), this can of course also be expressed as a ppm (volume) or mg/L value in the system under investigation, the stronger the attenuation of the fluorescence will be—there is thus a direct correlation between the oxygen content and the fluorescence intensity observed (which is passed on via a light guide system).

Behind this—it was already mentioned in Part III—is the transfer of energy from the fluorescence system (which is currently in an *excited state*) to the oxygen. Let us look at the whole thing in the form of (by now familiar) orbital representations:

- If the ruthenium complex is *not* excited, a *singlet* state is present: ^{1}Ru. The HOMO is doubly occupied spin-parallel, the LUMO is vacant: $_{(↑↓)}(\ )$.
- After excitation by light of suitable wavelength, the system is in an excited *triplet* state: ^{3}Ru*. Here, an electron has changed from the HOMO to the (now former) LUMO under spin reversal: $_{(↑)}^{(↑)}$.
- In the case of dioxygen, the two energetically degenerate π*-orbitals, i.e., the HOMO set, are single-occupied spin-parallel in the ground state: a diradical in a triplet state is present: $^3O_2$. This leads to the stylised orbital diagram $_{(↑)(↑)}$.
- If there is a radiationless energy transfer from ^{3}Ru* to $^3O_2$, the previously electronically excited ruthenium complex returns to its electronic ground state ^{1}Ru—schematic: $_{(↑↓)}(\ )$—(but it may *still* be rotationally/vibrationally excited in the process), while in the dioxygen molecule there is a spin reversal of one of the two electrons in the π*-orbital set: (↑)(↓) (or even (↓)(↑), that is meaningless). After the transfer of energy, the dioxygen molecule is in an electronically *excited singlet* state: $^1O_2^*$.

Concrete calculations on fluorescence (or more generally: luminescence) quenching, and how the different reaction rate constants of processes with luminous effects and of *deactivation processes* affect the **quantum yield**, as already discussed in Part III, can be found in Harris.

Harris, Section 18.6: Sensors based on fluorescence quenching

Such systems are extremely sensitive: Minimal sample volumes (on the order of magnitude of less than a picoliter (just a reminder: that's 10^{-12} L) are sufficient; in current systems (as of August 2023), the detection limit for diacid has dropped below 10 *attomoles* (1 amol = 10^{-18} mol, this is of course just supposed to be a reminder of the SI prefixes).

> **Relative Sensitivity**
> It is advisable to keep in mind that at a detection limit of <10 amol, the *absolute* number of oxygen atoms required to be present in order to be detected is still quite considerable: Since 1 mol corresponds to 6.022×10^{23} particles (as is well known), one attomole of oxygen still contains 6.022×10^5 O_2 molecules, i.e. more than 60,000. So, as admirable as such a low detection limit undoubtedly is for an artificially generated system: In the course of evolution, far more sensitive "sensors" have emerged.
>
> The **sensilla** of males of the butterfly species *Bombyx mori*, for example, respond (i.e. provide a "measuring signal") when they come into contact with just *one single molecule of* the sexual attractant bombycol (see below; the IUPAC designation is (10*E*,12*Z*)-10,12-hexadecadien-1-ol) produced by the female silkmoth. 200 molecules in contact with the sensilla receptors are sufficient to be interpreted as a "clear signal" by these animals. So far, such low detection limits have been sought in vain in instrumental analytics—but progress is constantly being reported.

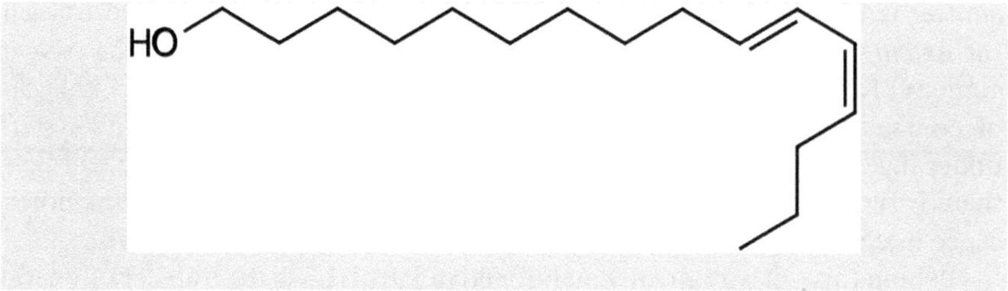

15.2 A Bio-Organic Example

In Part II we briefly discussed blood glucose meters in which the enzyme glucose *oxidase* catalyses the oxidation of glucose to gluconolactone and the hydrogen peroxide released in the process is quantified, allowing direct inference of the glucose content of the analyte sample. In 1996, the first miniaturised glucose sensor was presented, which is also based on this reaction, but in the form of a combination of an immobilised enzyme with an optode. Here, the quenching caused by the elemental oxygen released from the hydrogen peroxide is quantified; the detection limit here is one femtomol of glucose (kindly remember: 1 fmol = 10^{-15} mol).

15.3 An Inorganic Example in Living Cells

Synthetic dyes can also be used to obtain information in living organisms: For example, porous nanoparticles based on silicon dioxide (with a diameter of 100–500 nm) can be specifically introduced into cells (how exactly this works would again lead too far here). Important: Into the cavities of these nanoparticles, two different fluorescent dyes have previously been introduced:
- The dye tris-(4,7-diphenyl-1,10-phenanthroline)ruthenium chloride (◘ Fig. 15.2a; the counter ions are not shown) emits fluorescence photons with a wavelength of λ = 610 nm (red) upon appropriate excitation.

The patented pigment designated as "Oregon Green 488" (2′,7′-difluorofluorescein, ◘ Fig. 15.2b—yes, it is possible to obtain a patent even for such a small molecule!), on the other hand, fluoresces at $\lambda \cong$ 520 nm (= *green*, which should not be surprising given the trade name of this pigment). By the way, the name of this fluorescent dye comes from the fact that, due to its absorption maximum at λ_{abs} = 490 nm, it can be excited very well using an argon ion laser, which, among other things, produces radiation in the VIS range, with λ = 488 nm (turquoise).

As in the example from ▶ Sect. 15.1, this method of analysis is based on the quenching of fluorescence by elemental oxygen: Depending on the partial

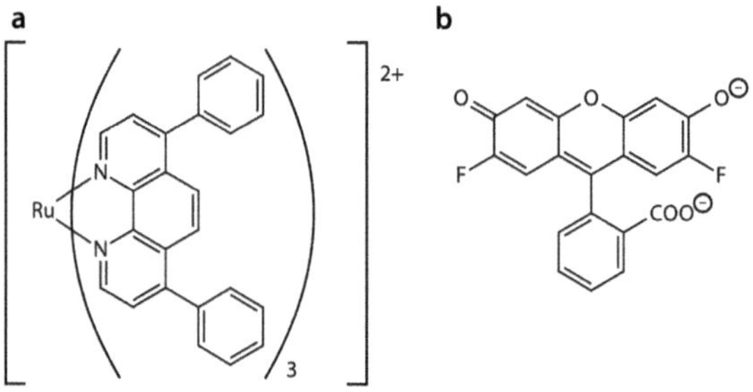

◘ Fig. 15.2 Tris-(4,7-diphenyl-1,10-phenanthroline)ruthenium(II) **a** and Oregon Green 488 **b**

15.3 An Inorganic Example in Living Cells

pressure of oxygen or the value of ppm(O_2) inside the cells treated this way, the fluorescence originating from the ruthenium complex decreases, whereas the Oregon-Green fluorescence *does not* change even at high p(O_2). The curves of the respective fluorescence intensities as a function of oxygen concentration are shown in Fig. 18.23 of ▶ Harris.

Harris, Section 18.6: Sensors based on fluorescence quenching

With increasing oxygen content, the red light emitted by the Ruthenium(II) complex becomes weaker and weaker and is therefore increasingly masked by the oxygen-*independent* green fluorescence, so that the change can even be seen with the naked eye. Recording an emission spectrum, the effect can of course also be quantified. Accordingly, this system then allows the oxygen content in the cells concerned to be determined; even the oxygen *consumption* over a period of several minutes can be determined in this way. What something like this actually looks like is shown in Colour Plate 18 from Harris.

Harris, Colour plates

Newer Developments and More Specific Applications

In the meantime, in addition to Oregon Green 488 from ◘ Fig. 15.2b, numerous derivatives of this parent compound are also used for such analyses, such as the succinimidyl ester of its 5-carboxylic acid (a) or even more complex compounds in which the succinimide unit is spatially separated from the actual system responsible for the fluorescence, the **fluorophore**, by incorporation of a longer hydrocarbon chain, as shown in (b). The emission wavelength is hardly changed by this, but such *"molecular spacers"* can avoid or at least reduce unwanted interaction of the fluorophore with the biomolecule to be labelled (which can lead to quenching)—and that can significantly facilitate such measurements. In the dye shown in (b), ε-aminocaproic acid (i.e., 6-aminohexanoic acid) serves as a spacer; the trade name of this fluorescent dye is "Oregon Green 488-X". (Not that it needs to be memorised, but because this dye has also been patented, it should at least be mentioned here.)

Instead of succinimide, much larger molecules or molecular moieties can be used, such as phalloidin, the main toxin of the green button mushroom (*Amanita phalloides*). This cyclic heptapeptide with a sulfide bridge attaches quite tightly to the structural protein actin, which occurs in all eukaryotic cells; therefore, the resulting fluorescent dye (c) allows staining of cytoskeletal parts. However, this is already *far* into the field of "more specific applications".

■ **Investigation of Further Metabolic Activities**

If fluorescent sensor molecules similar to the dyes shown in ◘ Fig. 15.2 (and in the box "Recent developments and more specific applications") are fixed in sensor foils, several substances produced or consumed in metabolic processes can also be quantified simultaneously, for example oxygen and carbon dioxide; the local pH value can also be determined in this way (since physiological conditions must ultimately prevail in living organisms, it will not fluctuate excessively, but remain be in the range pH = 6–8). Various fluorescence systems are used:

– On the one hand, systems responsive to the analyte in question are required (the analysis may well be based again on fluorescence quenching, i.e. quenching by the analyte under consideration),
– on the other hand, it needs, as an internal standard, a reference system that can also be excited to fluorescence (with the same excitation wavelength).

In this way, for example, the blood flow and oxygen supply of tissue can be checked in clinical analytics; such fluorescence sensors are also used in the increasingly important field of tissue cultivation.

❓ Questions

3. In this chapter, sensor systems were addressed in which not only individual molecules but also whole cells function as "detectors". What other problem arises in addition to the lack of pH stability of such sensors?

Flow Injection Analysis (FIA)

Contents

16.1 Summary – 214

Further Reading – 216

The most important component of flow injection analysis is a capillary with a diameter <1 mm and a length of about 1–2 m, which is continuously flowed through by a flow medium that has been mixed in advance with a reagent for the detection of the desired analyte. When an analyte solution is then injected into the capillary, during the migration of the analyte through the capillary a chemical reaction occurs between the analyte and the detection reagent; when the (now chemically modified) analyte finally reaches the end of the capillary, it is detected there by an appropriate detector and quantified on the basis of the (physico-)chemical properties of the reaction product. The actual detection/quantification can be

- photometrically (e.g. in the case of reaction products that exhibit an absorption maximum at a precisely defined wavelength or can be excited to fluorescence) *or*
- electrochemically (i.e. in the usual way potentiometrically, amperometrically, etc.).

Harris, Section 18.4: Flow injection analysis and sequential injection analysis

So far, this experimental set-up, which is shown schematically in Fig. 18.10 from ▶ Harris, still bears striking resemblance to already known analytical methods, such as capillary electrophoresis from Part III of "Analytical Chemistry I". However, this similarity is only superficial, as there are some differences. Let us first look at what exactly happens after the injection of the analyte solution:

A very elongated "analyte plug" is obtained (in comparison with the diameter of the capillary), in front of and behind which the flow medium (mixed with the appropriate reagent) is located; clearly illustrated in Fig. 18.11 (*above*) in the same chapter of Harris. Immediately after injection, this plug consists exclusively of the analyte solution, but its front and rear parts are in immediate contact with the flow medium, so that dispersion of the analyte, mixing with the medium, and hence the desired reaction between analyte and detection reagent occur almost immediately in these two parts of the "plug". (This partial mixing is shown graphically in the *centre* of ▶ Harris, Fig. 18.11), whereas the central part of this plug hardly comes into contact with the medium (and thus with the reagent) at all.

However, the flow medium and also the "analyte plug", which are in principle subject to laminar flow, do not move completely homogeneously within the capillary: Different velocities occur.

- The flow is particularly fast in the middle of the capillary (viewed in a cross-section of the flow pipe).
- At the edge of the capillary, the flow velocity is almost zero (due to friction effects).

In this partially retained "edge zone", due to dispersion of the analytes further mixing with the flow medium occurs, which was originally located *behind* the—now increasingly spread—plug, so that the amount of unreacted analyte there continues to decrease with increasing delay time in the capillary; this results in a second "mixing and reacting zone".

Thus, while there is more and more reacted analyte in front of and behind the (as mentioned, increasingly elongated) analyte plug, the middle part, which is hardly in contact with the medium and thus the detection reagent, still contains mainly unreacted analyte; quantitative analyte conversion is *not* achieved (this is shown graphically in ▶ Harris in Fig. 18.11 *below*).

This problem can be mitigated somewhat by the design of the apparatus: If the capillary is not used in an elongated form but bent into the shape of a spiral, the (more or less continuous) curvature results in turbulence/vortexes that ensure increased mixing of the as-yet unconverted analyte with the flow medium. But even in this case, the quantitative conversion of the analyte would only be achieved after a considerable amount of time—in most cases: too much of it, because in an average flow injection analysis with average flow velocities, the analyte reaches the detector after less than one minute. The decisive factor here is, therefore:

> **Important**
> In flow injection analysis, there is *no* quantitative turnover of the analyte; *no* equilibrium is reached.

Nevertheless, flow injection analysis allows quantitative investigations with remarkable precision. This is made possible by extremely precisely reproducible measurement conditions. The concentration profile of the analyte depends on many factors, in particular:
- the flow velocity,
- the temperature (and thus also the viscosities of analyte solution and flow medium; it should be understood that a higher degree of mixing is achieved at higher temperatures, leading to reduced viscosity),
- the sample volume, *and*
- the rate at which the analyte and detection reagent react with one another.

Therefore, if reproducible assay conditions are provided, equal amounts of analyte will result in equal signals. (The deviation here lies in the low single-digit percentage range; we will return to such considerations in Part V.)

The actual special feature of flow injection analysis, however, is something else: The flow of the medium mixed with the detection reagent is *continuously* maintained, so that once the analyte plug has moved far enough away from the injection site, immediately another sample can be injected.

> **Important**
> Of course, care must be taken to ensure that the respective "analyte plugs" are far enough apart to avoid **cross-contamination**.

In this way, a whole series of measurements can be carried out within a comparatively short time—which is particularly advantageous for routine analysis (e.g. of drinking water or seawater or similar): More than 100 measurements per hour are achievable; with the aid of **autosamplers**, which independently initiate injections following a predefined computer program, this method can also be automated quite well, so that a large number of measurements can be carried out without excessive personnel expenditure.

- **A Photometric Example from Organic Chemistry**

In Part III of "Analytical Chemistry I", dealing with liquid chromatography we mentioned that aromatic compounds can be easily detected photometrically due to their absorption in the UV range; of course, this is even easier with dyes whose absorption maximum is in the range of *visible* light. From *organic chemistry* you know that many aromatic compounds can be converted to dyes with diazonium salts via **azo coupling.** For reproducible quantification of corresponding aromatic analytes, a (stable) aryl diazonium salt can be used as a detection reagent dissolved in the flow medium, which then reacts with the analyte to form an **azo dye** with an absorption maximum in the VIS range (with a general analyte Ar-R shown in ◘ Fig. 16.1).

- **A Colorimetric Example from Inorganic Chemistry**

Inorganic analytes can also be quantified with the aid of flow injection analysis: Let us take the quantification of chloride ions as an example. If the flow medium (here: water) is mixed with (acceptably water-soluble) mercury(II)

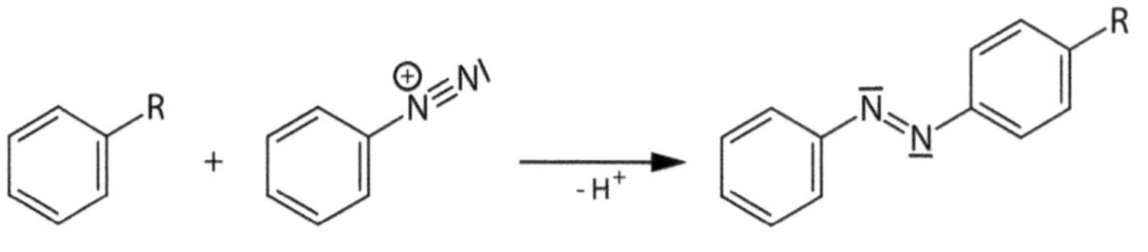

Fig. 16.1 Azo coupling of aromatic analytes to a detectable azo dye

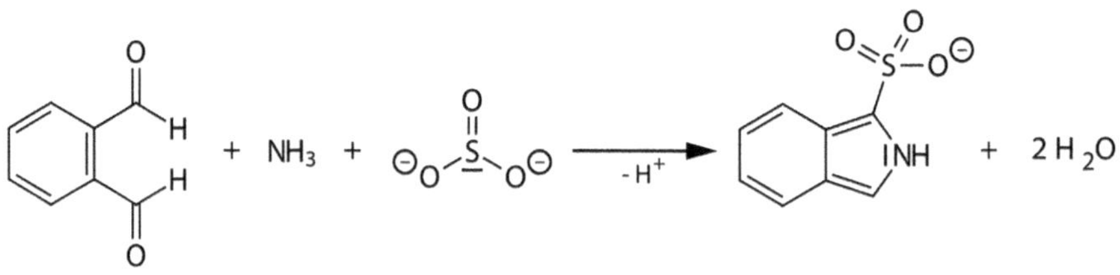

Fig. 16.2 Conversion of ammonia to 2H-isoindole-1-sulfonic acid

thiocyanate and additionally with iron(III) ions, the following reactions occur after injection of the analyte solution containing chloride due to the interaction of the "analyte plug" and the flow medium:

(a) First, the chloride ions quantitatively displace the thiocyanate ions ($HgCl_2$ is much more water soluble than $Hg(SCN)_2$):

$$Hg(SCN)_{2(aq)} + 2Cl^- \rightarrow HgCl_2 + 2SCN^-$$

(b) The thiocyanide ions released in this manner then react with the ferric ions (which are present in aqueous solution as hexaquaferric complex cations) by ligand displacement to form the pentaquathiocyanato ferric complex cation:

$$[Fe(H_2O)_6]^{3+} + SCN^- \rightarrow [Fe(H_2O)_5(SCN)]^{2+} + H_2O$$

This complex ion leads to a deep red colouration of the solution, which can now be quantified colorimetrically (using a reference scale) or (more precise:) photometrically (absorption maximum at $\lambda = 482$ nm) and thus allows conclusions to be drawn about the analyte content of the solution.

A Fluorometric Example

If ammonia is to be detected as part of a routine analysis (for example, in water analyses), it is possible to take advantage of the fact that benzene-1,2-dicarbaldehyde (*o-phthaldialdehyde*) reacts with ammonia in the presence of sulfite ions (SO_3^{2-}) to form 2H-isoindole-1-sulfonic acid (◘ Fig. 16.2), which emits fluorescent radiation with $\lambda = 423$ nm when suitably excited—and can thus be easily detected fluorometrically.

The emitted fluorescence radiation is then passed on again—as described in ▶ Chap. 12—via a fibre optic to the corresponding measuring instrument.

> **More Complex Detection Reactions**
> If the analyte can only be detected after undergoing *several* reaction steps, it is also possible to add further reagents "downstream" of the analyte injection site, which then finally react to form the detectable form of the analyte; here, for example, biomolecules—such as enzymes or antigens/antibodies—can also be used, as you are already familiar with from ▶ Sect. 14.2. Here too, of course, strict reproducibility of the reaction conditions must be ensured.

However, as mentioned at the beginning of this chapter, detection methods are not limited to the quantification of absorbed or emitted electromagnetic radiation. If an analyte is oxidisable or reducible, voltammetric detection is also possible:

■ A Voltammetric Example

Very small cells are used (with a sample volume that is usually less than one microliter), in the wall of which a working electrode is embedded. This then measures, in conjunction with a reference electrode (often the familiar silver/silver chloride electrode, known from Part II), the potential, which of course is dependent on the concentration of the analyte. The detection limits here might require an amount of substance of analyte even below the nanomole range.

■ Variation: The Sequential Injection Analysis

In **sequential** injection analysis, the continuous flow is dispensed with: This variant (usually computerised) is based on the fact that the flow can be stopped selectively, and the direction of flow can also be *reversed* by using appropriate pumps. With the aid of the associated apparatus, working with numerous valves and pumps (shown schematically in Fig. 18.14 of Harris), the analyte is first introduced (this is shown schematically in Fig. 18.15A, also from Harris) and then—via a second valve—mixed with the required detection reagent (Fig. 18.15B). Then the flow is first stopped so that the reaction product required for detection is formed (Fig. 18.15C); dispersion processes can even cause *complete* conversion of the analyte if necessary (Fig. 18.15D)—provided a sufficiently long waiting time is kept. However, this is not absolutely necessary, after all: With completely reproducible experimental conditions, usable measurement results are obtained even with incomplete conversion, as has already been described. Finally, the flow is reversed; a pump conveys the (fully or reproducibly partially) converted analyte to the detector so that quantification can be performed (▶ Harris, Fig. 18.15E).

Harris, Section 18.4: Flow injection analysis and sequential injection analysis

The advantage of this method is that significantly smaller quantities of reaction solution (flow agent and detection reagent!) are required, so that less waste is also produced. (Keyword: sustainability!)

The equipment required for this purpose has been miniaturised considerably in the course of technical development (which is still progressing) and is increasingly being used in routine analysis of solutions that are to be checked continuously (e.g. in technical chemistry for quality control of the mixtures obtained).

■ In Combination with Sensors

As already mentioned in this chapter, detection is not limited to photometric/fluorometric or electrochemical methods: *all* sensors described in this part—be they "inorganic" or biological/biochemical—can be combined with flow injection analysis or sequential injection analysis. In this way, the versatility of the detection methods is combined with a fast, automatable and comparatively cost-effective practical implementation (especially in the case of sequential injection analysis with regard to the consumption of chemicals). Whereas such equipment was the exception only a few years ago, it is increasingly developing into genuine "standard equipment" in all analytical laboratories.

❓ Questions

4. Why is it not possible to increase the number of analyses carried out within any given time window indefinitely, even in the case of computer-controlled flow injection analysis?
5. Why is meticulous calibration of the equipment used essential in flow injection analysis?

16.1 Summary

- **Electrochemical Sensors**

Electrochemical sensors allow conclusions to be drawn about the analyte content on the basis of potentiometric or voltammetric measurements. Important examples of such sensors are the Clark electrode (for determining the oxygen content of an aqueous solution) or the lambda probe. The latter is a solid-state electrode that determines the current flow resulting from the different oxygen partial pressuren of the analyte sample and a reference substance.

In the case of biosensors, it is often not only electrochemical processes that are important, but also other interactions; the changes induced in the system under consideration are then ultimately converted into electrochemical signals via transducers. Biosensors based on biomolecules such as antigens/antibodies, proteins/enzymes, DNA/RNA fragments, cell organelles or even whole, biologically active cells are often not only very specific but also extremely sensitive; remarkably low detection limits can often be achieved. However, such sensors can usually only be used under very moderate conditions, since the systems interacting specifically with the analyte cannot survive more rigid conditions (higher temperature, pH fluctuations, extreme concentrations of analytes or even matrix components). In particular, meticulous buffering is required here.

- **Optical Sensors**

Optodes can be used in combination with suitable transducers to quantify light phenomena. Fluorescence can be observed as well as its quenching by corresponding analytes. Such sensors can be of inorganic or organic origin and can also be based on biochemical reactions (in the presence of enzymes or similar). Optodes are important in the quantification of oxygen or other biologically significant molecules (e.g. in clinical analysis); optical sensors are also playing an increasingly important role in the elucidation of biochemical processes in living systems.

- **Flow Injection Analysis**

In flow injection analysis, the analyte solution to be examined is injected into a capillary through which a solvent (usually water) flows continuously. Reagents are dissolved in the solvent and convert the analyte into the detection form. The latter has properties that permit photometric or fluorometric quantification of the analyte; for example, aromatic analytes can be converted by dissolved diazonium salts to azo dyes with a precisely known absorption maximum. Similarly, colour-intensive complexes allow the quantification of inorganic analytes. If necessary, after injection of the analyte in a point of the capillary behind the injection site, it is possible to add further reagents to effect the required chemical modification of the analyte.

Quantitative conversion of the analyte usually does not take place due to the short delay time in the capillary, which is why calibration with analyte solutions of known concentration is essential. Due to these short delay times, a three-digit number of analyses can be performed per hour, often automated and using an autosampler. Characteristic for FIA is the comparatively high consumption of solvent and detection reagent (or reagents).

In the technically more complex sequential injection analysis, which is almost exclusively computer-controlled, the direction of flow within the capillary can be reversed with the aid of pumps and valves and the quantified solution can also be removed from the capillary again. In this way, the consumption of solvents as well as detection reagents is drastically reduced in some cases.

16.1 · Summary

✓ **Answers**
1. In addition to the usual factors for Nernst's equation z (number of electrons to be transferred) and the obligatory Faraday's constant (F = 96,485 C/mol) as well as, of course, the partial pressure of oxygen p(O$_2$) compared to the reference partial pressure p$_{Reference}$, four further factors play an important role:
   - the *thickness of the membrane* (d, usually given in cm, even if this is not quite in the SI sense), through which said diffusion must first occur; the thicker the membrane, the more the effect is attenuated, so the resulting sensor current is inversely proportional to the membrane thickness,
   - the (again temperature-dependent) *diffusion coefficient of oxygen* (D, again not quite SI-compliant with the unit cm^2/s) in the membrane material used (in this case: siloxane); the larger it is, the stronger the resulting current flow,
   - the *solubility of the oxygen* (S) in the membrane material: the more dissolves, the stronger the current will be, *and*
   - surface A (also here in cm^2), at which all these processes occur.

Overall, therefore, the quantitative calculation results in the following relationship:

$$I = z \cdot F \cdot A \cdot D_{O_2}^{Siloxan} \cdot S_{O_2}^{Siloxan} \cdot \frac{p_{O_2}}{p_{Reference}} \cdot \frac{1}{d}$$

(Don't worry, we are not going to calculate this now. In addition to concrete values for the partial pressures, we would also need the values for D and S to be taken from tables and/or data bases, and to conjure up only exemplary values out of the proverbial hat seems to me to make little sense.)

2. The fact that the glass electrode also responds to ammonium ions does not, of course, change the influence of pH fluctuations, which would significantly falsify the measurement result. Corresponding fluctuations must therefore be prevented.
3. The question already touches the border to biology: You probably know (and if not, this will change at the latest when you come to *biochemistry*) that cells have a certain electrolyte content, which may only change within very narrow limits if the functionality of the cell is to be maintained. For this reason, sensors that use living cells are extremely unsuitable for analyte solutions with an electrolyte content that is too high or too low, because *osmosis* would occur: The concentration gradient between the contents of the cell and the analyte solution leads to solvent molecules (i.e.: Water) being transported out of the cell or flowing into the cell:
   - If the ion concentration of the analyte solution is too high, water would leave the cell due to osmosis in order to reduce the "excessive" concentration outside the cell and at the same time increase the electrolyte concentration inside the cell, so that balancing of the concentration gradient takes place. The cells would accordingly continue to shrink and at some point would no longer be able to function in the way we call "life".
   - If the analyte solution is too low in concentration, the opposite phenomenon would occur: Water would continue to flow into the cell until the internal and external concentrations are balanced, but this uncontrolled influx of water almost inevitably leads to bursting of the cell, which also severely restricts its functionality.
4. Even if the chemical modifications of the analyte required for detection take place within the capillary used, i.e. do not have to be carried out in advance, there is always the risk of cross-contamination, as already men-

tioned: If the "analyte plugs" follow each other too closely, they might come into contact with each other and thus falsify the measurement results.

5. Of course, you might now say: "Because in analytics you actually *always* have to calibrate!"—and you wouldn't even be entirely wrong. But here the main reason for the necessity to calibrate is the fact that in most cases the reaction of the analyte with the required reagents to obtain the "detection form" does not take place complete because the (dwelling) time within the capillary is not sufficient for a chemical equilibrium to be established. However, if reproducible measurement conditions are meticulously ensured, this "problem" applies to each individual "analyte plug": Thus, it can be assumed that for all analyte samples, regardless of their individual concentrations, the same percentage of analyte molecules will always be converted to the detection form. However, this percentage is *not known*, so it must first be determined, using reference samples of known concentration, which actual measured values result in each case. The evaluation of the readings/measurement values belonging to the samples of unknown concentration is then carried out in the form of **interpolation**. If things go badly, **extrapolation** beyond the measured values of the highest or lowest known concentration may occasionally be required—in which case, of course, one must trust that even at these very high or very low concentrations there is still a linear relationship between the concentration and the measured value obtained. (You might remember that the problem arising with extrapolation has already been dealt with in Part I of "Analytical Chemistry I".)

Further Reading

Berg JM, Tymoczko JL, Stryer L (2013) Stryer: Biochemie. Springer Spektrum, Heidelberg
Brown TL, LeMay HE, Bursten BE (2007) Chemie: Die zentrale Wissenschaft. Pearson, München
Cammann K (2010) Instrumentelle Analytische Chemie. Spektrum, Heidelberg
Gründler P (2004) Chemische Sensoren. Springer, Berlin
Harris DC (2014) Lehrbuch der Quantitativen Analyse. Springer, Heidelberg
Holleman/Wiberg (2007) Lehrbuch der Anorganischen Chemie. DeGruyter, Berlin
Lottspeich F, Engels JW (2012) Bioanalytik. Springer, Heidelberg
Murphy K, Travers P, Walport M (2009) Janeway: Immunologie. Spektrum, Heidelberg
Skoog DA, Holler FJ, Crouch SR (2013) Instrumentelle Analytik. Springer, Heidelberg
If you want to enlarge upon the topic of chemical sensors, I suggest the very well accessible and yet detailed book by Cammann; for the topic "bio sensors", particularly Stryer and Lottspeich are to be mentioned.

Statistics

Requirements

In this part we close the circle in a way, because we will refer again and again to the statistical basics that were covered in Part V of "Analytical Chemistry I". So please do not resort to the "can be checked off and that's it" method, you really need the mathematics and in this case the statistics.

You probably already suspect what is coming now: We naturally fall back on the contents that you will certainly have learned in courses on the topic of mathematical concepts in the natural sciences. In particular, we will use functions and their derivatives—including partial ones—and the calculation of extrema will play a role. To simplify some mathematical formulas, we will use series expansions—the Taylor series and its special case of MacLaurin's series. Integration also comes up frequently in statistics when calculating probabilities, but you are certainly already familiar with determining integrals from basic mathematics. We will even refer to methods of solving systems of linear equations in this part. You can see that mathematics is really needed if we want to do statistics.

And of course, the usual symbols of mathematics will be used again and again. Accordingly, it will be teeming with sum signs (Σ), and the product sign (Π) will also appear occasionally; this abbreviated notation simply saves space (and is also very common, so you should be familiar with it). But don't panic: It won't be that hard in this part either, and most of the time even the most complex mathematical expression at first glance can ultimately be broken down to a fairly to very simple formula.

Official Learning Target

(This "breaking down" does, however, require a little more complex mathematical thinking now and then, for which one should take a little more time. To that extent, we don't consider it an "official learning objective" of this chapter, and have presented it, shaded in gray, in a slightly smaller font, so that readers in a hurry know they may skip it for now if necessary. But only for now.)

Learning Objectives

You will learn the various techniques for the statistical evaluation of measured values, including different methods for determining the mean and calculating standard deviations and confidence intervals. The extent to which errors can add up or otherwise mutually reinforce each other is considered in the context of error propagation; using linear regression, you will learn to identify, quantify, and graph linear dependencies.

The various methods and types of measurement value distribution are understood in principle, and you also know how to assess their respective areas of application and how to use them sensibly.

Furthermore, you know estimation methods for the parameters and characteristic values of these distributions of measured values on the basis of samples. You are able to estimate confidence intervals for such parameters with the help of Student t-distribution or chi-square distribution. Through the application of parameter tests, you have learned to evaluate measurement series, such as those known from Part II of "Analytical Chemistry I" (keyword: multiple titrations).

For a quantitative determination of analyte contents—see Part I of "Analytical Chemistry I"—method validation by calibration is essential. You will learn the associated mathematics—linear regression and determination of compensation lines—in this part and will then be able to apply this to procedures such as standard addition and internal standard.

In addition, you have understood the purpose of outlier tests and can apply them to measurement series.

Contents

Chapter 17 Experimental Errors – 219

Chapter 18 Statistical Analysis – 221

Chapter 19 Gaussian Error Propagation – 235

Chapter 20 Measured Value Distribution – 245

Chapter 21 Parameter Estimates – 277

Chapter 22 Validation of Methods – 309

Chapter 23 Outlier Tests – 319

Glossary – 333

Experimental Errors

All measurements, no matter how carefully and scientifically they are carried out, are subject to certain uncertainties. These uncertainties are investigated and evaluated in the context of **error analysis**. The two most important functions are:
- Estimation of the size of the measurement uncertainties
- Reduction of uncertainties

The final measured value must therefore always be provided with an **error indication in order** to be able to recognise its accuracy. A distinction is made here between
- **Systematic** errors
- **Statistical** (random) errors

Systematic errors or systematic deviations are, as the name suggests, errors of the measuring system itself: They are *reproducible* and affect every single measured value in the *same way*.

> Systematic errors can, for example, be the result of incorrect calibration of the measuring instrument used, so that measured values are constantly too high or too low.

Statistical errors, on the other hand, which are often also referred to as **random errors** (for example, also in Part I of "Analytical Chemistry I"—remember that?), are errors that result from a scattering of the measured values.

> Statistical errors occur, for example, when you have to pipette the same amount of a solution several times: You will not succeed in always measuring exactly 25,000 mL, even if you always use the same bulb pipette. Once, the meniscus in the pipette may not be visible to the naked eye, but may have been a little too high, the second time a little too low, etc. This results in certain statistical (and unavoidable) fluctuations.

While systematic errors can be compensated for or at least largely minimised by corrective measures, for example on the measuring device or other components, statistical fluctuations of the individual measured values cannot be avoided in principle. The more measured values that are available, the less significant the fluctuations—including the frequently cited **outliers**—are. These must then be statistically evaluated.

Skoog, Annex A.1: Precision and accuracy of measurement

▶ Skoog addresses this issue in a very comprehensible way.

Statistical Analysis

Contents

18.1 Mean Value ($\bar{x}$) – 222

18.2 Standard Deviation (s) – 226

18.3 Confidence Interval – 232

A statistical evaluation of the measured values obtained only makes sense if a certain number of comparable measured values is available; if the number of samples is too small, outliers—which are statistically possible—can significantly distort the overall picture. Three key figures here are:
- the mean value ($\bar{x}$)
- the standard deviation (s)
- the confidence interval

Skoog, Annex A.2.1: Populations and samples

You have already encountered the first two terms in Part I of "Analytical Chemistry I". Nevertheless, we will look at all three again in turn, and Skoog also help you here.

18.1 Mean Value ($\bar{x}$)

The mean value is nothing other than the *arithmetic mean of* all measured values. With the help of the mean value, we can make statements about quantities that are measured not only once, but several times. In this way, we obtain the value that should/should not occur "on average" in each measurement, usually in combination with the expected deviations from this mean value ($\pm\sigma_x$); we will discuss deviations in ▶ Sect. 18.2.

Before we determine a measurand X, we minimise, if possible, the systematic errors (i.e. everything that is associated with a systematic falsification of the measured values due to the measurement setup, such as calibration errors, device-related errors and the like), so that the resulting measurement results are essentially subject to *statistical* fluctuations.

In general, all measured values follow a distribution and then fluctuate around this mean value, whereby the width of the distribution can be regarded as a measure of the quality of the measurements. A typical histogram of the measured values and the Gaussian normal distribution fitted to the data is shown in ◘ Fig. 18.1. On the abscissa (as the *x-axis* is called in "good mathematical" terms) lie the measured values x_i, and on the ordinate (as the *y-axis* is called in technical terms) we read of the assigned frequency $h(x_i)$ with which the respective measured values occurred in the course of the experiment.

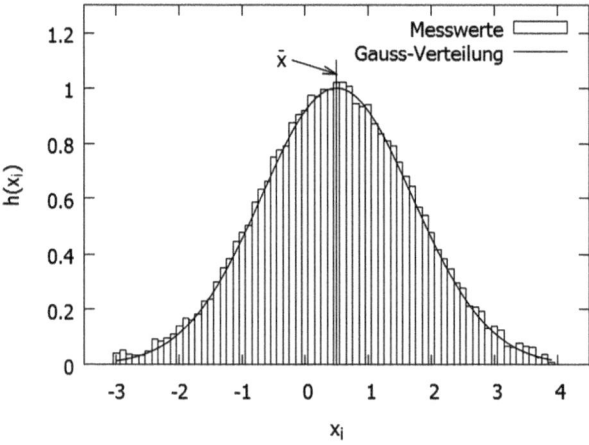

◘ **Fig. 18.1** Histogram with measured values and the fitted Gaussian normal distribution

18.1 · Mean Value ($\bar{x}$)

> **Important**
> The value that *we can influence ourselves* is plotted on the *x-axis* (i.e. the volume of the titrant added during a titration, etc.).
> The *y-axis* shows the measured values dependent on the *x-value*.

Thus the continuous solid line does not represent the measurement situation correctly, because for a solid line we would actually need an *infinite* number of measured values, but we only have a finite number of them. The actual number of measured values is given as N in the statistics.

Thus, if we take N measurements, the totality of the measured values together represents a **sample**: A random partial survey from a **population** (also called an **ensemble**) to determine a characteristic X.

The individual measurement results x_i (with $i = 1, 2, 3, ..., N$) are called **sample values**. (What is really meant here are the concrete measured values of each individual measurement).

To calculate the *mean value* of all measurement results, we use a formula that you already know:

$$\text{Mean value } \bar{x} = \frac{\sum_i x_i}{N} \tag{18.1}$$

$\bar{x}$ is therefore, as already mentioned above, nothing other than the arithmetic mean of the measured values obtained.

If the individual measurements are subject to statistical fluctuations, these can—as is well known—occur "in both directions", i.e. be "too large" or "too small". However, the more measured values are available, the more extreme outliers will "average out". (We will deal with outliers separately in 23.)

Please consider what is actually to be found out here: Under ideal conditions (i.e. if one could measure completely error-free), each measurement would yield exactly the same value—each sample value would then correspond to the **true value** (in statistics usually indicated with μ). For this reason, as the number of measured values (the samples, i.e.: N) increases, the resulting mean value approaches $\bar{x}$ the value μ more and more.

> **Attention**
> Please keep in mind that with almost no measurement—even under (anyway unachievable) *ideal conditions*—measured values are obtained which correspond to the theoretical (and factually unachievable) "true value" (which we do not know at all!), because the system under investigation is itself subject to fluctuations:
>
> Let us imagine, for example, that we wanted to determine the position of an atom within a molecule or another multi-atom compound: As you know, there are (above the also unattainable) absolute zero (0 K) always certain oscillations (which we even specifically excite in the context of IR spectroscopy, remember Part IV of "Analytical Chemistry I"?), i.e. the "true" location of an atom *cannot be* determined at all. The same applies to other measurements, which are **in principle** subject to certain fluctuations—which may also be purely metrological.

> **Important**
> Each Gaussian distribution is described by two parameters:
> - $\bar{x}$—the mean value *and*
> - s—the measured standard deviation
>
> Both parameters originate from the *measurement*.

Their theoretical counterparts are then the abstract and *actually unattainable* values
- μ—the *true* mean and
- σ—the mathematically determined, idealised standard deviation

Please do not confuse these two sets of parameters.

For Those Who Would Like to Know More
In addition to the arithmetic mean, there is also the
(a) geometric mean and the
(b) harmonic mean

The *geometric* mean is calculated according to Eq. (18.2):

$$\bar{x}_g = \sqrt[N]{x_1 \cdot x_2 \cdot \ldots \cdot x_N} = \sqrt[N]{\prod_{i=1}^{N} x_i} \qquad (18.2)$$

The mean value obtained in this way is particularly important in economic statistics, e.g. when the mean value of interest rate changes over several years is to be calculated.

For the *harmonic* mean, on the other hand, Eq. (18.3) holds:

$$\bar{x}_h = \frac{N}{\dfrac{1}{x_1} + \dfrac{1}{x_2} + \ldots + \dfrac{1}{x_N}} = \frac{N}{\sum_{i=1}^{N} \dfrac{1}{x_i}} \qquad (18.3)$$

This is used if characteristic X has a dimension consisting of a numerator and a denominator and the averaging is to be carried out over the dimension in the *denominator*.

In Eq. (18.3) each measured value x_i *occurs* exactly once. If, on the other hand, in a measurement, the different measured values occur in clusters with weighting factors w_i, then the corresponding weighted harmonic mean value is:

$$\bar{x}_h = \frac{w_1 + w_2 + \ldots + w_N}{\dfrac{w_1}{x_1} + \dfrac{w_2}{x_2} + \ldots + \dfrac{w_N}{x_N}} = \frac{\sum_{i=1}^{N} w_i}{\sum_{i=1}^{N} \dfrac{w_i}{x_i}} \qquad (18.4)$$

▶ **Example**

As an example, consider an alloy whose average density is to be calculated. This alloy consists of 1.7 cm³ of iron (with density $\rho(Fe) = 7.87$ g/cm³), 2.1 cm³ of cobalt ($\rho(Co) = 8.89$ g/cm³), and 1.8 cm³ of nickel ($\rho(Ni) = 8.91$ g/cm³). Using the harmonic mean, we obtain for the density:

$$\rho(\text{Alloy}) = \frac{1.7 + 2.1 + 1.8}{\dfrac{1.7}{7.87} + \dfrac{2.1}{8.89} + \dfrac{1.8}{8.91}} = 8.56 \text{ g/cm}^3$$

◀

However, the mean value is now determined: Since it is not possible to perform an infinite number of measurements, the true value μ generally remains unknown. Under real conditions (with a finite number of samples), only a **best value can be** achieved: We will come back to this in a moment.

18.1 · Mean Value ($\bar{x}$)

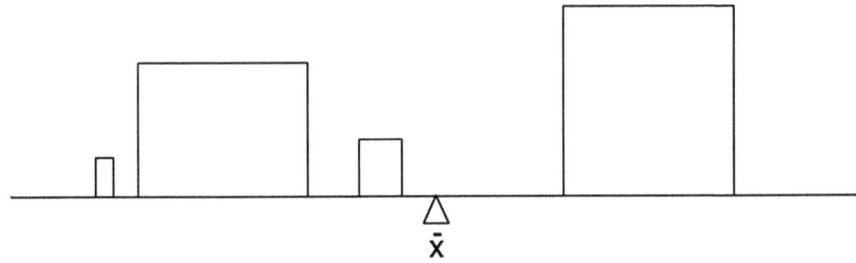

■ **Fig. 18.2** The arithmetic mean—"balanced"

- **Absolute Error of the Mean Value**

Please note: The mean $\bar{x}$ always subject to an error ε compared to the true mean μ. This error ε is called the **absolute error** of the arithmetic mean. It corresponds to the difference between the true value and the mean value:

$$\varepsilon = \bar{x} - \mu \tag{18.5}$$

- ■ **What Do You Need to Know About the Mean?**

The formula for calculating the mean value $\bar{x}$ from N measurements (Eq. 18.1), which you are familiar with (by now at the latest), does not fall from the sky, and deriving it yourself would perhaps be a bit of a stretch. In the following, therefore, we will briefly consider how a parameter $\bar{x}$ must be structured as a mean value so that it can serve as a representative variable for our measurement. The mean is a well-considered combination of the sample values xi and is now determined in such a way that the sum S of the squares of all deviations $(\bar{x} - x_i)$ becomes minimal, that is, that holds:

$$S = \sum_{i=1}^{N} (\bar{x} - x_i)^2 = \text{Minimum} \tag{18.6}$$

This sum is based on the **least-squares method** and goes back to Carl Friedrich Gauss, probably one of the most important mathematicians. All measured values x_i and the still unknown mean value are included $\bar{x}$ this sum. If you imagine that the measured values—weighted according to the frequency—are placed on a "ruler", then the equilibrium point with which you balance the ruler lies exactly where the *arithmetic mean is*; illustrated in ■ Fig. 18.2.

At this optimal point the deviations cancel each other out—the (arithmetic) mean is thus similar to the centre of gravity of a mass distribution—and *related to this centre of gravity* the squares of the deviations are minimal. What does this mean for our sum S?—Since the measured values x_i are known from the measurements made and are available as numerical values, the value of the sum depends only on $\bar{x}$. For the sum to be minimal, we need the first derivative of S with respect to the unknown mean $\bar{x}$. (Yes, this is really just like in the curve discussion!) For the derivative must then hold:

$$\frac{dS}{d\bar{x}} = 0 \quad \text{and} \quad \frac{dS}{d\bar{x}} = 2 \cdot \sum_{i=1}^{N} (\bar{x} - x_i) = 0$$

> **For Those Who Really Want to Do the Math**
> This is where the chain rule comes into play. Have fun with it.

From this you get as an average value:

$$\bar{x} = \frac{1}{N}\sum_{i=1}^{N} x_i \qquad (18.7)$$

The arithmetic mean of all measurements (which you already know from Eq. 18.1). Another notation for the representation of the mean value is often the symbol $E(X)$—it stands for the term **expected value of** the measurand X and provides the mean value with the **realisations/measured values** x_i determined by sampling $\bar{x}$ and, in the case of a **complete survey** or knowledge of the probability function, the true value μ.

18.2 Standard Deviation (s)

Now that we have established that the mean is $\bar{x}$ our best value for the quantity X, *we need to* look at $x_i - \bar{x}$ the difference.

This difference, which is often called the *deviation of the individual measurements xi from the mean value* or $\bar{x}$ **standard deviation** (usually with the formula symbol **s**), indicates how much the ith measured value x_i differs from the mean value: If only very small deviations occur, i.e. if all measured values are close to each other, this initially indicates that the measurements are probably very accurate. (Cave: If there is a *systematic* error, it will not be noticed here!)

The standard deviation is then the value that can be determined from measurements:

$$s = \sqrt{\frac{1}{N-1}\sum_{i=1}^{N}(\bar{x} - x_i)^2} \qquad (18.8)$$

But what exactly does this value s actually mean, and why is $1/(N-1)$ used as the prefactor instead of $1/N$?

- **Meaning of the Standard Deviation**

s characterises the uncertainty of the individual measured values. Later, we will deal in more detail with the probability distributions of measured quantities and the associated measured values. As a small anticipation, we would like to refer to a probability function typical for measured quantities: the **Gaussian distribution** or **normal distribution** with a mean value $\mu = \bar{x}$ and a standard deviation $\sigma = s$. The graph of this probability function is reminiscent of the silhouette of a bell—hence the name **bell curve**. The mean value of the measured variable X of this function lies at the point $x = \mu$, i.e. where the *expected value is* located, and the dispersion of the possible measured values is characterised by the standard deviation $s = \sigma$. (Of course, Harris also covers this important topic).

Harris, Section 4.1: Gaussian distribution

The total area enclosed by the function and abscissa is 1 (or alternatively expressed as 100%), this means that **all** numerical values x_i (measured values) possible for the quantity X have been recorded—in the case of the Gaussian curve this range extends from $-\infty < x_i < \infty$.

You can easily imagine that with such a large **definition range** (and thus also: **value range**), a complete survey will not be possible. With a partial survey or even a sample, we will only catch a small section of the totality of possible measured values. The good news is that our measured values, the data of a complete survey, are centered around the expected value.

So let's consider an area that is symmetrical about the expected value $\mu = \bar{x}$. The area spanning the interval from $\bar{x} - s$ to $\bar{x} + s$ is about 68.3%, i.e. about 68.3% of all measured values are in the range of the simple standard deviation around the mean, i.e. $\bar{x} \pm s$. This is shown graphically in ◘ Fig. 18.3.

18.2 · Standard Deviation (s)

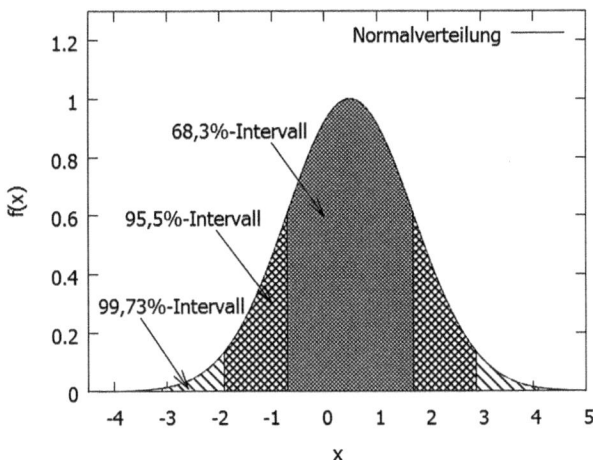

☐ **Fig. 18.3** Confidence intervals in the Gaussian distribution

If a measurement with the same experimental set-up and otherwise unchanged measurement conditions delivers the mean value $\bar{x}$ and the associated standard deviation s, then with a new measurement/sample with the same measurement set-up again about 68.3% of the measured values will $\bar{x} \pm s$ lie in the range, after all nothing has changed in the distribution of the measured values for our measurand X on which the experiment is based.

- **Variance and Standard Deviation (Formerly: Mean Error of the Individual Measured Values)**

The **variance (s^2) is** given as a measure of the dispersion of the respective individual values x_i around the mean value. This is the square of the standard deviation of the respective individual measurements.

$$s^2 = \frac{1}{N-1} \sum_{i=1}^{N} (x_i - \bar{x})^2 \qquad (18.9)$$

⚠ It is important to divide by $N - 1$ and not just by N.

■■ But Why $N - 1$?

Let us first consider another form for calculating the variance, which is very often used when the recording of the measured values is very lengthy and/or the size of the sample is large. In such a case, one wants to determine the (current) mean $\bar{x}$ and the variance s^2 *in* between. This is done with the **theorem of Huygens-Steiner**.

$$s^2 = \frac{1}{N-1} \sum_{i=1}^{N} x_i^2 - \frac{N}{N-1} \bar{x}^2 \qquad (18.10)$$

This formula allows the calculation of mean value and standard deviation quasi *en passant*. For this purpose it is necessary to continuously calculate the respective square of the measured value in addition to the measured values x_i and to add these values to the previously measured ones.

▶ Example

To illustrate, here's a quick example.
　　Application of the theorem of Huygens-Steiner for the calculation of mean and variance

i	x_i	$\sum x_i$	$\bar{x} = \frac{1}{N}\sum x_i$	x_i^2	$\sum x_i^2$	$s^2 = \frac{1}{N-1}\sum x_i^2 - \frac{N}{N-1}\bar{x}^2$
1	3.5	3.5	3.5	12.25	12.25	–
2	3.8	7.3	3.65	14.44	26.69	0.045
3	2.4	9.7	3.23	5.76	32.45	0.576
4	4.2	13.9	3.475	17.64	50.09	0.596
5	2.7	16.6	3.32	7.29	57.38	0.567
6	3.9	20.5	3.417	15.21	72.59	0.507
7	3.6	24.1	3.443	12.96	85.55	0.428

Something like this table can be easily created in a spreadsheet program and allows the calculation of mean and variance even if it is not specified beforehand how large the sample should be.

For Those Who Want to Know Exactly
Here is the derivation of the displacement theorem:

$$s^2 = \frac{1}{N-1}\sum_{i=1}^{N}(x_i - \bar{x})^2$$

$$= \frac{1}{N-1}\sum_{i=1}^{N}(x_i^2 - 2\bar{x}x_i + \bar{x}^2) = \frac{1}{N-1}\left[\sum_{i=1}^{N}x_i^2 - 2\cdot\bar{x}\sum_{i=1}^{N}x_i + \sum_{i=1}^{N}\bar{x}^2\right]$$

$$s^2 = \frac{1}{N-1}\left[\sum_{i=1}^{N}x_i^2 - 2\cdot\bar{x}\sum_{i=1}^{N}x_i + \sum_{i=1}^{N}\bar{x}^2\right] = \frac{1}{N-1}\left(\sum_{i=1}^{N}x_i^2 - N\cdot\bar{x}^2\right)$$

It follows that:

$$s^2 = \frac{1}{N-1}\sum_{i=1}^{N}x_i^2 - \frac{N}{N-1}\bar{x}^2$$

Let us now return to the question of why in Eq. (18.8) the sum over the squared deviations is divided by $N - 1$:

First, we must remember that the expected value is $\bar{x}$ the average of N measurements. If we measure a measurand X whose expected value $\bar{x}$ we know, the first $N - 1$ measurements are independent, and the Nth measurement is then $\bar{x}$ determined by. The number of **degrees of freedom** with knowledge of is $\bar{x}$ thus $N - 1$ instead of N—or in other words: The determination $\bar{x}$ of $N - 1$ instead of N, so we can also only divide by $N - 1$.

For Those Interested in Theory
In an explanation formulated with mathematical terminology, we use the fact that the variance σ^2 is $E\left((x-\bar{x})^2\right)$ the expected value of the distance squares.
For this expected value holds:

$$\sigma^2 = E\left((x-\bar{x})^2\right) = E\left(x^2 - 2x\bar{x} + \bar{x}^2\right) = E\left(x^2\right) - 2\bar{x}E(x) + E\left(\bar{x}^2\right)$$

$$= E\left(x^2\right) - 2\bar{x}^2 + \bar{x}^2 = E\left(x^2\right) - \bar{x}^2$$

$$\sigma^2 = E\left(x^2\right) - \left(E(x)\right)^2 = E\left(x^2\right) - \mu^2$$

18.2 · Standard Deviation (s)

Thus, the expected value for the squared measured values:

$$E(x^2) = \sigma^2 + \mu^2$$

For the variance of the expected value ($V(\bar{x})$ coming up in the next section):

$$V(\bar{x}) = \sigma^2 / N = E(\bar{x}^2) - \mu^2$$

Thus, the following applies to the expected value:

$$E(\bar{x}^2) = \sigma^2 / N + \mu^2$$

As N increases, the mean $\bar{x}$ converge.

If we now calculate the expected value of the variance, then:

$$E(s^2) = E\left(\frac{1}{N-1}\sum_i x_i^2 - \frac{N}{N-1}\bar{x}^2\right) = \frac{1}{N-1}\sum_i E(x_i^2) - \frac{N}{N-1}E(\bar{x}^2)$$

Note that we have used the variance formula here, where the sum is divided by $N - 1$.

$$E(s^2) = \frac{1}{N-1}\sum_i(\sigma^2 + \mu^2) - \frac{N}{N-1}(\sigma^2/N + \mu^2) = \frac{N}{N-1}\left(\sigma^2 + \mu^2 - \frac{\sigma^2}{N} - \mu^2\right) = \sigma^2$$

This result means that the formula we used for the variance is **expectation-true**. **This is** also referred to as the *unbiased* variance.

If we mistakenly use N (instead of $N - 1$, *as is* correct) to calculate the variance, we get:

$$E\left(\tilde{s}^2\right) = E\left(\frac{1}{N}\sum_i x_i^2 - \bar{x}^2\right) = \frac{1}{N}\sum_i E(x_i^2) - E(\bar{x}^2) = \frac{1}{N}\sum_i(\sigma^2 + \mu^2) - \left(\frac{\sigma^2}{N} + \mu^2\right)$$

If we evaluate the sum, we get:

$$E\left(\tilde{s}^2\right) = \sigma^2 + \mu^2 - \left(\frac{\sigma^2}{N} + \mu^2\right)$$

The two μ^2 cancel out because of the sign and fall away. What remains, converted again to the number of measurements N:

$$E\left(\tilde{s}^2\right) = \frac{N-1}{N}\sigma^2$$

This result reflects that the formula

$$\tilde{s}^2 = \frac{1}{N}\sum_i x_i^2 - \bar{x}^2 \tag{18.11}$$

has *no* expectation-true behavior: the formula we now use is a *biased* estimate for the variance.

Granted: With increasing sample sizes (large N), both formulas for the variance converge more and more. Whether you ultimately divide by 9999 or by 10,000 when estimating the variance of a measurement with, say, ten thousand values, then hardly matters. But nevertheless only Eq. (18.9) is *correct*.

The standard deviation s is now nothing more than the square root of the variance s^2 just discussed in detail:

$$s = \sqrt{s^2} = \sqrt{\frac{1}{(N-1)}\sum_{i=1}^{N}(x_i - \bar{x})^2} \tag{18.12}$$

This standard deviation is also known as the mean error/uncertainty of the individual measurement.

- **Standard Deviation of the Mean (Formerly Known As: Mean Error of the Mean, Standard Error)**

If $x_1, x_2, ..., x_N$ are the results of N measurements of the same quantity X, then, as we saw in ▶ Sect. 18.1, the best value for the quantity X (i.e.: x_{Best}) is its mean value $\bar{x}$. The standard deviation s then characterises—as discussed in the last paragraph—the **mean uncertainty of** the individual measured values $x_1, x_2, ..., x_N$.

The use of this mean value leads to more reliable statements than each individual measurement considered on its own: The uncertainty of the mean value ultimately becomes smaller and smaller as the number of measured values increases.

The measure of certainty (and hence accuracy) of the mean is its **mean error** (σ_x). It is defined as:

$$\sigma_x = \frac{s}{\sqrt{N}} = \sqrt{\frac{1}{N(N-1)}\sum_{i=1}^{N}(x_i - \bar{x})^2} \qquad (18.13)$$

The specification of the measured value x for the experimentally determined measurand X is then in the form:

$$x = \bar{x} \pm \sigma_x$$

This completely defines the final result.

> ▶ **Example**
>
> In one measurement the measured value x *is* recorded. A total of 10 measurements ($N = 10$) are made, each resulting in the measured values x_i (with $i = 1, ..., 10$); a systematic error is not known. This results in the measured values x_i given in the table:
> Sample of size $N = 10$ for the measurand X with measured values x_i.

N	x_i
1	73.2
2	74.6
3	78.5
4	74.3
5	72.1
6	75.0
7	75.5
8	77.1
9	71.3
10	74.4

What statements can be made here?—From Eq. (18.1) it follows first of all:

$$\bar{x} = (73.2 + 74.6 + 78.5 + 74.3 + 72.1 + 75.0 + 75.5 + 77.1 + 71.3 + 74.5)/10 = 74.6$$

And now, using the following table, let's look at what the individual values are for s and s^2 *at* each individual x_i.

Calculation of mean value and squared deviations. The sum of the deviations is formed to check the centre of gravity property

18.2 · Standard Deviation (s)

i	x_i	$x_i - \bar{x}$	$(x_i - \bar{x})^2$
1	73.2	−1.4	1.96
2	74.6	0.0	0.00
3	78.5	3.9	15.21
4	74.3	−0.3	0.09
5	72.1	−2.5	6.25
6	75.0	0.4	0.16
7	75.5	0.9	0.81
8	77.1	2.5	6.25
9	71.3	−3.3	10.89
10	74.4	−0.2	0.04
$\bar{x} = 74.6$		$\sum_{i=1}^{10}(x_i - \bar{x}) = 0$ (it has to cancel each other out)	$\sum_{i=1}^{10}(x_i - \bar{x})^2 = 41.66$

Overall, the parameters are as follows:
(a) Mean value $\bar{x} = 74.6$
(b) For the standard deviation s holds:

$$s = \sqrt{\frac{\sum_{i=1}^{10}(x_i - \bar{x})^2}{N-1}} = \sqrt{\frac{41.66}{9}} \approx 2.15$$

(c) As variance s^2 results:

$$s^2 = \frac{1}{N-1}\sum_{i=1}^{N}(x_i - \bar{x})^2 = (2.15)^2 = 4.63$$

Accordingly, the standard deviation of the mean σ_x is calculated to:

$$\sigma_x = \frac{s}{\sqrt{N}} = \sqrt{\frac{1}{N(N-1)}\sum_{i=1}^{N}(x_i - \bar{x})^2} = \sqrt{\frac{41.66}{10 \cdot 9}} = 0.680$$

The measured value x must therefore be specified as follows:

$$x = \bar{x} \pm \sigma_x = 74.6 \pm 0.7$$

So we have two standard deviations:
1. the standard deviation of the individual measured values (s) *and*
2. the standard deviation of the mean (σ_x)

Since the mean is, in a sense, a well-considered combination of the individual measurements, the standard deviation of the mean (i.e. σ_x) is smaller by a factor $\frac{1}{\sqrt{N}}$ than the standard deviation of the individual measurements. (So you can, if you like, gloss over the standard deviation of a sample a bit—at least for people who don't really know statistics).

Please note the **significant figures** (known, for example, from Part I of "Analytical Chemistry I"): Since an addition is performed here, only *one* decimal place can be specified for the *deviation* (i.e.: two significant figures).

18.3 Confidence Interval

The specification of the above measured value does not, of course, correspond to the "true" (unattainable!) mean µ. However, we can specify intervals within which the mean lies with a certain (high) probability (think back to ◘ Fig. 18.3). These so-called confidence regions, or confidence intervals, specify ranges, symmetric about the mean $\bar{x}$, within which the (unknown) "true" mean µ of a measured quantity lies $\bar{x}$ with a chosen **confidence level** γ. This confidence level indicates the probability that the "true" mean µ lies within the **confidence interval**. (Again, Harris and Skoog also help you).

As the probability increases, the width of the confidence **interval** increases. The following applies to the confidence interval:

$$\bar{x} - t\frac{s}{\sqrt{N}} \leq \mu \leq \bar{x} + t\frac{s}{\sqrt{N}} \tag{18.14}$$

The parameter t (which we encounter again in ▶ Sect. 21.2, but—unfortunately—has no name of its own) depends on
- the number of measured values (N) and
- the confidence level (γ).

The corresponding values for t can be taken from ◘ Table 18.1 or can be determined with a spreadsheet program—company names should be avoided here.

(Increasingly, such statistical calculations are left to spreadsheet programs mentioned above. Harris even goes into this topic quite extensively; we will dispense with it here).

> ▶ **Example**
> For confidence levels of γ = 68.3%, γ = 90%, γ = 95%, and γ = 99%, t for each of N = 9 readings, the parameter is 1.07; 1.86; 2.31; and 3.36, respectively, indicating that the mean µ has a probability of
>
> 68.3% in the interval $[\bar{x} - 1.07\sigma_x, \bar{x} + 1.07\sigma_x]$
>
> 90% in the interval $[\bar{x} - 1.86\sigma_x, \bar{x} + 1.86\sigma_x]$
>
> 95% in area $[\bar{x} - 2.31\sigma_x, \bar{x} + 2.31\sigma_x]$

lies. If, on the other hand, the probability that the true mean lies within the confidence interval is to be 99%, then the confidence interval is $[\bar{x} - 3.36\sigma_x, \bar{x} + 3.36\sigma_x]$.

As the number of readings increases ($N \to \infty$), the parameter $t = 1$ for a confidence level γ = 68.3%, for γ = 90% the parameter t = 1.65; if the confidence level is γ = 95% then t = 1.96, and if the confidence level is chosen to be 99%, then for a number of readings approaching infinity t = 2.58.

The figure shows the single σ-environment comprising 68.3% of the measured values, the double σ-environment (95.5% of the measured values) and triple σ-environment (99.73% of the measured values) in the form of a normal distribution with characteristic values µ = 0.5 and σ = 1.2.

Harris, Section 4.2: Confidence intervals
Skoog, Appendix A.2.2: Statistical treatment of random errors

On the use of spreadsheet programs for various aspects of statistics, *see also*
Harris, Section 4.1: Gaussian distribution
Harris, Section 4.3: Comparison of means with Student's *t-test*
Harris, Section 4.5: *t-tests* with spreadsheets
Harris, Section 4.7: The method of least squares
Harris, Section 4.9: Least squares worksheet

18.3 Confidence Interval

Table 18.1 Quantiles of the t-distribution for one-sided/two-sided confidence intervals

Confidence level degrees of freedom	0.9/0.95	0.95/0.975	0.975/0.9875	0.99/0.995	0.995/0.9975	0.999/0.9995
1	3.077684	6.313752	12.706205	31.820516	63.656741	318.308839
2	1.885618	2.919986	4.302653	6.964557	9.924843	22.327125
3	1.637744	2.353363	3.182446	4.540703	5.840909	10.214532
4	1.533206	2.131847	2.776445	3.746947	4.604095	7.173182
5	1.475884	2.015048	2.570582	3.364930	4.032143	5.893430
6	1.439756	1.943180	2.446912	3.142668	3.707428	5.207626
7	1.414924	1.894579	2.364624	2.997952	3.499483	4.785290
8	1.396815	1.859548	2.306004	2.896459	3.355387	4.500791
9	1.383029	1.833113	2.262157	2.821438	3.249836	4.296806
10	1.372184	1.812461	2.228139	2.763769	3.169273	4.143700
11	1.363430	1.795885	2.200985	2.718079	3.105807	4.024701
12	1.356217	1.782288	2.178813	2.680998	3.054540	3.929633
13	1.350171	1.770933	2.160369	2.650309	3.012276	3.851982
14	1.345030	1.761310	2.144787	2.624494	2.976843	3.787390
15	1.340606	1.753050	2.131450	2.602480	2.946713	3.732834
16	1.336757	1.745884	2.119905	2.583487	2.920782	3.686155
17	1.333379	1.739607	2.109816	2.566934	2.898231	3.645767
18	1.330391	1.734064	2.100922	2.552380	2.878440	3.610485
19	1.327728	1.729133	2.093024	2.539483	2.860935	3.579400
20	1.325341	1.724718	2.085963	2.527977	2.845340	3.551808
21	1.323188	1.720743	2.079614	2.517648	2.831360	3.527154
22	1.321237	1.717144	2.073873	2.508325	2.818756	3.504992
23	1.319460	1.713872	2.068658	2.499867	2.807336	3.484964
24	1.317836	1.710882	2.063899	2.492159	2.796940	3.466777
25	1.316345	1.708141	2.059539	2.485107	2.787436	3.450189
26	1.314972	1.705618	2.055529	2.478630	2.778715	3.434997
27	1.313703	1.703288	2.051831	2.472660	2.770683	3.421034
28	1.312527	1.701131	2.048407	2.467140	2.763262	3.408155
29	1.311434	1.699127	2.045230	2.462021	2.756386	3.396240
30	1.310415	1.697261	2.042272	2.457262	2.749996	3.385185
40	1.303077	1.683851	2.021075	2.423257	2.704459	3.306878
50	1.298714	1.675905	2.008559	2.403272	2.677793	3.261409
60	1.295821	1.670649	2.000298	2.390119	2.660283	3.231709
70	1.293763	1.666914	1.994437	2.380807	2.647905	3.210789
80	1.292224	1.664125	1.990063	2.373868	2.638691	3.195258
90	1.291029	1.661961	1.986675	2.368497	2.631565	3.183271
100	1.290075	1.660234	1.983972	2.364217	2.625891	3.173739
200	1.285799	1.652508	1.971896	2.345137	2.600634	3.131480
500	1.283247	1.647907	1.964720	2.333829	2.585698	3.106612
∞	1.281552	1.644853	1.959964	2.326348	2.575829	3.090232

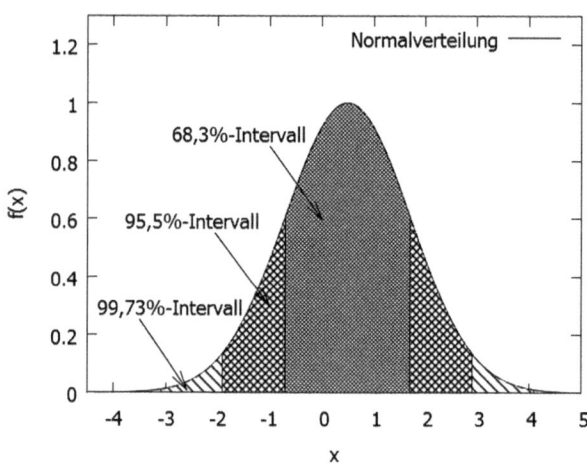

Normal distribution with µ = 0.5 and σ = 1.2 and confidence intervals

The fine checkered region marks the *1σ-environment* containing 68.3% of the distribution, the coarse checkered interval contains the *2σ-environment* with 95.5% of the distribution, and the dashed region includes the *3σ-environment* with 99.73% of the distribution. ◄

? Questions

1. What are the resulting values $\bar{x}$, s, s^2, and σ_x with the following (abstract) measured values: x_i = 23.1; 24.0; 23.8; 22.9; 24.1; 23.4; and 23.6? How should the resulting total measured value be stated?

2. Multiple titrations of 20.00 mL each of a sodium hydroxide solution of unknown concentration will consume 24.24; 24.33; 24.48; 24.52; and 24.53 mL of a 1000-molar hydrochloric acid until the neutral point is reached. What is the total concentration of the sodium hydroxide solution? What is the standard deviation?

Gaussian Error Propagation

Contents

19.1 Linear Regression/Equilibrium Line – 237

19.2 Adjustment of Fit Parameters for Equalisation Curves/Parabolas – 241

© Springer-Verlag GmbH Germany, part of Springer Nature 2025
U. Ritgen, *Analytical Chemistry II*, https://doi.org/10.1007/978-3-662-68710-9_19

How does one determine the accuracy of a physical quantity, which is not amenable to direct measurement? There are plenty of examples of this:
- The density of a substance, which is determined indirectly as the quotient of the two measurands mass and volume
- The acceleration, which is calculated (just as indirectly) from the measurement of distances and times, and much more.

Skoog, Annex A.2.3: Propagation of measurement inaccuracies (error propagation)

Here, too, a look at Skoog certainly help if necessary.

Let us assume that the quantities $x, y, z, \ldots$ have been measured with uncertainties $\Delta x, \Delta y, \Delta z, \ldots$ (these are positive numbers) and the measured values have been used to calculate the function $f(x, y, z, \ldots)$. If the uncertainties of $x, y, z, \ldots$ are random, then the uncertainty of f is:

$$f = f(x,y,z,\ldots)$$

From the total differential

$$df = \frac{\partial f}{\partial x}dx + \frac{\partial f}{\partial y}dy + \frac{\partial f}{\partial z}dz$$

we can derive Δf an approximate formula for the error:

$$\Delta f \approx \frac{\partial f}{\partial x}\Delta x + \frac{\partial f}{\partial y}\Delta y + \frac{\partial f}{\partial z}\Delta z$$

From this it is easy to see that the errors in a composite measurand may well "cancel" each other out.

This "mutual cancellation" can be avoided by using the *amounts of* the partial derivatives:

$$|\Delta f| = \left|\frac{\partial f}{\partial x}\Delta x\right| + \left|\frac{\partial f}{\partial y}\Delta y\right| + \left|\frac{\partial f}{\partial z}\Delta z\right| \tag{19.1}$$

The uncertainty for the quantity (f formula symbol u_f) is obtained from the **Gaussian error propagation law**:

$$u_f = \sqrt{\left(\frac{\partial f}{\partial x}\Delta x\right)^2 + \left(\frac{\partial f}{\partial y}\Delta y\right)^2 + \left(\frac{\partial f}{\partial z}\Delta z\right)^2 + \ldots} \tag{19.2}$$

Here, $\frac{\partial f}{\partial x}$ the partial derivative of the function is f according to the variable x (all other variables are considered as parameters, i.e. as constant with respect to the variation of the variable x).

▶ **Example**

Let's take a look at the moment of inertia J as an example. The following applies here:

$$J = \frac{1}{2}mr^2$$

with the total differential:

$$\Delta J = mr\Delta r + \frac{1}{2}r^2\Delta m$$

and the error propagation:

$$u_J = \sqrt{(mr\Delta r)^2 + \left(\frac{1}{2}r^2\Delta m\right)^2}$$

19.1 · Linear Regression/Equilibrium Line

Thus, if we know the errors in the determination of the radius (Δr) and the mass determination (Δm), we are able to determine the resulting total errors. If the radius of the cylinder $r = 56.0$ cm and the measurement error $\Delta r = 0.1$ cm and the mass is at $m = 2.26$ kg with an error $\Delta m = 0.01$ kg, then we get for the error of the moment of inertia:

$$\Delta J = mr\Delta r + \frac{1}{2}r^2\Delta m = 2.26 \text{ kg} \cdot 56.0 \text{ cm} \cdot 0.1 \text{ cm} + \frac{1}{2}(56.0 \text{ cm})^2 \cdot 0.01 \text{ kg}$$
$$= 28.3 \text{ kg cm}^2$$

Here, note that we have only three significant figures. The error propagation u_J is accordingly:

$$u_J = \sqrt{(mr\Delta r)^2 + \left(\frac{1}{2}r^2\Delta m\right)^2} = \sqrt{\frac{(2.26 \text{ kg} \cdot 56.0 \text{ cm} \cdot 0.1 \text{ cm} +)^2}{+\left(\frac{1}{2}(56.0 \text{ cm})^2 \cdot 0.01 \text{ kg}\right)^2}}$$
$$= 15.7 \text{ kg cm}^2$$

The measurement result is then given in $J \pm \Delta J$ the form. In this example, the moment of inertia of the cylinder is $J = 3.70 \cdot 10^3 \pm 28.3$ kg cm^2. ◄

■ Regression Calculation

Equilibrium calculus is intended to establish relationships between experimentally available measurements and theoretical models (usually in the form of functions). As examples we will consider linear, quadratic and exponential functions/relationships. Let us start with the simplest case.

19.1 Linear Regression/Equilibrium Line

Assume that there is a linear relationship between two physical or chemical quantities y and x:

$$y = mx + b$$

Now there is a series of measured values: $y_i(x_i)$ with $i = 1, \ldots, N$. Then inevitably the question arises: How exactly can the constants m and b be determined from pairwise measurements of these quantities (of y at different values of x)? And how can a straight line $y = mx + b$ be laid through these measuring points in such a way that the errors of m and b become as small as possible?

It often happens that the quantity x can be measured more accurately than y.

> In free fall, for example, the time t elapsed in each case can be measured much more accurately by electronically controlled clocks than the corresponding distance covered or the speed reached in each case.

In such cases the errors of the quantity x can be neglected compared to those of the quantity y. The compensation line that we use in such cases always passes through the center of gravity $S(\bar{x}, \bar{y})$.

The derivation of the formulae follows from the *least square method* (known from ► Sect. 18.1), according to which the deviation of the measured value from the theoretical value is minimised:

$$S = \sum_{i=1}^{N}\left(y_i - (mx_i + b)\right)^2 = \sum_{i=1}^{N}\left(y_i^2 - 2y_i(mx_i + b) + m^2 x_i^2 + 2mx_i b + b^2\right) \quad (19.3)$$

S here is the sum of the squared deviations between the measured value y and the theoretical value obtained for y when the experimental x-values are substituted into the straight line equation y = mx + b. (This is again an aspect of statistics discussed quite extensively in both ▶ Harris and ▶ Skoog).

Harris, Section 4.7: The method of least squares
Skoog, Annex A.4: The method of least squares

From the zeros of the partial derivatives

$$\left(\frac{\partial S}{\partial m} = 0\right) \quad \text{and} \quad \left(\frac{\partial S}{\partial b} = 0\right)$$

one obtains determination equations for the coefficients m and b:

$$\frac{\partial S}{\partial b} = \sum_{i=1}^{N}\left(-2y_i + 2mx_i + 2b\right) = 0 \quad (19.4)$$

and

$$\frac{\partial S}{\partial m} = \sum_{i=1}^{N}\left(-2y_i x_i + 2mx_i^2 + 2x_i b\right) = 0 \quad (19.5)$$

From Eq. (19.4) one obtains:

$$2Nb = 2\sum_{i=1}^{N} y_i - 2m\sum_{i=1}^{N} x_i \quad \text{and thus} \quad b = \frac{1}{N}\sum_{i=1}^{N} y_i - m\frac{1}{N}\sum_{i=1}^{N} x_i = \bar{y} - m\cdot\bar{x}$$

The intercept b thus results from the mean values for the measured variables x and y. However, b still depends on the slope measure m—and we want to determine this now.

From Eq. (19.5) we now obtain:

$$2m\sum_{i=1}^{N} x_i^2 = 2\sum_{i=1}^{N} y_i x_i - 2b\sum_{i=1}^{N} x_i$$

With the result for the intercept b, we thus get:

$$m\sum_{i=1}^{N} x_i^2 = \sum_{i=1}^{N} y_i x_i - \left(\frac{1}{N}\sum_{i=1}^{N} y_i - m\frac{1}{N}\sum_{i=1}^{N} x_i\right)\sum_{i=1}^{N} x_i \quad \text{resp.}$$

$$m\sum_{i=1}^{N} x_i^2 = \sum_{i=1}^{N} y_i x_i - (\bar{y} - m\bar{x})\sum_{i=1}^{N} x_i$$

❗ These two calculations lead to exactly the same result. The expression after the "or" is always the shorter variant, in which we have already included known information.

It follows:

$$m\sum_{i=1}^{N} x_i^2 - m\frac{1}{N}\sum_{i=1}^{N} x_i \sum_{i=1}^{N} x_i = \sum_{i=1}^{N} y_i x_i - \left(\frac{1}{N}\sum_{i=1}^{N} y_i\right)\sum_{i=1}^{N} x_i \quad \text{resp.} \quad m\sum_{i=1}^{N} x_i^2 - m\bar{x}\sum_{i=1}^{N} x_i$$

$$= \sum_{i=1}^{N} y_i x_i - \bar{y}\sum_{i=1}^{N} x_i$$

19.1 · Linear Regression/Equilibrium Line

So it goes:

$$m\left(\sum_{i=1}^{N} x_i^2 - \frac{1}{N}\left(\sum_{i=1}^{N} x_i\right)^2\right) = \sum_{i=1}^{N} y_i x_i - \frac{1}{N}\sum_{i=1}^{N} y_i \sum_{i=1}^{N} x_i \quad \text{resp.} \quad m\left(\sum_{i=1}^{N} x_i^2 - N\bar{x}^2\right) = \sum_{i=1}^{N} y_i x_i - N \cdot \bar{y} \cdot \bar{x}$$

The **slope measure** m and the **ordinate intercept** b of the compensation line are obtained from:

$$m = \frac{\sum_i (x_i \cdot y_i) - \frac{\sum_i x_i \cdot \sum_i y_i}{N}}{\sum_i x_i^2 - \frac{(\sum_i x_i)^2}{N}} \qquad b = \bar{y} - m \cdot \bar{x} \qquad (19.6)$$

respectively

$$m = \frac{\sum_i (x_i \cdot y_i) - N \cdot \bar{x} \cdot \bar{y}}{\sum_i x_i^2 - N(\bar{x})^2} \qquad b = \bar{y} - m \cdot \bar{x} \qquad (19.7)$$

Linear Regression with Different Measured Values

The investigation of the solubility L of $NaNO_3$ in water as a function of temperature T resulted in the following pairs of measurements:

I	1	2	3	4	5	6
Ti	0	20	40	60	80	100
Li	70.7	88.3	104.9	124.7	148.0	176.0

Determine the balance line

- Step: Plotting the measured values in a diagram

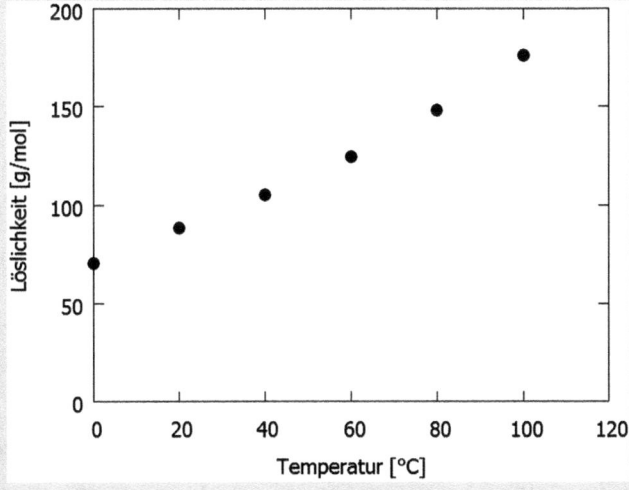

Solubility of $NaNO_3$ as a function of temperature in °C

- Step: Calculating the regression line using Eqs. (19.6) and (19.7). To do this, we create a table with the required sums:

$$m = \frac{\sum_i (x_i \cdot y_i) - N\overline{xy}}{\sum_i x_i^2 - N(\overline{x})^2} = \frac{42,884 - 6 \cdot 50 \cdot 118.7\overline{6}}{22,000 - 6 \cdot 50^2} = 1.0363$$

$$b = \overline{y} - m \cdot \overline{x} = 118.7\overline{6} - 1.0363 \cdot 50 = 66.952$$

Calculation of sums and averages for Eqs. (19.6) and (19.7), respectively

i	1	2	3	4	5	6	Totals	Mean values
Ti	0	20	40	60	80	100	300	50
Li	70.7	88.3	104.9	124.7	148.0	176.0	712.6	118.77
Ti$_2$	0	400	1600	3600	6400	10,000	22,000	
TiLi	0.0	1766.0	4196.0	7482.0	11,840.0	17,600.0	42,884	

- Step: Drawing the regression line

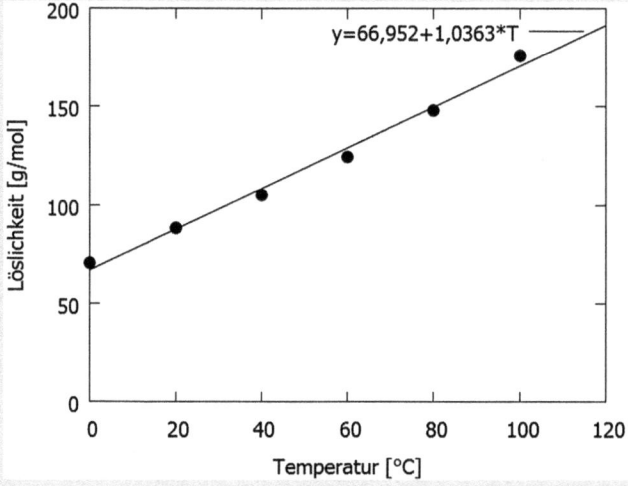

Measured values of the solubility and the corresponding compensation line

A Tip for Everyday Life
Human optics is quite excellent for inspecting straight lines: A typical physicist's or statistician's hand movement is therefore to hold the data points drawn on paper in a coordinate system obliquely in front of the eyes—keyword: grazing incidence—and to turn the paper in such a way that the perspective on the data points is *shortened*. Starting from the eye, we then have an—imaginary—straight line and can easily see the deviations of the points or just the non-deviation from this—still imaginary—line. If you feel like it, try it out, and if the imaginary line doesn't work so well, then use your index finger—as an extension of the eye and in *any case* visible as the opposite of the imaginary line—to help you. You can also do this with a pencil, but then you should be careful not to jam it into your eye. That can go into the eye—and exactly that has actually already happened occasionally …

19.2 Adjustment of Fit Parameters for Equalisation Curves/Parabolas

Of course, there are such mathematical tools not only for balancing lines, but also for—in principle—all conceivable functions/curves, i.e. also exponential and power functions. Because we do not want to go too far into the nitty-gritty in this part, we will limit ourselves to *parabolas in the* following section; however, *it* should be noted that other curves can also be treated in this way.

19.2 Adjustment of Fit Parameters for Equalisation Curves/Parabolas

Consider the case where the data points lie approximately on a parabola. The functional relationship between N measurement points (x_i, y_i) with $i = 1, \ldots, N$ is a general parabola of the form: $y_i = ax_i^2 + bx_i + c$, where the parameters a, b and c are to be determined from the measured values. As in the case of linear regression, we again minimise the sum of the squares of the deviations:

$$S(a,b,c) = \sum_{i=1}^{N}\left(y_i - \left(ax_i^2 + bx_i + c\right)\right)^2 \to \text{Minimum}$$

To do this, we calculate the zeros of the partial derivatives according to the parameters a, b and c in each case:

$$\frac{\partial S}{\partial a} = \sum_{i=1}^{N}\left(y_i - \left(ax_i^2 + bx_i + c\right)\right)x_i^2 = 0 \tag{19.8}$$

$$\frac{\partial S}{\partial b} = \sum_{i=1}^{N}\left(y_i - \left(ax_i^2 + bx_i + c\right)\right)x_i = 0 \tag{19.9}$$

$$\frac{\partial S}{\partial c} = \sum_{i=1}^{N}\left(y_i - \left(ax_i^2 + bx_i + c\right)\right) = 0 \tag{19.10}$$

To calculate the solution, we present the above problem in the form of a system of equations:

$$\begin{pmatrix} \sum_{i=1}^{N} x_i^4 & \sum_{i=1}^{N} x_i^3 & \sum_{i=1}^{N} x_i^2 \\ \sum_{i=1}^{N} x_i^3 & \sum_{i=1}^{N} x_i^2 & \sum_{i=1}^{N} x_i \\ \sum_{i=1}^{N} x_i^2 & \sum_{i=1}^{N} x_i & N \end{pmatrix} \cdot \begin{pmatrix} a \\ b \\ c \end{pmatrix} = \begin{pmatrix} \sum_{i=1}^{N} y_i x_i^2 \\ \sum_{i=1}^{N} y_i x_i \\ \sum_{i=1}^{N} y_i \end{pmatrix}$$

The elements of the coefficient matrix result directly from sums over powers of the *x-values* and the number of individual measurements, the inhomogeneity is obtained from the calculation of the sums over products of the *y-values* with powers of the *x-values*. With suitable methods for solving systems of equations (e.g. **Gaussian algorithm, Cramer's rule**) the solution vector and thus the parameters a, b and c can be determined.

> ▶ **Example**
> The temperature dependence of the volume of water can be described in the range from −20 to 100 °C by a parabola: $V(T) = aT^2 + bT + c$.
> The readings are:
> Measured values for the temperature dependence of the volume. The other columns contain the quantities T^2, T^3, T^4, VT and VT^2 necessary for the determination of the parabola parameters together with the sums.

$$\begin{pmatrix} 156{,}800{,}256 & 1{,}792{,}064 & 22{,}416 \\ 1{,}792{,}064 & 22{,}416 & 284 \\ 22{,}416 & 284 & 8 \end{pmatrix} \begin{pmatrix} a \\ b \\ c \end{pmatrix} = \begin{pmatrix} 23{,}133.72 \\ 291.9109 \\ 8.105976 \end{pmatrix}$$

	T/°C	V	T²	T³	T⁴	VT	VT²
	−20	1.006580	400	−8000	160,000	−20.1316	402.63
	0	1.000160	0	0	0	0	0
	4	1.000028	16	64	256	4.0001	16.00
	20	1.001797	400	8000	160,000	20.0359	400.72
	40	1.007842	1600	64,000	2,560,000	40.3137	1612.55
	60	1.017089	3600	216,000	12,960,000	61.0253	3661.52
	80	1.029027	6400	512,000	40,960,000	82.3222	6585.77
	100	1.043453	10,000	1,000,000	100,000,000	104.3453	10,434.53
Totals	284	8.105976	22,416	1,792,064	156,800,256	291.9109	23,113.72
	T/°C	V	T²	T³	T⁴	VT	VT²

The solution of this system of equations can be determined using Cramer's rule. For this we need the determinant of the coefficient matrix:

$$\det A = \begin{vmatrix} 156{,}800{,}256 & 1{,}792{,}064 & 22{,}416 \\ 1{,}792{,}064 & 22{,}416 & 284 \\ 22{,}416 & 284 & 8 \end{vmatrix} = 1{,}333{,}397{,}094{,}400.00$$

In addition, there are the auxiliary determinants, where the columns are successively replaced by the inhomogeneity:

$$\det A_1 = \begin{vmatrix} 23{,}113.72165 & 1{,}792{,}064 & 22{,}416 \\ 291.910932 & 22{,}416 & 284 \\ 8.105976 & 284 & 8 \end{vmatrix} = 6{,}478{,}122.65$$

$$\det A_2 = \begin{vmatrix} 156{,}800{,}256 & 23{,}113.72165 & 22{,}416 \\ 1{,}792{,}064 & 291.910932 & 284 \\ 22{,}416 & 8.105976 & 8 \end{vmatrix} = -74{,}764{,}971.01$$

$$\det A_3 = \begin{vmatrix} 156{,}800{,}256 & 1{,}792{,}064 & 23{,}133.72165 \\ 1{,}792{,}064 & 22{,}416 & 291.920932 \\ 22{,}416 & 284 & 8.105976 \end{vmatrix} = 1{,}335{,}563{,}062{,}516.12$$

The coefficients are obtained from the ratio of the auxiliary determinants to the determinants of the coefficient matrix. Since this has a very large value, we have considered more decimal places in the inhomogeneity when determining the auxiliary determinants in order to avoid rounding errors at this point when calculating the determinants.

At this point it should be pointed out that strong *rounding errors* can occur in the numerical solution of linear systems of equations with strongly varying coefficients due to error propagation.

Thus, for the parabola parameters we obtain:
A = det A_1/det A = 0.000004858360,
b = det A_2/det A = −0.000056071047 and
c = det A_3/det A = 1.001624398407.

19.2 Adjustment of Fit Parameters for Equalisation Curves/Parabolas

The parabola for the volume is:

$$V(T) = 0.000004858360\ T^2 - 0.000056071047\ T + 1.001624398407$$

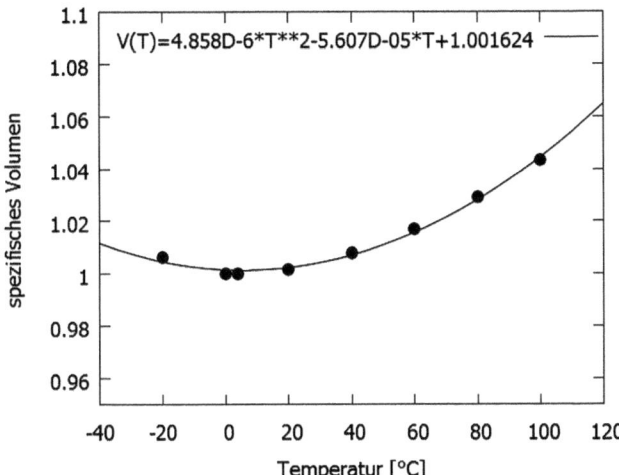

Volume change of water as a function of temperature and compensation parabola

A nice example of the approximation of a correlation which is no longer quite linear—such as in the case of solutions which are too highly concentrated in photometry—can be found in Fig. 4.12 from Harris. Here, the construction of a corresponding calibration curve is also dealt with extensively.

Harris, Section 4.8: Calibration curves

❓ Questions

3. For a reaction at two temperatures $T_1 = 290$ K and $T_2 = 300$ K the rate constants are $k_1 = 1.1 \times 10^{-3}$ s^{-1} and $k_2 = 2.3 \times 10^{-3}$ s^{-1}, the absolute error of the temperature measurement is ± 0.3 K, for the rate constants the relative error is 5%. The activation energy E_a is determined according to the Arrhenius equation:

$$E_a = R \frac{T_1 T_2}{T_1 - T_2} \ln\left(\frac{k_1}{k_2}\right) = 8.314 \frac{J}{\text{mol} \cdot K} \frac{290\ K \cdot 300\ K}{290\ K - 300\ K}$$

$$\ln\left(\frac{1.1 \cdot 10^{-3}\ s^{-1}}{2.3 \cdot 10^{-3}\ s^{-1}}\right) \approx 53.4\ \text{kJ mol}^{-1}$$

   Calculate the absolute and relative error for the activation energy.

4. For a gas—in adiabatic process control—the temperature depends on the volume according to the formula $T = k \cdot V^{1-\kappa}$; in this equation k and κ are constants

You have received the following series of measurements:

Volume dependence of the temperature of a gas

V/L	1	2	3	4	5
T/K	401	302	259	229	211

Determine the two constants (k and κ). Solution: Here a suitable transformation is recommended—keyword logarithmise—in order to then be able to exploit the straight line equation ($y = mx + b$) and its parameters m and b.

Measured Value Distribution

Contents

20.1 Discrete Uniform Distributions – 246

20.2 Two-Point Distribution – 247

20.3 Binomial Distribution – 249

20.4 Hypergeometric Distribution – 254

20.5 Poisson Distribution – 259

20.6 Continuous Uniform Distribution – 264

20.7 Exponential Distribution – 267

20.8 Gaussian Normal Distribution – 269

20.9 Logarithmic Normal Distribution – 274

When performing measurements, the expected results are subject to a certain distribution. A distinction must be made between **discrete** and **continuous** distributions for the measured values.

Let us start with the quantities important in discrete distributions. A random variable X describing quantitative discrete characteristics is usually represented by the realisations $x_1, x_2, x_3 \ldots$ whose probabilities are associated with the values $p_i = P(X = x_i) = f(x_i)$. Here $f(x_i)$ is a function with the property

$$\sum_i f(x_i) = \sum_i p_i = 1$$

We refer to this function $f(x_i)$ as the **probability function**. $F(x_i)$ is called the associated distribution function of the probability distribution of a random variable and is defined as

$$F(x) = P(X \leq x) = \sum_{i, x_i \leq x} f(x_i) = \sum_{i, x_i \leq x} p_i$$

We calculate the mean $E(x) = \bar{x}$ for discrete distributions using this equation:

$$\bar{x} = \sum_i x_i \cdot f(x_i) = \sum_i x_i \cdot p_i \tag{20.1}$$

We obtain the variance $D^2(X) = \sigma^2$ using:

$$\sigma^2 = \sum_i (x_i - \bar{x})^2 \cdot f(x_i) = \sum_i (x_i - \bar{x})^2 \cdot p_i \tag{20.2}$$

The standard deviation is obtained from the square root of the variance, as seen in ▶ Sect. 18.2:

$$s = \sqrt{\sigma^2} \tag{20.3}$$

In the immediate aftermath, we first introduce discrete uniform distribution, and then in further sections we come to the **Bernoulli experiments**, which are fundamental to the **two-point distribution** (more on this in ▶ Sect. 20.2), the **binomial distribution** (which comes in ▶ Sect. 20.3), the **hypergeometric distribution** (▶ Sect. 20.4), and the **Poisson distribution** (more on this in ▶ Sect. 20.5).

20.1 Discrete Uniform Distributions

Examples of discrete uniform distributions are the classical dice experiments. Here, there is a uniform discrete distribution with identical individual probabilities $f(x_i) = 1/N$, this means that when N different outcomes x_i occur, but the possible realisations x_i are all equally likely. The associated distribution function is: $F(x)$

$$F(x) = P(X \leq x) = \sum_{i, x_i \leq x} f(x_i) = \sum_{i, x_i \leq x} \frac{1}{N} \tag{20.4}$$

It is shown in ◘ Fig. 20.1.

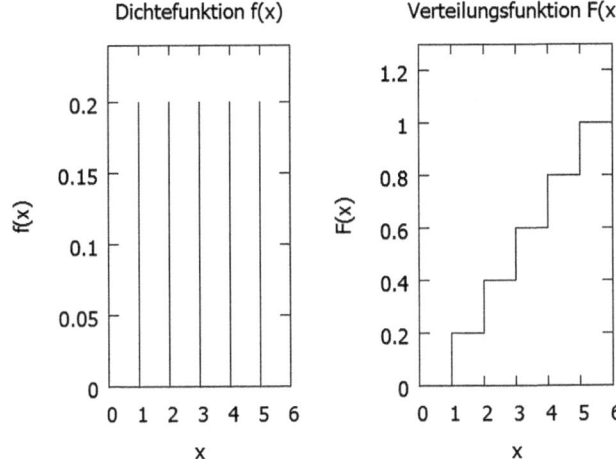

◘ **Fig. 20.1** Probability function and distribution function of a discrete uniform distribution with the parameter $f(x_i) = 1/5$

> ▶ **Example**
> A simple example: You look at the clock (with second hand) at any given time. What is the probability that you have just caught exactly one full minute?
>
> How many different possibilities are there for the second hand?—60 pieces, so: $N = 60$. For only one of these possibilities "exactly x:00 seconds" is true, so: $x = 1$. The summation according to Eq. (20.4) is then very simple: $F(x) = 1/60$, so the probability for the occurrence of this event is one 60th or 0.01666...
>
> The same applies, of course, to any other pointer position: the probability for the event x:23 is also $P = 1/60$.
>
> And the probability for a pointer position between x:40 and x:45? Here it also becomes only slightly more complicated. First the number of possible positions must be determined: x:40, x:41 ... x:45, that is 6 pieces. The total number of possible pointer positions does not change, of course: $N = 60$. This results in $P = 6/60$, i.e. 10%. ◀

20.2 Two-Point Distribution

If a random variable X can only take on two values (a and b), we speak of a **Bernoulli experiment**. The distribution of the two possible outcomes is also called a two-point distribution. The probability of event b occurring is $P(X = b) = p$; the probability of value *a being* accepted is $P(X = a) = q = 1 - p$.

The sum of the two probabilities necessarily results in 1, since one of the two possible events for the quantity X is inevitably realised: The system under consideration *must* take one of the two values; "neither-nor" is just as inadmissible as "both-as well as".

We obtain the mean $E(X) = \mu$ of a two-point distribution by:

$$\mu = a \cdot q + b \cdot p = a + p \cdot (b - a) \tag{20.5}$$

For the variance $D^2(X) = s_x^2$ holds:

$$\begin{aligned}
s_x^2 &= (a-\mu)^2 \cdot q + (b-\mu)^2 \cdot p \\
&= (a^2 - 2\cdot a\cdot\mu + \mu^2)\cdot q + (b^2 - 2\cdot b\cdot\mu + \mu^2)\cdot p \\
&= (a^2\cdot q + b^2\cdot p) - 2\cdot(a\cdot q + b\cdot p)\mu + \mu^2 \\
&= (a^2\cdot q + b^2\cdot p) - (a\cdot q + b\cdot p)^2 \\
&= (a^2\cdot q + b^2\cdot p) - \left((a\cdot q)^2 + 2\cdot a\cdot q\cdot b\cdot p + (b\cdot p)^2\right) \\
&= a^2\cdot q\cdot p + b^2\cdot p\cdot q - 2\cdot a\cdot q\cdot b\cdot p \\
&= p\cdot q\cdot(a-b)^2
\end{aligned}$$

Thus, the variance $D^2(X) = s_x^2$ of the two-point distribution is calculated according to:

$$D^2(X) = s_x^2 = p\cdot q\cdot(a-b)^2 \tag{20.6}$$

and the standard deviation:

$$s_x = \sqrt{s_x^2} = \sqrt{p\cdot q\cdot(a-b)^2} \tag{20.7}$$

> ▶ **Example**
>
> In amorphous solids such as glasses, so-called two-level systems exist in which atoms relax by tunnel effect from a local minimum A into a neighbouring minimum B separated by an energy barrier and return to A again. The probability of atoms staying in minimum A is $P(X = A) = q = 1 - p$; the probability of staying in minimum B is $P(X = B) = p$. The potential energy in minimum A is E_A, in minimum B it is E_B. The mean potential energy $\bar{E}_{pot}$ and its standard deviation s_E are to be calculated.
>
> We calculate the expected value for the potential energy $\bar{E}_{pot}$ in a two-level system with:
>
> $$\bar{E}_{pot} = q\cdot E_A + p\cdot E_B = E_A + p\cdot(E_B - E_A)$$
>
> We then obtain the variance of s_E^2 the potential energy E_{pot} according to
>
> $$s_E^2 = p\cdot q\cdot(E_A - E_B)^2 \quad \blacktriangleleft$$

If the random variable X has only the value 0 with a probability $P(X = 0) = q = 1 - p$ and the value 1 with a probability $P(X = 1) = p$, such a two-point distribution is also called a *Bernoulli distribution*, and the corresponding experiment is called a *Bernoulli experiment*.
- The mean value of a Bernoulli experiment is $\mu = 0\cdot q + 1\cdot p = p$.
- The following then applies to the associated variance $s_x^2 = p\cdot q$.

> ▶ **Example**
>
> Let us take the above example and say that one energy level (x_1) has an energy content of 3.0 eV, the second (x_2) an energy content of 2.3 eV. The first state is taken with a probability $P(x_1) = p = 0.80$ (i.e. 80%), for the second state $P(x_2) = q = 1 - p = 0.20$ (i.e. 20%). What is the total energy content of our system? What is the variance and standard deviation?
> - For the energy content, the following applies: The energy level x_1 with the energy content ($= a$) 3.0 eV is occupied to 80% (p), the energy level x_2 (with the energy content $b = 2.3$ eV) to 20% (q). According to Eq. (20.5), we get: $\mu = 3.0$ eV $\times 0.80 + 2.3$ eV $\times 0.20 = 2.86$ eV

20.3 · Binomial Distribution

- For the variance, we need Eq. (20.6):
  $s_x^2 = 0.80 \times 0.20 \times (3.0 \text{ eV} - 2.3 \text{ eV})^2 = 0.0784 \text{ eV}^2$ (Don't be surprised about the unit: That's still the *variance*. When we are done, the unit will be correct again).
- For the standard deviation, we then need only take the square root according to Eq. (20.7): $s_x = 0.28$ eV. (See?)
  The result is for the total energy is $\mu = 2.9 \pm 0.3$ eV (because of the moderate accuracy of the measurement we have only two significant figures). ◄

20.3 Binomial Distribution

If a Bernoulli experiment is repeated N times without the experiments being interdependent, we speak of a multistage experiment or a Bernoulli experiment of scope N.

Examples include coin tosses or drawing a ball from a bag containing multiple balls that are two—*and only two*—different colours, with discarding. In each case, the probability $P(A) = p$ (*P* here stands for *probability*) for event *A* to occur and $P(B) = q = 1 - p$ *for event B to* occur does not change. The distribution used to describe such a multistage experiment is the binomial distribution. If we consider the possible outcomes of an *N-stage* experiment using a tree diagram, we get a total of 2 N possible combinations/paths of outcomes, as shown in ◘ Fig. 20.2:

> **As a Reminder**
> Does this look familiar to you?—This is the same principle we already had with the tree diagrams in NMR spectroscopy (Part I). (Cue the quartet with a relative signal intensity of 1:3:3:1, the quintet with 1:4:6:4:1, etc. ...)

Let us be interested in a particular outcome in this multistage experiment. Let us consider the case where the event A occurs k times. Note that k can of course only take values between 0 and N, i.e. there are two "extreme possibilities":
- Event *A* does not occur at all.
- Event *A* occurs every time.
- In addition, there are all the "in between" results.

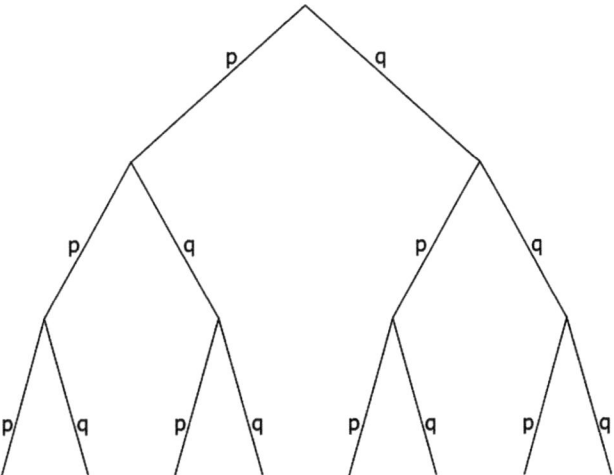

◘ **Fig. 20.2** Tree diagram of a Bernoulli experiment of scope 3: Binomial distribution

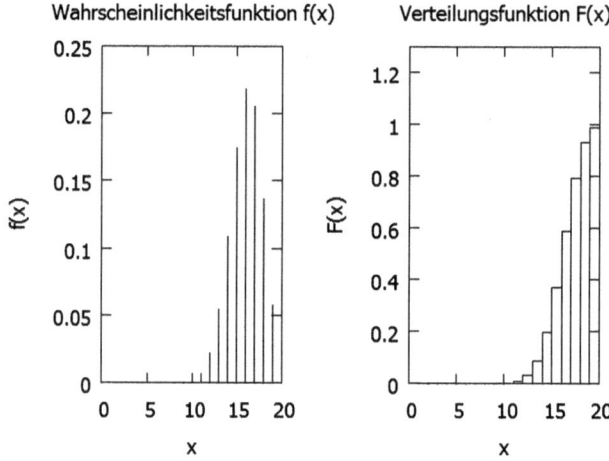

Fig. 20.3 Probability function and distribution function of a binomial distribution with the single probability $P = 0.8$ with $N = 20$-fold repetition

Each of these possible outputs (i.e. each possible "measurement result") can now be realised by at least one path. We obtain the number of possible paths/outputs in each case by the associated binomial coefficient—hence the term for the distribution. The coefficient results in:

$$\binom{N}{k} = \left(\frac{N!}{k!(N-k)!} \right)$$

The probability of event A *occurring k times* along such a path is then calculated by $p^k \cdot q^{N-k}$.

Thus, the probability for a *k-fold* realisation of A considering all possible combinations is

$$P(X = k) = \binom{N}{k} p^k \cdot q^{N-k} \tag{20.8}$$

Figure 20.3 shows the binomial distributions that result when an event A *occurs* with probability $P(A) = 0.8$ and the experiment is repeated $N = 20$ times.

The mean $E(X) = \mu$ for the occurrence of event A is given by in the binomial distribution:

$$\mu = \sum_{k=0}^{N} k \cdot \binom{N}{k} p^k \cdot (1-p)^{N-k} = N \cdot p \tag{20.9}$$

Since a one-step Bernoulli experiment has mean $\mu = p$, if it is repeated N times, the mean is $\mu = N \cdot p$.

> **To Follow Up: A Side Calculation**
> From the definition for the mean, we get:
>
> $$\mu = \sum_{k=0}^{N} k \cdot \binom{N}{k} p^k \cdot (1-p)^{N-k} = \sum_{k=1}^{N} k \cdot \binom{N}{k} p^k \cdot (1-p)^{N-k}$$
>
> The summation does not start until $k = 1$, since the term for $k = 0$ takes the value 0 and can be omitted. Using the binomial coefficient further results:

20.3 · Binomial Distribution

$$\mu = \sum_{k=1}^{N} k \cdot \frac{N!}{k! \cdot (N-k)!} p^k \cdot (1-p)^{N-k} = \sum_{k=1}^{N} \frac{N!}{(k-1)! \cdot (N-k)!} p^k \cdot (1-p)^{N-k}$$

$$\mu = \sum_{k=1}^{N} \frac{N \cdot (N-1)!}{(k-1)! \cdot (N-k)!} p \cdot p^{k-1} \cdot (1-p)^{N-k}$$

$$= N \cdot p \cdot \sum_{k=1}^{N} \frac{(N-1)!}{(k-1)! \cdot (N-k)!} p^{k-1} \cdot (1-p)^{N-k}$$

Now let's introduce a new variable (saves a bit of typing):

$$\tilde{k} = k - 1$$

Thus the equation is:

$$\mu = N \cdot p \cdot \sum_{\tilde{k}=0}^{N-1} \frac{(N-1)!}{\tilde{k}! \cdot (N-1-\tilde{k})!} p^{\tilde{k}} \cdot (1-p)^{N-1-\tilde{k}}$$

The sum in this expression is nothing more than the binomial expansion for the term $(q + p)^{N-1}$, and since this term has the value 1, as seen above: $\mu = N \cdot p$.

The variance $D^2(X) = \sigma^2$ of a binomially distributed quantity is given by

$$\sigma^2 = N \cdot p \cdot q = N \cdot p \cdot (1-p). \tag{20.10}$$

Now, since the variance of a simple Bernoulli experiment is at $\sigma^2 = p \cdot q = p \cdot (1-p)$ when repeated $\sigma^2 = N \cdot p \cdot q = N \cdot p \cdot (1-p)$ N times it is. The standard deviation of the Bernoulli distribution is then

$$\sigma = \sqrt{N \cdot p \cdot q} = \sqrt{N \cdot p \cdot (1-p)} \tag{20.11}$$

And Here's a Side Calculation

From the definition for variance we get:

$$\sigma^2 = \sum_{k=0}^{N} (k - Np)^2 \cdot \binom{N}{k} p^k \cdot (1-p)^{N-k}$$

$$= \sum_{k=0}^{N} (k^2 - 2Npk + N^2 p^2) \cdot \binom{N}{k} p^k \cdot (1-p)^{N-k}$$

Let's look at the individual totals:

$$\sigma^2 = \sum_{k=0}^{N} k^2 \cdot \binom{N}{k} p^k \cdot (1-p)^{N-k} - 2Np \sum_{k=0}^{N} k \binom{N}{k} p^k \cdot (1-p)^{N-k}$$

$$+ N^2 p^2 \sum_{k=0}^{N} \binom{N}{k} p^k \cdot (1-p)^{N-k}$$

We already know the second sum in this expression, it corresponds to the mean $\mu = N \cdot p$, and the last sum is the binomial formula for $(p + q)^N$ necessarily has the value 1 because $p + q = 1$. This inserted gives:

$$\sigma^2 = \sum_{k=0}^{N} k^2 \cdot \binom{N}{k} p^k \cdot (1-p)^{N-k} - 2NpNp + N^2 p^2$$

$$= \sum_{k=0}^{N} k^2 \cdot \binom{N}{k} p^k \cdot (1-p)^{N-k} - N^2 p^2$$

If we use the formula for the binomial coefficient here, we get:

$$\sigma^2 = \sum_{k=0}^{N} k^2 \cdot \frac{N!}{k! \cdot (N-k)!} p^k \cdot (1-p)^{N-k} - N^2 p^2$$

$$= \sum_{k=0}^{N} k \cdot \frac{N!}{(k-1)! \cdot (N-k)!} p^k \cdot (1-p)^{N-k} - N^2 p^2$$

$$\sigma^2 = \sum_{k=0}^{N} k \cdot \frac{N \cdot (N-1)!}{(k-1)! \cdot (N-k)!} p \cdot p^{k-1} \cdot (1-p)^{N-k} - N^2 p^2$$

$$= N \cdot p \cdot \sum_{k=0}^{N} k \cdot \frac{(N-1)!}{(k-1)! \cdot (N-k)!} p^{k-1} \cdot (1-p)^{N-k} - N^2 p^2$$

Since in the sum the first summand takes the value 0, we start the summation at $k = 1$.

$$\sigma^2 = N \cdot p \cdot \sum_{k=1}^{N} k \cdot \frac{(N-1)!}{(k-1)! \cdot (N-k)!} p^{k-1} \cdot (1-p)^{N-k} - N^2 p^2$$

So it goes:

$$\sigma^2 = N \cdot p \cdot \sum_{k=1}^{N} (k-1+1) \cdot \frac{(N-1)!}{(k-1)! \cdot (N-k)!} p^{k-1} \cdot (1-p)^{N-k} - N^2 p^2$$

That leads us to:

$$\sigma^2 = N \cdot p \cdot \sum_{k=1}^{N} (k-1) \cdot \frac{(N-1)!}{(k-1)! \cdot (N-k)!} p^{k-1} \cdot (1-p)^{N-k}$$

$$+ N \cdot p \cdot \sum_{k=1}^{N} \frac{(N-1)!}{(k-1)! \cdot (N-k)!} p^{k-1} \cdot (1-p)^{N-k} - N^2 p^2$$

If we now introduce a new variable $\tilde{k} = k - 1$, the equation is:

$$\sigma^2 = N \cdot p \cdot \sum_{\tilde{k}=0}^{N-1} \tilde{k} \cdot \frac{(N-1)!}{\tilde{k}! \cdot (N-1-\tilde{k})!} p^{\tilde{k}} \cdot (1-p)^{N-1-\tilde{k}}$$

$$+ N \cdot p \cdot \sum_{\tilde{k}=0}^{N-1} \frac{(N-1)!}{\tilde{k}! \cdot (N-1-\tilde{k})!} p^{\tilde{k}} \cdot (1-p)^{N-1-\tilde{k}} - N^2 p^2$$

The second sum in this expression is nothing more than the binomial expansion for the term $(p + q)^{N-1}$, and since this term has the value 1, we can start $\tilde{k} = 1$ again with in the first sum and it holds—as seen above:

$$\sigma^2 = N \cdot p \cdot \sum_{\tilde{k}=1}^{N-1} \tilde{k} \cdot \frac{(N-1)!}{\tilde{k}! \cdot (N-1-\tilde{k})!} p^{\tilde{k}} \cdot (1-p)^{N-1-\tilde{k}} + N \cdot p - N^2 p^2 \sigma^2$$

$$= N \cdot p \cdot \left(1 + \sum_{\tilde{k}=1}^{N-1} \frac{(N-1)!}{(\tilde{k}-1)! \cdot (N-1-\tilde{k})!} p^{\tilde{k}} \cdot (1-p)^{N-1-\tilde{k}} \right) - N^2 p^2$$

After renaming the variables using $k = \tilde{k} - 1$, we get:

20.3 · Binomial Distribution

$$\sigma^2 = N \cdot p \cdot \left(1 + \sum_{k=0}^{N-2} \frac{(N-1)(N-2)!}{k! \cdot (N-2-k)!} pp^k \cdot (1-p)^{N-2-k}\right) - N^2 p^2 \sigma^2$$

$$= Np\left(1 + (N-1)p \sum_{k=0}^{N-2} \frac{(N-2)!}{k! \cdot (N-2-k)!} p^k \cdot (1-p)^{N-2-k}\right) - N^2 p^2$$

Since the sum gives $(p + q)^{N-2}$ the binomial (again with the value 1), we get the variance after a longer calculation with:

$$\sigma^2 = Np(1 + (N-1)p) - N^2 p^2 = Np + N(N-1)p^2 - N^2 p^2 = Np - Np^2$$
$$= Np(1-p) = Npq$$

▶ **Example**

As an example of a Bernoulli distributed quantity, consider the probability of failure for several stirred boilers in operation. Let the probability that such a boiler fails be $p = 0.20$, and hence $q = 1 - p = 1 - 0.20 = 0.80$. In the plant, $N = 12$ such boilers are in operation. The probability that at least 9 boilers are in operation is to be calculated.

For this, of course, we need the *failure probabilities* for 0 to 3 boilers.
The probability of failure of 0 boilers is:

$$P(X=0) = \binom{12}{0} \cdot 0.20^0 \cdot 0.80^{12} = 0.80^{12} \approx 0.06871$$

We obtain the failure of *a* boiler with a probability:

$$P(X=1) = \binom{12}{1} \cdot 0.20^1 \cdot 0.80^{11} = 12 \cdot 0.20 \cdot 0.80^{11} \approx 0.20616$$

Two boilers fall with a probability of:

$$P(X=2) = \binom{12}{2} \cdot 0.20^2 \cdot 0.80^{10} = \frac{12 \cdot 11}{2} \cdot 0.20^2 \cdot 0.80^{10} \approx 0.28347 \text{ from}$$

Finally, for the failure of *three* boilers, we obtain a probability of:

$$P(X=3) = \binom{12}{3} \cdot 0.20^3 \cdot 0.80^9 = \frac{12 \cdot 11 \cdot 10}{6} \cdot 0.20^3 \cdot 0.80^9 \approx 0.23622$$

The associated distribution function is then:

$$F(3) = P(X \leq 3) = \sum_{k=0}^{3} P(X=k) = P(X=0) + P(X=1)$$
$$+ P(X=2) + P(X=3) \approx 0.79456$$

So the probability of having at least nine boilers in operation is:
$P(X \geq 9) = 1 - 0.79456 = 0.2054545 \approx 20.5\%$. ◀

> **A Little Note That Simplifies Life**
> The binomial distribution transitions to the normal distribution for large values of N, for which the following parameters then apply:
>
> $$\mu = N \cdot p \quad \text{and} \quad \sigma = \sqrt{N \cdot p \cdot q} = \sqrt{N \cdot p \cdot (1-p)}$$
>
> **As a rule of thumb, the binomial distribution can be replaced by the normal distribution if:** $N \cdot p \cdot (1-p) \geq 9$.
>
> Experiments based on a binomial distribution are also referred to as *sampling with re-sampling*, since *all* objects to be examined are always available for each repetition of the Bernoulli experiment.

20.4 Hypergeometric Distribution

If the objects to be examined are *not* placed back in a sample, the experiment is based on a *hypergeometric* distribution. Such samples are typical for examinations that are carried out in the context of quality controls at manufacturers or acceptance inspections on the customer side. The probability function of the hypergeometric distribution is:

$$f(x) = P(X = x) = \frac{\binom{M}{x} \cdot \binom{N-M}{n-x}}{\binom{N}{n}} \tag{20.12}$$

Out of a total of N objects, M pieces possess property A, this feature is encountered with probability $p = M/N$; property $\bar{A}$ is found in the remaining pieces and therefore occurs with probability

$$q = \frac{N-M}{N} = 1 - p$$

before. Taking a sample in the size of n objects, the hypergeometric distribution is used to compute the probability that x (with $x = 1, 2, ..., n$) objects possess property A. The distribution function $F(x)$ of the hypergeometric distribution is given by

$$F(x) = \sum_{k \leq x} \frac{\binom{M}{k} \cdot \binom{N-M}{n-k}}{\binom{N}{n}} \quad \text{mit } x = 0, 1, 2, ..., n \tag{20.13}$$

For $x < 0$, $F(x) = 0$. N, M, and n are parameters of this distribution.
The following applies:
(a) $N = 1, 2, 3, ...$ is the total number of objects
(b) $M = 1, 2, 3, ..., N$ (can therefore run to N) is number of objects with property A
(c) $n = 1, 2, 3, ... N$ is the number of objects sampled (sample size).

The hypergeometric distribution $f(x)$ comes about by taking n objects—without putting them back—from the ensemble of N parts. The **ensemble** contains M objects that have the characteristic A. Conversely, $N - M$ objects then possess the characteristic $\bar{A}$ (non-A).

20.4 · Hypergeometric Distribution

If x of the randomly sampled n objects have characteristic A, then the remaining $n - x$ objects have characteristic $\bar{A}$. The number of ways in which the x objects can be taken from the (partial) totality is calculated as:

$$\binom{M}{x}$$

For the number of ways in which the remaining $n - x$ parts of the sample can be realised, it then follows:

$$\binom{N-M}{n-x}$$

The number of possibilities for taking n pieces, x of which have property A, from the set of all objects is thus:

$$\binom{M}{x} \cdot \binom{N-M}{n-x}$$

Thus, if we have N parts in total, and we take n parts from this total, this can be realised in a variety of different ways, which can be denoted by the following term:

$$\binom{N}{n}$$

Thus, the probability that among the n parts there are exactly x parts with property A results from the

ratio of $\binom{M}{x} \cdot \binom{N-M}{n-x}$ to the total $\binom{N}{n}$.

Thus we obtain:

$$f(x) = P(X = x) = \frac{\binom{M}{x} \cdot \binom{N-M}{n-x}}{\binom{N}{n}}.$$

● Figure 20.4 shows the course of the hypergeometric probability function $f(x)$ and the corresponding distribution function $F(x)$.

The mean $E(X) = \mu$ of the hypergeometric distribution is calculated by:

$$\mu = n \cdot \frac{M}{N} = n \cdot p \qquad (20.14)$$

Since the probability of an object with property A is $p_A = M/N$, the mean value multiplies accordingly for n pieces.

The variance $D^2(X) = \sigma^2$ of this distribution is given by:

$$\sigma^2 = \frac{nM(N-M)(N-n)}{N^2(N-1)} = n \cdot p \cdot q \cdot \frac{(N-n)}{(N-1)} \qquad (20.15)$$

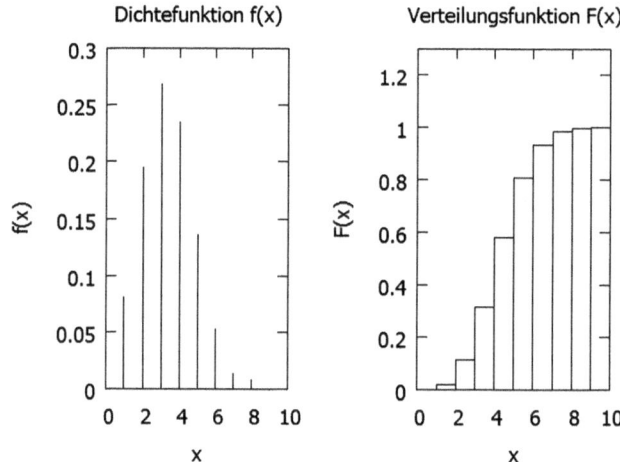

Fig. 20.4 Probability function and distribution function of a hypergeometric distribution with the characteristic values $N = 150$, $M = 50$ and $n = 10$

Therefore, the standard deviation σ of the hypergeometric distribution is:

$$\sigma = \sqrt{\frac{nM(N-M)(N-n)}{N^2(N-1)}} = \sqrt{n \cdot p \cdot q \cdot \frac{(N-n)}{(N-1)}} \qquad (20.16)$$

> **So That Such Formulas Do Not "Fall from the Sky", Here Is the Corresponding Derivation**
>
> From the definition for the expected value
>
> $$\mu = \sum_{x=0}^{n} x \cdot f(x)$$ we obtain with the hypergeometric probability function according to Eq. (20.12):
>
> $$\mu = \sum_{x=0}^{n} x \frac{\binom{M}{x}\binom{N-M}{n-x}}{\binom{N}{n}} = \sum_{x=1}^{n} x \frac{\binom{M}{x}\binom{N-M}{n-x}}{\binom{N}{n}} = \sum_{x=1}^{n} x \frac{\frac{M}{x}\binom{M-1}{x-1}\binom{N-M}{n-x}}{\frac{N}{n}\binom{N-1}{n-1}}$$
>
> Since in the sum the first summand is equal to 0, we can start the sum with $x = 1$. For the binomial coefficient we use its property
>
> $$\binom{k}{l} = \frac{k}{l}\binom{k-1}{l-1}$$ both for the binomial coefficient of the denominator and in the numerator. x can now be truncated, and we subtract the constant factors before the sum. Then we get:
>
> $$\mu = \frac{nM}{N} \sum_{x=1}^{n} \frac{\binom{M-1}{x-1}\binom{N-M}{n-x}}{\binom{N-1}{n-1}} = \frac{nM}{N} \sum_{\tilde{x}=0}^{n-1} \frac{\binom{M-1}{\tilde{x}}\binom{N-1-(M-1)}{n-1-\tilde{x}}}{\binom{N-1}{n-1}} = n\frac{M}{N} = np$$
>
> By renaming the running index, $\left(\tilde{x} = x - 1\right)$ the new index takes all values from 0 to $n - 1$. The sum is the hypergeometric function $f\left(\tilde{x}\right)$; this sum has the value 1, because the sum of the probabilities—and that's what it's all about—is 1. (We *have to* find something).

20.4 · Hypergeometric Distribution

For the variance, we obtain by definition Eq. (20.2) by substituting the hypergeometric probability function $f(x)$:

$$\sigma^2 = \sum_{x=0}^{n}(x-\mu)^2 \frac{\binom{M}{x}\binom{N-M}{n-x}}{\binom{N}{n}} = \sum_{x=0}^{n}(x^2 - 2\mu x + \mu^2)\frac{\binom{M}{x}\binom{N-M}{n-x}}{\binom{N}{n}}$$

By multiplying out, three sums are obtained in the following:

$$\sigma^2 = \sum_{x=0}^{n} x^2 \frac{\binom{M}{x}\binom{N-M}{n-x}}{\binom{N}{n}} - 2\mu \sum_{x=0}^{n} x \frac{\binom{M}{x}\binom{N-M}{n-x}}{\binom{N}{n}} + \mu^2 \sum_{x=0}^{n} \frac{\binom{M}{x}\binom{N-M}{n-x}}{\binom{N}{n}}$$

The second sum gives the *expected value*, the third sum again the *sum value* (i.e. 1, which we just had), since we have again added the probabilities of *all* possible outcomes. (We will come to the *first sum in* a moment).

Thus, the overall result is:

$$\sigma^2 = \sum_{x=0}^{n} x^2 \frac{\binom{M}{x}\binom{N-M}{n-x}}{\binom{N}{n}} - 2\mu\mu + \mu^2 = \sum_{x=0}^{n} x \frac{\binom{M}{x}\binom{N-M}{n-x}}{\binom{N}{n}} - \mu^2$$

Using the properties for binomial coefficients, we get:

$$\sigma^2 = \sum_{x=0}^{n} x^2 \frac{\frac{M}{x}\binom{M-1}{x-1}\binom{N-M}{n-x}}{\frac{N}{n}\binom{N-1}{n-1}} - \mu^2 = \frac{nM}{N}\sum_{x=1}^{n} x \frac{\binom{M-1}{x-1}\binom{N-M}{n-x}}{\binom{N-1}{n-1}} - \mu^2$$

In the sum, we replace x with $(x-1) + 1$—adding a "clever 0" $(-1 + 1)$ is a trick often used in mathematics—and create two sums:

$$\sigma^2 = \frac{nM}{N}\sum_{x=1}^{n}(x-1)\frac{\binom{M-1}{x-1}\binom{N-M}{n-x}}{\binom{N-1}{n-1}}$$

$$+\frac{nM}{N}\sum_{x=1}^{n}\frac{\binom{M-1}{x-1}\binom{N-M}{n-x}}{\binom{N-1}{n-1}} - \mu^2$$

Here we replace the expression $x - 1$ with a new variable x, running from 0 to $n - 1$:

$$\sigma^2 = \frac{nM}{N} \sum_{x=0}^{n-1} x \cdot \frac{\binom{M-1}{x}\binom{N-M}{n-1-x}}{\binom{N-1}{n-1}}$$

$$+ \frac{nM}{N} \sum_{x=0}^{n-1} \frac{\binom{M-1}{x}\binom{N-M}{n-1-x}}{\binom{N-1}{n-1}} - \mu^2$$

The first sum thus gives the *mean of* the hypergeometric distribution:

$$(n-1)\frac{(M-1)}{(N-1)}$$

The second sum has—once again—the value 1.

$$\sigma^2 = \frac{nM}{N}\frac{(n-1)(M-1)}{(N-1)} + \frac{nM}{N} - \mu^2 = \frac{nM}{N}\frac{(n-1)(M-1)}{(N-1)} + \frac{nM}{N} - \left(\frac{nM}{N}\right)^2$$

$$\sigma^2 = \frac{nM}{N}\left(\frac{(n-1)(M-1)}{(N-1)} + 1 - \frac{nM}{N}\right) = \frac{nM}{N}\left(\frac{(n-1)(M-1)}{(N-1)} + \frac{N-nM}{N}\right)$$

$$\sigma^2 = \frac{nM}{N}\left(\frac{N(n-1)(M-1)}{N(N-1)} + \frac{(N-1)(N-nM)}{N(N-1)}\right)$$

$$= \frac{nM}{N}\left(\frac{N(n-1)(M-1) + (N-1)(N-nM)}{N(N-1)}\right)$$

$$\sigma^2 = \frac{nM}{N}\left(\frac{N(nM-n-M+1) + N^2 - NnM - N + nM}{N(N-1)}\right)$$

$$= \frac{nM}{N}\left(\frac{N^2 - nN - NM + nM}{N(N-1)}\right)$$

$$\sigma^2 = \frac{nM}{N}\left(\frac{N-M}{N}\right)\left(\frac{N-n}{N-1}\right) = \frac{nM}{N}\left(1 - \frac{M}{N}\right)\left(\frac{N-n}{N-1}\right)$$

Thus:

$$\sigma^2 = \frac{nM}{N}\left(1 - \frac{M}{N}\right)\left(\frac{N-n}{N-1}\right)$$

▶ Example

Let's assume we have 100 items that are produced with a reject probability of 8%. For quality control, we take a sample of 4 pieces and test it. For the hypergeometric distribution with parameters $N = 100$, $M = 8$ and $n = 4$, we obtain for the hypergeometric probability function $f(x)$ and the corresponding distribution function $F(x)$:

X	0	1	2	3	4
$f(x)$ in %	71.26	25.62	2.99	0.13	0.00
$F(x)$ in %	71.26	96.88	99.87	100.00	100.00

20.5 Poisson Distribution

A reasonable agreement for the acceptance of the delivery can be the presence of at most *one* defective piece in the sample. The probability of this is 96.88%.

Since the number of test pieces taken is much smaller than the total, their distribution hardly changes due to the fact that the objects are just *not* put back. The above test approximates an experiment with binomial distribution with parameters $n = 4$ and $p = 0.08$. The corresponding values of $f(x)$ and $F(x)$ for the binomial distribution are:

X	0	1	2	3	4
$f(x)$ in %	71.64	24.92	3.25	0.19	0.00
$F(x)$ in %	71.64	96.56	99.81	100.00	100.00

The probability for the occurrence of at most one faulty piece within the sample is 96.56%. We therefore obtain a similar result to the hypergeometric distribution.

> **So That It Is Clear When Now What Is Needed**
> 1. In the case of a random sample, where the drawn objects are *laid back*, the *binomial distribution* is to be applied.
> 2. Without laying back, the *hypergeometric* distribution is applied. But:
>   (a) For large values of N and under the condition $N \gg n$, the hypergeometric distribution transitions to the binomial distribution (with parameters n and $p = M/N$), since the probability p (= M/N) hardly changes even when the drawn objects are *no longer* returned.
>   (b) As a rule of thumb, if $n < 0.05\ N$, the (mathematically simpler) binomial distribution can be used instead of the hypergeometric distribution.

20.5 Poisson Distribution

For Bernoulli experiments, where the probability p for the occurrence of an event is very small and at the same time the size N is very large, we use the Poisson distribution. An important example of an event with such a small probability and at the same time a large ensemble is the radioactive decay of chemical elements (we dealt with this in Part III, among other things—think of age determination).

The characteristics of the elements are:
- *Decay* =: A with probability p and
- *Non-decay* =: $\bar{A}$ with probability $q = 1 - p$.

The probability function $f(x)$ of the binomial distribution for this experiment is:

$$f(x) = \binom{N}{x} p^x \cdot q^{N-x} = \frac{N!}{x! \cdot (N-x)!} p^x (1-p)^{N-x}$$

$$= \frac{N!}{x! \cdot (N-x)!} p^x \sum_{k=0}^{N-x} \binom{N-x}{k} (-p)^k$$

In the last part of the equation we use the binomial theorem for natural exponents and replace the binomial $(1-p)^{N-x}$ by a sum with exponents k. Since the number of decayed elements (x) is very small compared to the total number (N), the expression

$$\frac{N!}{(N-x)!} = N \cdot (N-1) \cdot \ldots \cdot (N-x+1) = N^x + O(N^{x-1})$$

can be approximated by the term N^x: When multiplying out the x factors, we get a sum of terms ordered by the powers of N. The highest power we get for N when multiplying out—and thus the most important term—is N^x, all other terms have a lower power. The next term is of order (hence the capital letter O) N^{x-1}, but we do not care about its prefactor and also about the further terms, since these terms are small compared to the first contribution, so the following approximations result:

$$f(x) \approx \frac{N^x}{x!} p^x \sum_k \binom{N-x}{k} (-p)^k \approx \frac{N^x}{x!} p^x \sum_{k=0}^{N-x} \frac{(N-x)!}{k!(N-x-k)!} (-p)^k$$

For the term

$$\frac{(N-x)!}{(N-x-k)!} = (N-x) \cdot (N-x-1) \cdot \ldots \cdot (N-x-k+1) = N^k + O(N^{k-1})$$

we can use the approximation N^k. Then we get:

$$f(x) \approx \frac{N^x}{x!} p^x \sum_k \frac{N^k}{k!} (-p)^k \approx \frac{(Np)^x}{x!} \sum_k \frac{(-Np)^k}{k!} \approx \frac{1}{x!} (Np)^x e^{-Np}$$

The sum represents the $\sum_k \frac{(-Np)^k}{k!}$ **MacLaurin's series expansion of** the e-function with the argument $-Np$. Thus, we obtain the probability function of the Poisson distribution:

$$f(x) = \frac{1}{x!} (Np)^x e^{-Np}$$

For the distribution function of $F(x)$ the Poisson distribution holds:

$$F(x) = e^{-Np} \sum_{k \le x} \frac{(Np)^k}{k!}$$

For the expected value we get: μ

$$\mu = e^{-Np} \sum_{x=0}^{N} x \frac{(Np)^x}{x!} = e^{-Np} \sum_{x=1}^{N} x \frac{(Np)^x}{x!} = e^{-Np} \sum_{x=1}^{N} \frac{(Np)^x}{(x-1)!}$$

$$\mu = e^{-Np} \sum_{x=1}^{N} \frac{(Np)^{x-1+1}}{(x-1)!} = Np e^{-Np} \sum_{x=1}^{N} \frac{(Np)^{x-1}}{(x-1)!}$$

So, for the *expected value of* a Poisson distribution:

$$\mu = N \cdot p \tag{20.17}$$

20.5 Poisson Distribution

With this result, the probability function and the distribution function of the Poisson distribution can be compactly represented:

$$f(x) = \frac{1}{x!}\mu^x e^{-\mu} \quad \text{und} \quad F(x) = e^{-\mu}\sum_{k \le x}\frac{\mu^k}{k!} \qquad (20.18)$$

The parameter of the Poisson distribution is the expected value ($E(X) = \mu = N \cdot p$ known from Eq. (20.17)).

For the variance $D^2(X) = \sigma^2$ of the Poisson distribution, we first obtain:

$$\sigma^2 = e^{-\mu}\sum_{x=0}^{N}(x-\mu)^2\frac{\mu^x}{x!} = e^{-\mu}\sum_{x=0}^{N}(x^2 - 2x\mu + \mu^2)\frac{\mu^x}{x!} = e^{-\mu}\sum_{x=0}^{N}\left(x^2\frac{\mu^x}{x!} - 2x\mu\frac{\mu^x}{x!} + \mu^2\frac{\mu^x}{x!}\right)$$

$$\sigma^2 = e^{-\mu}\sum_{x=0}^{N}x^2\frac{\mu^x}{x!} - 2\mu e^{-\mu}\sum_{x=1}^{N}x\frac{\mu^x}{x!} + \mu^2 e^{-\mu}\sum_{x=0}^{N}\frac{\mu^x}{x!} = e^{-\mu}\sum_{x=0}^{N}x^2\frac{\mu^x}{x!} - 2\mu e^{-\mu}\sum_{x=1}^{N}\frac{\mu^{x-1+1}}{(x-1)!} + \mu^2 e^{-\mu}\sum_{x=0}^{N}\frac{\mu^x}{x!}$$

$$\sigma^2 = e^{-\mu}\sum_{x=0}^{N}x^2\frac{\mu^x}{x!} - 2\mu^2 e^{-\mu}\sum_{x=1}^{N}\frac{\mu^{x-1}}{(x-1)!} + \mu^2 e^{-\mu}\sum_{x=0}^{N}\frac{\mu^x}{x!} = e^{-\mu}\sum_{x=0}^{N}x^2\frac{\mu^x}{x!} - 2\mu^2 e^{-\mu}\sum_{\tilde{x}=0}^{N-1}\frac{\mu^{\tilde{x}}}{\tilde{x}!} + \mu^2 e^{-\mu}\sum_{x=0}^{N}\frac{\mu^x}{x!}$$

For all sums:

$$\sum_{x=0}^{N}\frac{\mu^x}{x!} \approx e^{\mu}$$

Therefore, the last two sums add up to $-\mu^2$. Putting this in, we get:

$$\sigma^2 = e^{-\mu}\sum_{x=0}^{N}x\frac{\mu^x}{(x-1)!} - \mu^2 = e^{-\mu}\sum_{x=1}^{N}x\frac{\mu^{x-1+1}}{(x-1)!} - \mu^2 = \mu e^{-\mu}\sum_{x=1}^{N}x\frac{\mu^{x-1}}{(x-1)!} - \mu^2 = \mu e^{-\mu}\sum_{\tilde{x}=0}^{N-1}(\tilde{x}+1)\frac{\mu^{\tilde{x}}}{\tilde{x}!} - \mu^2$$

In the last transformation, we use as new variable $\tilde{x} = x-1$. Then we get:

$$\sigma^2 = \mu e^{-\mu}\sum_{x=0}^{N-1}(x+1)\frac{\mu^x}{x!} - \mu^2 = \mu e^{-\mu}\sum_{x=1}^{N-1}x\frac{\mu^x}{x!} + \mu e^{-\mu}\sum_{x=0}^{N-1}\frac{\mu^x}{x!} - \mu^2 = \mu e^{-\mu}\sum_{x=0}^{N-1}\frac{\mu^{x-1+1}}{(x-1)!} + \mu e^{-\mu}\sum_{x=0}^{N-1}\frac{\mu^x}{x!} - \mu^2$$

$$\sigma^2 = \mu^2 e^{-\mu}\sum_{x=0}^{N-1}\frac{\mu^{x-1}}{(x-1)!} + \mu e^{-\mu}\sum_{x=0}^{N-1}\frac{\mu^x}{x!} - \mu^2$$

The sums result—as before—in each case

$$\sum_{x=0}^{N-1}\frac{\mu^x}{x!} = e^{\mu}$$

This simplifies things, because then the *variance* $D2(X) = \sigma 2$ of the Poisson distribution holds:

$$\sigma^2 = \mu^2 e^{-\mu}\sum_{x=0}^{N-1}\frac{\mu^{x-1}}{(x-1)!} + \mu e^{-\mu}\sum_{x=0}^{N-1}\frac{\mu^x}{x!} - \mu^2 = \mu^2 + \mu - \mu^2 = \mu \qquad (20.19)$$

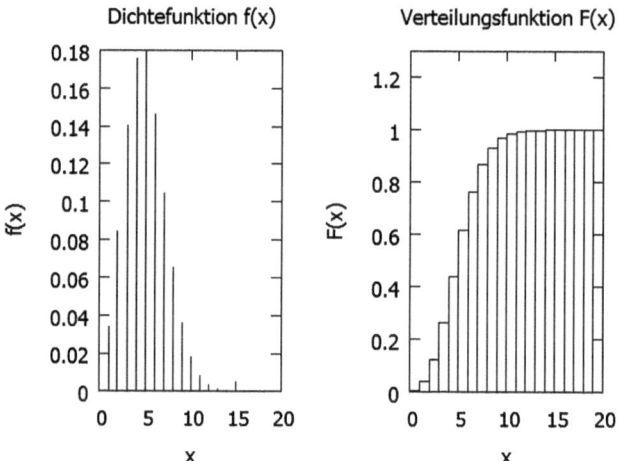

Fig. 20.5 Probability function and distribution function of a Poisson distribution with characteristic value $\mu = 5$

> Please note: With the Poisson distribution, the *expected value is* actually equal to the *variance of* this very distribution! (This is different from all previous measured value distributions).

Thus, the *standard deviation* of the Poisson distribution is:

$$\sigma = \sqrt{\mu}. \tag{20.20}$$

The probability function and the distribution function of the Poisson distribution are shown graphically in ◘ Fig. 20.5.

▶ Example

We have already referred to radioactive decay as an example of a Poisson distributed quantity. In the application we consider a radioactive substance with a decay constant (probability) of $k = 1.4 \cdot 10^{-11} \cdot s^{-1}$ and a quantity of $N = 1.5 \cdot 10^{11}$. The expected value is then $\mu = 1.4 \cdot 1.5\ s^{-1} = 2.1\ s^{-1}$, the corresponding probability**Error! Bookmark not defined.** function is $f(x) = (1/x!) \cdot 2.1^x \cdot e^{-2,1}$.

Listed below are some values for the probability**Error! Bookmark not defined.** function $f(x)$ and the distribution function $F(x)$ for various values of x:

20.5 Poisson Distribution

x	0	1	2	3	4	5
$f(x)$	$e^{-2.1} \approx 0.122\,46$	$2.1 \cdot e^{-2.1} \approx 0.257\,16$	$\dfrac{2.1^2}{2} \cdot e^{-2.1} \approx 0.270\,02$	$\dfrac{2.1^3}{6} \cdot e^{-2.1} \approx 0.189\,01$	$\dfrac{2.1^4}{24} \cdot e^{-2.1} \approx 0.099\,23$	$\dfrac{2.1^5}{120} \cdot e^{-2.1} \approx 0.041\,68$
$F(x)$	0.122 46	0.379 61	0.649 63	0.838 64	0.937 87	0.979 55

We can use this to address various problems: The question of how likely $n \geq 4$ atoms decay per second is solved by determining the distribution function $F(3)$ that determines the event that *fewer* than 4 atoms decay in 1 s:

$$F(3) = \sum_{k=0}^{3} \frac{1}{k!} 2.1^k e^{-2.1} = e^{-2.1} \sum_{k=0}^{3} \frac{1}{k!} 2.1^k = e^{-2.1} \cdot \left(1 + 2.1 + \frac{1}{2} \cdot 2.1^2 + \frac{1}{6} \cdot 2.1^3\right)$$
$$= e^{-2.1} \cdot 6.8485 = 0.8386$$

The probability that $n \geq 4$ atoms decay in 1 s is therefore
$P(X \geq 4) = 1 - F(3) = 1 - 0.8386 = 0.1614 = 16.14\,\%$.

> There are two things to consider about the Poisson distribution:
> - Characteristically, mean µ and variance σ^2 have the same value.
> - The binomial distribution may be replaced by the Poisson distribution if the conditions $Np < 10$ and $N > 1500p$ are satisfied.

20.6 Continuous Uniform Distribution

In this section, we introduce the uniform continuous distribution, the exponential distribution, the Gaussian normal distribution**Error! Bookmark not defined.**, and the log normal distribution.

Let us start with the quantities important in continuous**Error! Bookmark not defined.** distributions. A random variable X describing quantitatively continuous characteristics is usually represented by realisations $x \in B$. As in the discrete case, the probability distribution results from the distribution function $F(x)$ of a random variable and is defined as:

$$F(x) = P(X \leq x) = \int_{-\infty}^{x} f(t)\,dt$$

In the continuous case integrals are used and replace the sums of the discrete cases.

The probability that the random variable X takes a $B = [x_1, x_2]$ value in the interval is determined by:

$$P(x_1 \leq X \leq x_2) = \int_{x_1}^{x_2} f(x)\,dx = F(x_2) - F(x_1)$$

Thus, the probability of the event $X = a$ occurring is:

$$P(a \leq X \leq a) = \int_{a}^{a} f(x)\,dx = F(a) - F(a) = 0$$

This fact is an essential and fundamental difference to *discrete* distributions, where a value $f(a) = P(X = a) \geq 0$ can be assigned to a particular (single) event. Thus, this particular function $f(x)$ does *not* represent probability (i.e., $f(x)$ is not a probability**Error! Bookmark not defined.** function, since probability in the continuous case takes positive values only as an integral of $f(x)$ over a finite range). In the continuous uniform distribution, we therefore refer to $f(x)$ as the **probability density**, **density function** or *density of a continuous random variable X*.

Here $f(x)$ is a function with the properties

$$\int_{-\infty}^{\infty} f(x)\,dx = 1 \text{ (\textbf{completeness relation}) and } f(x) \geq 0,\ x \in R$$

20.6 Continuous Uniform Distribution

We calculate the mean value for continuous distributions with the equation

$$E(X) = \bar{x} = \int_{-\infty}^{\infty} x \cdot f(x) dx \qquad (20.21)$$

We obtain the variance $D^2(X) = \sigma^2$ using:

$$D^2(X) = \sigma^2 = \int_{-\infty}^{\infty} (x - \bar{x})^2 \cdot f(x) dx \qquad (20.22)$$

These integrals are typically performed over the set of real numbers. However, if the distribution differs from 0 only in a certain interval, the integrals are also calculated only over the interval in question.

Let us now turn to the individual continuous distributions:

- **Uniformly Continuous Distribution**

For a uniformly continuous distribution, the probability density $f(x)$ in an interval takes $B = [x_1, x_2]$ only a constant value $c > 0$ and otherwise has the value 0. From the completeness relation

$$\int_{x_1}^{x_2} f(x) dx = \int_{x_1}^{x_2} c \cdot dx = c \cdot (x_2 - x_1) = 1$$ we get the value for this constant:

$$f(x) = c = \frac{1}{(x_2 - x_1)}$$

The corresponding distribution function is

$$F(x) = \frac{(x - x_1)}{(x_2 - x_1)} \quad \text{for } x_1 \leq x \leq x_2$$

In the range $x < x_1$, $F(x) = 0$; for $x > x_2$, $F(x) = 1$.

An example of a uniformly continuous probability distribution with probability density including distribution function is shown in ◘ Fig. 20.6.

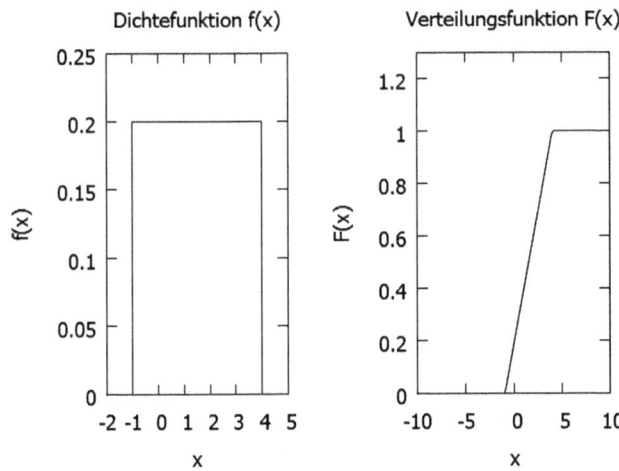

◘ **Fig. 20.6** Probability density and distribution function of a uniformly continuous distribution with the characteristic value $c = 0.2$

The mean value of this distribution lies—as one can easily imagine—exactly in the middle of the interval boundaries, i.e. from the integral for the calculation of the mean value

$$E(X) = \bar{x} = \int_{x_1}^{x_2} x \cdot f(x) dx$$

follows:

$$\bar{x} = \int_{x_1}^{x_2} x \cdot \frac{1}{x_2 - x_1} dx = \frac{1}{x_2 - x_1} \int_{x_1}^{x_2} x \cdot dx = \frac{1}{2} \cdot \frac{1}{x_2 - x_1} \cdot x^2 \bigg|_{x_1}^{x_2}$$

$$= \frac{1}{2} \cdot \frac{x_2^2 - x_1^2}{x_2 - x_1} = \frac{1}{2}(x_2 + x_1).$$

The variance $D^2(X) = \sigma^2$ is determined by calculating the following integral:

$$D^2(X) = \sigma^2 = \int_{x_1}^{x_2} (x - \bar{x})^2 \cdot f(x) dx = \int_{x_1}^{x_2} (x - \bar{x})^2 \cdot \frac{1}{x_2 - x_1} dx$$

Using the second binomial formula, we get for the variance:

$$\sigma^2 = \frac{1}{x_2 - x_1} \int_{x_1}^{x_2} \left(x^2 - 2\bar{x}x + \bar{x}^2 \right) dx = \frac{1}{x_2 - x_1} \left(\frac{1}{3} x^3 - \bar{x} x^2 + \bar{x}^2 x \right) \bigg|_{x_1}^{x_2}$$

The evaluation of the primitive function then yields:

$$\sigma^2 = \frac{1}{x_2 - x_1} \left(\frac{1}{3} x^3 - \bar{x} x^2 + \bar{x}^2 x \right) \bigg|_{x_1}^{x_2} = \frac{1}{x_2 - x_1} \left(\frac{x_2^3 - x_1^3}{3} - \bar{x}(x_2^2 - x_1^2) + \bar{x}^2 (x_2 - x_1) \right)$$

The last two summands simplify to:

$$-\bar{x}^2 \left(= -2\bar{x}^2 + \bar{x}^2 \right)$$

The first term results in:

$$\frac{1}{3} \cdot \frac{x_2^3 - x_1^3}{x_2 - x_1} = \frac{1}{3} \left(x_2^2 + x_2 x_1 + x_1^2 \right)$$

Together, then, we get:

$$\sigma^2 = \frac{1}{3}\left(x_2^2 + x_2 x_1 + x_1^2\right) - \frac{1}{4}(x_2 + x_1)^2 = \frac{1}{12}\left(4\left(x_2^2 + x_2 x_1 + x_1^2\right) - 3\left(x_2^2 + 2x_2 x_1 + x_1^2\right) \right)$$

This calculates to:

$$\sigma^2 = \frac{x_2^2 - 2x_2 x_1 + x_1^2}{12}$$

Overall, then, for the variance:

$$\sigma^2 = \frac{(x_2 - x_1)^2}{12} \tag{20.23}$$

20.7 Exponential Distribution

> Let's assume that you regularly travel a certain route by train (or use another means of transport belonging to the public transport system). Furthermore, you only know that the train runs there every 20 min—so not in the sense of wrong, but with the railway you never know. Let's also assume that you can't remember the departure times because they change regularly—as you know with the railway. Then your waiting time at the platform is a continuous, equally distributed random variable. If you are lucky, your waiting time is 00:00 min, because you just catch the train at the final spurt on the platform. If, however, the train doors are locked at the very moment of your arrival, then you are unlucky and the waiting time is 20:00 min. If you arrive on another occasion at an even different time, then *all waiting times in the* interval from 0 to 20 min (e.g. 12:42 min, 0:23 min, etc.) can occur, because the (waiting) time is a continuous variable—we want to disregard relativistic effects such as time dilation, which can be observed very often at railway stations, at this point.

20.7 Exponential Distribution

The continuous distribution needed to describe the lifetime of technical devices is called exponential distribution.

The probability density is $f(x) = \lambda e^{-\lambda x}$ with $x \geq 0$. For the range $x < 0$, $f(x) = 0$.

The distribution function here is:

$$F(x) = \int_0^x \lambda e^{-\lambda t} dt = -e^{-\lambda t}\Big|_0^x = 1 - e^{-\lambda x}$$

Here λ is the parameter that characterises the exponential distribution. The course of the density and distribution function is shown as an example in ◘ Fig. 20.7.

The mean value for the exponential distribution is calculated with:

$$E(X) = \bar{x} = \int_0^\infty x \cdot f(x) dx \qquad (20.24)$$

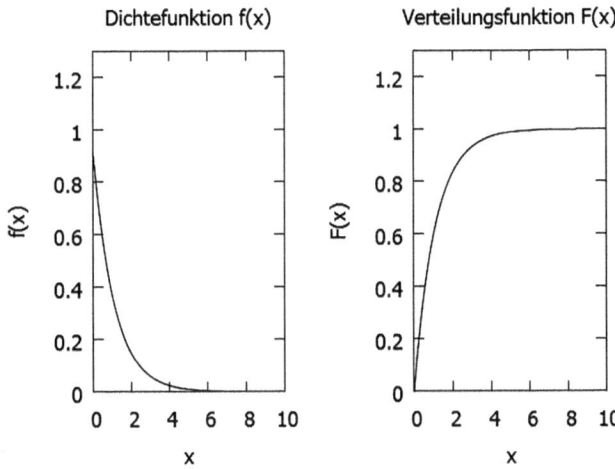

◘ **Fig. 20.7** Density and distribution function of an exponential distribution with the characteristic value $\lambda = 0.9$

This gives us:

$$\bar{x} = \int_0^\infty x \cdot \lambda e^{-\lambda x} dx = \left(-xe^{-\lambda x}\right)\Big|_0^\infty + \int_0^\infty e^{-\lambda x} dx = -xe^{-\lambda x} - \frac{1}{\lambda}e^{-\lambda x}\Big|_0^\infty = \frac{1}{\lambda}$$

For an *estimate of* the parameter λ, the reciprocal of the mean is used $\bar{x}$.

We calculate the variance $D^2(X) = \sigma^2$ of the exponential distribution from the definition, which you have here as Eq. (20.25):

$$\sigma^2 = \int_0^\infty (x - \bar{x})^2 \cdot \lambda e^{-\lambda x} dx = \int_0^\infty \left(x^2 - 2\bar{x}x + \bar{x}^2\right) \cdot \lambda e^{-\lambda x} dx \quad (20.25)$$

The evaluation of the integral yields:

$$\sigma^2 = -\left(x^2 - 2\bar{x}x + \bar{x}^2\right) e^{-\lambda x}\Big|_0^\infty + 2\int_0^\infty (x - \bar{x}) e^{-\lambda x} dx$$

$$\sigma^2 = -\left(x^2 - 2\bar{x}x + \bar{x}^2\right) e^{-\lambda x}\Big|_0^\infty + 2\left(\frac{x - \bar{x}}{-\lambda}\right) e^{-\lambda x}\Big|_0^\infty + \frac{2}{\lambda}\int_0^\infty e^{-\lambda x} dx$$

$$\sigma^2 = \left(-x^2 + 2\bar{x}x - \bar{x}^2 + 2\left(\frac{x-\bar{x}}{-\lambda}\right) - \frac{2}{\lambda^2}\right) e^{-\lambda x}\Big|_0^\infty = \bar{x}^2 - 2\frac{\bar{x}}{\lambda} + \frac{2}{\lambda^2} = \frac{1}{\lambda^2}$$

In short, once we have estimated the value of λ, the variance is calculated according to:

$$\sigma^2 = 1/\lambda^2 \quad (20.26)$$

▶ **Example**

The average useful life of an LED lamp is 50,000 h; the life of the luminaires is exponentially distributed. According to the above, we can determine the parameter/characteristic value λ of the exponential distribution from the reciprocal of the mean service life:

$$\lambda = \frac{1}{50,000} \frac{1}{h}$$

Using the exponential distribution, we can now calculate the percentage of lamps that operate for less than 30,000 h. From the distribution function for the exponential distribution
$F(x) = 1 - e^{-\lambda x}$ we get the probability
$P(X < 30,000) = F(30,000) = 1 - e^{-30,000/50,000} \approx 0.4512$.
Around 45% of lamps therefore fail after a service life of less than 30,000 h.
We obtain the probability that a lamp burns longer than 70,000 h from:
$P(X > 70,000) = 1 - F(70,000) = 1 - (1 - e^{-70,000/50,000}) \approx 0.2466$.
This means that only just under 25% of the luminaires burn for more than 70,000 h. ◀

So far in this chapter you have become acquainted with the basics, principles and most important parameters for the description of measured value distributions. Now we can turn to the two core elements for these very purposes, because as helpful (and hopefully comprehensible!) as the previous considerations have been, they serve primarily to ensure that you also understand the two most important methods, which are of incomparably greater importance than any of the previous ones. (The fact that they are mentioned comparatively late should not lead you to the erroneous assumption that they are mentioned here "only for the sake of completeness": You will encounter the contents of ▶ Sects. 20.8 and 20.9 again and again when it comes to statistical considerations).

20.8 Gaussian Normal Distribution

One of the most important and central distributions is the Gaussian normal distribution, also called the general **normal distribution.** (If you are wondering why, despite its importance, it is "only now" being treated, here is our explanation: First we wanted to give you an easy introduction to this topic with the simpler, discrete distributions and then steadily increase in complexity in the following sections. Now we have arrived at the first of the two highlights of statistics [which, of course, Harris and Skoog also discuss]—you will learn about the second in the immediately following ▶ Sect. 20.9. Back to the topic).

Harris, Section 4.1: Gaussian distribution
Skoog, Appendix A.2: Statistical treatment of random errors

The normal distribution is characterised by two parameters:
- Expected value $E(X) = \mu$ and
- Variance $D^2(X) = \sigma^2$.

The *density function* $f(x)$ of the normal distribution is given by:

$$f(x) = \frac{1}{\sqrt{2\pi}\sigma} e^{-\frac{1}{2}\left(\frac{x-\mu}{\sigma}\right)^2} \quad \text{for} \quad -\infty <; x < \infty \tag{20.27}$$

The *probability distribution of* $F(x)$ the normal distribution is the integral over it:

$$F(x) = \int_{-\infty}^{x} f(t) dt = \frac{1}{\sqrt{2\pi}\sigma} \int_{-\infty}^{x} e^{-\frac{1}{2}\left(\frac{t-\mu}{\sigma}\right)^2} dt \tag{20.28}$$

The course of this function is shown in Fig. ▶ 20.8 as an example for the values $\mu = 0.9$ and $\sigma = 1.0$.

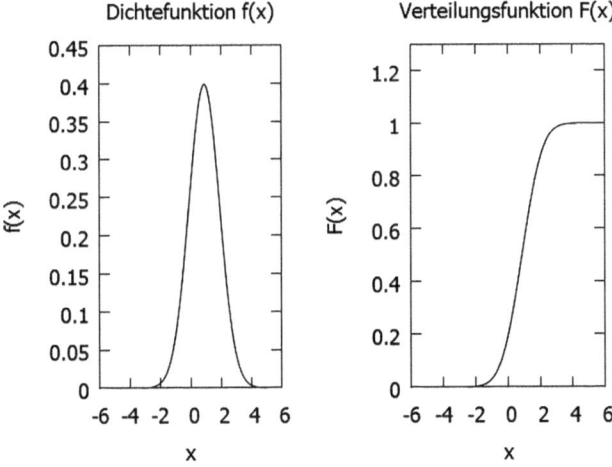

☐ **Fig. 20.8** Density function and distribution function of the Gaussian normal distribution with characteristic values $\mu = 0.9$ and $\sigma = 1.0$

The density function is also called a **bell curve distribution**—located axisymmetrically around μ—as can be easily seen from the graphical representation.

> The mean of this distribution is denoted by μ and the variance again by σ^2.

Through the transformation

$$t = \frac{x - \mu}{\sigma}$$

the density function $f(x)$ is transformed into the so-called *standard normal distribution*. This then has the mean $\mu = 0$ and the standard deviation $\sigma = 1$. The corresponding *density function is*:

$$f(x) = \frac{1}{\sqrt{2\pi}} e^{-\frac{1}{2}x^2}$$

The values of the associated *distribution function*

$$F(x) = \frac{1}{\sqrt{2\pi}} \int_{-\infty}^{x} e^{-\frac{1}{2}t^2} dt$$

are available in tabulated form (for mathematical interest, please refer to the box "Solution of the integral"), so that the integration does not have to be (explicitly) performed.

> **Solution of the Integral**
> To solve the integral, which cannot be represented in closed form, integration is performed by **series expansion** and the result is given by the error function
> $$F(x) = \frac{1}{2} + erf(x) \text{ is specified. Thereby is}$$
> $erf(x) = \frac{2}{\sqrt{\pi}} \int_0^x e^{-t^2} dt$ the so-called *error function*; alternatively, numerical integration methods can be used to calculate the integral.

For the total probability over all possible measured values (whose number and also their respective values are of course smaller than infinity, otherwise the measuring device burns us out), applies:

$$P(x < \infty) = \frac{1}{\sqrt{2\pi}} \int_{-\infty}^{\infty} e^{-\frac{1}{2}t^2} dt = 1 \tag{20.29}$$

Due to symmetry, the probability of negative arguments (i.e., with measured values <0) is:

$$F(-x) = \frac{1}{\sqrt{2\pi}} \int_{-\infty}^{-x} e^{-\frac{1}{2}t^2} dt = 1 - F(x) \tag{20.30}$$

So we only need to take the values for x from ◘ Table 20.1 and insert them. (If we know the value of the distribution function at the point x (with $x > 0$), we also know the value for its "negative counterpart", i.e. $-x$, via the relationship from Eq. (20.30)).

Explanation: The value of $\varphi(x)$ associated with a given *x-value* (with two decimal places) is obtained by first finding the relevant *row* (with the first decimal place) and then finding the relevant *column* (with the second decimal place)

20.8 Gaussian Normal Distribution

Table 20.1 Values of the standard normal distribution calculated with a spreadsheet program

	0	0.01	0.02	0.03	0.04	0.05	0.06	0.07	0.08	0.09
0.0	0.500000	0.503989	0.507978	0.511966	0.515953	0.519939	0.523922	0.527903	0.531881	0.535856
0.1	0.539828	0.543795	0.547758	0.551717	0.555670	0.559618	0.563559	0.567495	0.571424	0.575345
0.2	0.579260	0.583166	0.587064	0.590954	0.594835	0.598706	0.602568	0.606420	0.610261	0.614092
0.3	0.617911	0.621720	0.625516	0.629300	0.633072	0.636831	0.640576	0.644309	0.648027	0.651732
0.4	0.655422	0.659097	0.662757	0.666402	0.670031	0.673645	0.677242	0.680822	0.684386	0.687933
0.5	0.691462	0.694974	0.698468	0.701944	0.705401	0.708840	0.712260	0.715661	0.719043	0.722405
0.6	0.725747	0.729069	0.732371	0.735653	0.738914	0.742154	0.745373	0.748571	0.751748	0.754903
0.7	0.758036	0.761148	0.764238	0.767305	0.770350	0.773373	0.776373	0.779350	0.782305	0.785236
0.8	0.788145	0.791030	0.793892	0.796731	0.799546	0.802337	0.805105	0.807850	0.810570	0.813267
0.9	0.815940	0.818589	0.821214	0.823814	0.826391	0.828944	0.831472	0.833977	0.836457	0.838913
1.0	0.841345	0.843752	0.846136	0.848495	0.850830	0.853141	0.855428	0.857690	0.859929	0.862143
1.1	0.864334	0.866500	0.868643	0.870762	0.872857	0.874928	0.876976	0.879000	0.881000	0.882977
1.2	0.884930	0.886861	0.888768	0.890651	0.892512	0.894350	0.896165	0.897958	0.899727	0.901475
1.3	0.903200	0.904902	0.906582	0.908241	0.909877	0.911492	0.913085	0.914657	0.916207	0.917736
1.4	0.919243	0.920730	0.922196	0.923641	0.925066	0.926471	0.927855	0.929219	0.930563	0.931888
1.5	0.933193	0.934478	0.935745	0.936992	0.938220	0.939429	0.940620	0.941792	0.942947	0.944083
1.6	0.945201	0.946301	0.947384	0.948449	0.949497	0.950529	0.951543	0.952540	0.953521	0.954486
1.7	0.955435	0.956367	0.957284	0.958185	0.959070	0.959941	0.960796	0.961636	0.962462	0.963273
1.8	0.964070	0.964852	0.965620	0.966375	0.967116	0.967843	0.968557	0.969258	0.969946	0.970621
1.9	0.971283	0.971933	0.972571	0.973197	0.973810	0.974412	0.975002	0.975581	0.976148	0.976705
2.0	0.977250	0.977784	0.978308	0.978822	0.979325	0.979818	0.980301	0.980774	0.981237	0.981691
2.1	0.982136	0.982571	0.982997	0.983414	0.983823	0.984222	0.984614	0.984997	0.985371	0.985738
2.2	0.986097	0.986447	0.986791	0.987126	0.987455	0.987776	0.988089	0.988396	0.988696	0.988989
2.3	0.989276	0.989556	0.989830	0.990097	0.990358	0.990613	0.990863	0.991106	0.991344	0.991576
2.4	0.991802	0.992024	0.992240	0.992451	0.992656	0.992857	0.993053	0.993244	0.993431	0.993613
2.5	0.993790	0.993963	0.994132	0.994297	0.994457	0.994614	0.994766	0.994915	0.995060	0.995201

(continued)

Table 20.1 (continued)

	0	0.01	0.02	0.03	0.04	0.05	0.06	0.07	0.08	0.09
2.6	0.995339	0.995473	0.995604	0.995731	0.995855	0.995975	0.996093	0.996207	0.996319	0.996427
2.7	0.996533	0.996636	0.996736	0.996833	0.996928	0.997020	0.997110	0.997197	0.997282	0.997365
2.8	0.997445	0.997523	0.997599	0.997673	0.997744	0.997814	0.997882	0.997948	0.998012	0.998074
2.9	0.998134	0.998193	0.998250	0.998305	0.998359	0.998411	0.998462	0.998511	0.998559	0.998605
3.0	0.998650	0.998694	0.998736	0.998777	0.998817	0.998856	0.998893	0.998930	0.998965	0.998999
3.1	0.999032	0.999065	0.999096	0.999126	0.999155	0.999184	0.999211	0.999238	0.999264	0.999289
3.2	0.999313	0.999336	0.999359	0.999381	0.999402	0.999423	0.999443	0.999462	0.999481	0.999499
3.3	0.999517	0.999534	0.999550	0.999566	0.999581	0.999596	0.999610	0.999624	0.999638	0.999651
3.4	0.999663	0.999675	0.999687	0.999698	0.999709	0.999720	0.999730	0.999740	0.999749	0.999758
3.5	0.999767	0.999776	0.999784	0.999792	0.999800	0.999807	0.999815	0.999822	0.999828	0.999835
3.6	0.999841	0.999847	0.999853	0.999858	0.999864	0.999869	0.999874	0.999879	0.999883	0.999888
3.7	0.999892	0.999896	0.999900	0.999904	0.999908	0.999912	0.999915	0.999918	0.999922	0.999925
3.8	0.999928	0.999931	0.999933	0.999936	0.999938	0.999941	0.999943	0.999946	0.999948	0.999950
3.9	0.999952	0.999954	0.999956	0.999958	0.999959	0.999961	0.999963	0.999964	0.999966	0.999967
4.0	0.999968	0.999970	0.999971	0.999972	0.999973	0.999974	0.999975	0.999976	0.999977	0.999978
	0	0.01	0.02	0.03	0.04	0.05	0.06	0.07	0.08	0.09

20.8 Gaussian Normal Distribution

Example: For $x = 1.96$, you first look for the row with 1.9 (so you orient yourself to what is in the first *column*) and then the column with the corresponding second decimal place (which you find in the first *row*). For $x = 1.96$, the result is $\varphi(x) = 0.975002$

Conversely, for a given value of $\varphi(x)$, one can also determine the corresponding *x-value* (with two decimal places) by adding to the value from the *first column* of the row in which said value of $\varphi(x)$ is located the second decimal place belonging to the *first row*

Example: For $\varphi(x) = 0.99224$ the result is $x = 2.42$

Because of the great importance of the normal distribution, a single representation/designation for this function is not sufficient—*variatio delectat*. It has various names and is also often given its own formula symbol (φ) in numerous textbooks. The same applies to the density function:

$$\varphi(x;,\mu;,\sigma^2) = f\left(\frac{x-\mu}{\sigma}\right)$$

For the corresponding distribution function then holds:

$$\phi(x;,\mu;,\sigma^2) = F\left(\frac{x-\mu}{\sigma}\right) = \varphi\left(\frac{x-\mu}{\sigma}\right)$$

This representation is based on a transformation of the measurand X with $X = Z - \sigma + \mu X$, which allows to infer from the standard normal distribution of the quantity Z with characteristic values $\mu = 0$, $\sigma = 1$ to the normal distribution with values different from it for the parameters μ, σ. The probability for the quantity X is then given by:

$$P(x_1 \leq X \leq x_2) = P\left(\frac{x_1-\mu}{\sigma} \leq Z \leq \frac{x_2-\mu}{\sigma}\right) = \varphi\left(\frac{x_2-\mu}{\sigma}\right) - \varphi\left(\frac{x_1-\mu}{\sigma}\right)$$

We obtain the probability that the deviation of the measurand X from the mean μ is absolutely smaller than a given value $\varepsilon = k - \sigma$, (i.e., k *times the* standard deviation σ) from:

$$P(|X-\mu| \leq k \cdot \sigma) = P(-k \leq X \leq k) = \varphi(k) - \varphi(-k) = 2 \cdot \varphi(k) - 1$$

▶ **Example**

The probability of a measured variable X *being* within the simple standard deviation σ around the expected value μ is:

$P(|X - \mu| \leq 1 \cdot \sigma) = 2 \cdot \varphi(1) - 1 = 2 \cdot 0.841\ 345 - 1 = 0.682\ 69$

This means that about 68.27% of the readings are within the σ-surround of the mean.

Similarly, we can calculate that 95.45% of the readings occur in a 2σ environment and 99.73% of the results occur in a 3σ—environment. You already know these confidence intervals from ▶ Fig. 18.3.

μThen within what neighbourhood of are 50% of the readings to be found?—It must apply:

$P(|X - \mu| \leq k \cdot \sigma) = 2 \cdot \varphi(k) - 1 = 0.5$

It follows that: $\phi(k) = 0.75$. From the table we see that k must lie between 0.67 and 0.68. In other words: In an environment of slightly more than 0.67 σ, 50% of the measurement results are located. ◀

Such considerations are important, for example, in *statistical quality assurance*.

> **Example**
>
> The well-known active ingredient acetylsalicylic acid is often administered in tablet form with 500 mg. The figure of 500 mg is of course only the mean value of the amount of active ingredient, in fact there may be more or less of the active ingredient in a tablet.
>
> De facto, with a production of several million (or more) tablets, the quantity of an active ingredient follows a Gaussian distribution. However, the consumer only knows one parameter of this distribution, namely the mean value. (However, you do not need to worry about this now: The legislator precisely defines the limits for both tablet weights and the respective content via the pharmacopoeia—and these in Pharm. Eur. [the European Pharmacopoeia] are *based on the* Gaussian function. However, you will not find this information on the package insert—perhaps this would also be a little … confusing). ◄

For the Sake of Completeness: The Corresponding Derivation

$$F(x) = \frac{1}{\sqrt{2\pi}} \int_{-\infty}^{x} e^{-\frac{1}{2}t^2} dt = \frac{1}{2} + \frac{1}{\sqrt{2\pi}} \int_{0}^{x} e^{-\frac{1}{2}t^2} dt$$

The integral is calculated by series expansion of the *e-function*:

$$F(x) = \frac{1}{2} + \frac{1}{\sqrt{2\pi}} \int_{0}^{x} e^{-\frac{1}{2}t^2} dt = \frac{1}{2} + \frac{1}{\sqrt{2\pi}} \int_{0}^{x} \left(1 - \frac{t^2}{2} + \frac{t^4}{2^2 \cdot 2!} - \frac{t^6}{2^3 \cdot 3!} + \frac{t^8}{2^4 \cdot 4!} - + \ldots \right) dt$$

$$F(x) = \frac{1}{2} + \frac{1}{\sqrt{2\pi}} \left(t - \frac{t^3}{2 \cdot 3} + \frac{t^5}{2^2 \cdot 5 \cdot 2!} - \frac{t^7}{2^3 \cdot 7 \cdot 3!} + \frac{t^9}{2^4 \cdot 9 \cdot 4!} - + \ldots \right) \Big|_{0}^{x}$$

$$F(x) = \frac{1}{2} + \frac{1}{\sqrt{2\pi}} \left(x - \frac{x^3}{2 \cdot 3} + \frac{x^5}{2^2 \cdot 5 \cdot 2!} - \frac{x^7}{2^3 \cdot 7 \cdot 3!} + \frac{x^9}{2^4 \cdot 9 \cdot 4!} - + \ldots \right)$$

Gaussians are generally regarded as the ultimate, but in fact the logarithm is used whenever a distribution is *not symmetrical*. If, for example, we are interested in the particle size distribution in different grinding processes or in the distribution of metabolite concentrations in the blood, we cannot get any further with the Gaussian distribution and have to resort to the *logarithmic* normal distribution.

20.9 Logarithmic Normal Distribution

In numerous applications, the measured quantities have only positive values, i.e. $X > 0$. If it is also true that the logarithm of this quantity $Y = \ln(X)$ is normally distributed, the distribution of X is described by the logarithmic normal distribution. The density of the distribution is

$$f(x) = \frac{1}{\sqrt{2\pi}\sigma \cdot x} e^{-\frac{1}{2}\left(\frac{\ln(x)-\mu}{\sigma}\right)^2} \quad \text{for } x > 0, \text{ otherwise } f(x) = 0.$$

The corresponding distribution function $F(x)$ is:

$$F(x) = \int_{0}^{x} f(t) dt = \frac{1}{\sqrt{2\pi}\sigma} \int_{0}^{x} \frac{1}{t} e^{-\frac{1}{2}\left(\frac{\ln(t)-\mu}{\sigma}\right)^2} dt$$

20.9 Logarithmic Normal Distribution

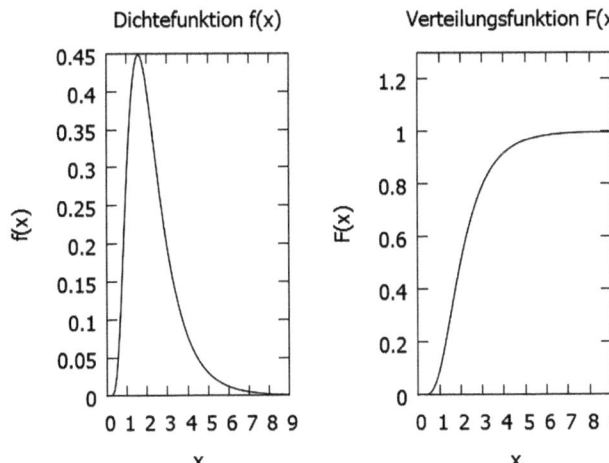

Fig. 20.9 Probability density and distribution function of the log normal distribution with characteristic values $\mu = 0.7$ and $\sigma = 0.5$

The characteristic values of the distribution are once again the parameters μ and σ. For the values $\mu = 0.7$ and $\sigma = 0.5$, the density and distribution functions are shown as an example in ◘ Fig. 20.9.

The expected value $E(X)$ is

$$E(X) = e^{\mu + \frac{\sigma^2}{2}} \tag{20.31}$$

The associated variance $D^2(X) = \sigma^2$ is given by:

$$D^2(X) = e^{2\mu + \sigma^2} \left(e^{\sigma^2} - 1 \right) \tag{20.32}$$

The logarithmic normal distribution is used, among other things, to investigate the distribution of particle size sizes in colloid chemistry, but also in material testing of building material properties or to determine the growths of crystals. Even astrophysicists use this approach: The distribution of galaxies in the universe can also be described with it.

Of course, this is by no means the end of the subject of measured value distributions; there would be a multitude of other distributions to mention, but once again the framework should not be blown up excessively. Only the **Weibull distribution** should be mentioned by name, because it is preferably used to determine and describe the stability (robustness) of modern materials such as (high-performance) ceramics and the like. (In corresponding textbooks of materials science and/or analysis you will stumble over the Weibull module of corresponding materials again and again).

❓ Questions

5. You toss 10 coins into the air, each of which will show heads or tails with exactly equal probability after bouncing. (No, you do not get any coins standing on the edge!) What is the probability of getting exactly 5 × heads, 5 × tails? What are the odds of 6 × heads 4 × tails? How (un)likely is the outcome 10 × heads, 0 × tails?
6. One unspecified measuring instrument was checked at different times to see whether it was in working order or not. It was checked a total of 443 times, and in 23 cases it was found to be inoperable. How likely is it that you can now work with it?

Parameter Estimates

Contents

21.1 Chi-Square Distribution (χ^2-Distribution) – 278

21.2 Student t-Distribution – 281

21.3 Estimation Methods – 284

21.4 Maximum Likelihood Method – 288

21.5 Confidence Intervals for the Unknown Parameters ϑ of a Distribution – 291

21.6 Parameter Tests – 299

As you will have noticed in ▶ Chapters 17–20, in statistics we are not always dealing with intermediate results or parameters that can be calculated in concrete terms. Now and then—because the calculation would be too costly or is simply impossible due to the large number of factors to be taken into account—we have no choice but to make reasonable *estimates with* regard to certain parameters. This applies in particular if these are to represent the bases of **estimation functions** for the determination of unknown parameters of measured value distributions.

However, before we get to know procedures with the help of which we can determine characteristics and parameters of statistical distributions, we direct our attention to two **test distributions**. We will need these for the actual parameter tests.

21.1 Chi-Square Distribution (χ^2-Distribution)

For the presence of a chi-squared distribution, it is necessary that the **stochastically** independent random variables $X_1, X_2, \ldots, X_n$ each have the standard normal distribution. From these random variables we form a new random variable $Z = \chi^2$ for which holds: $Z = \chi^2 = X_1^2 + X_2^2 + \ldots + X_n^2$.

Thus, by the symbol χ^2 it is indicated that the random variables X_i are squared. The values z of this new random variable $Z = \chi^2$ are continuously distributed; furthermore, $z \geq 0$. The density function $f(z)$ of this χ^2-distribution is given by:

$$f(z) = \begin{cases} A_n \cdot z^{\left(\frac{n-2}{2}\right)} \cdot e^{-\frac{z}{2}}, & z > 0 \\ 0 & z \leq 0 \end{cases} \tag{21.1}$$

In this formula, the parameter n is called the degree of freedom of the distribution (you already know this term from ▶ Sect. 18.2) and indicates how many random variables X are included in the distribution. A_n is a **normalisation constant** so that the area under the distribution (i.e. the distribution function of the χ^2-distribution) takes the value 1. For A_n holds:

$$A_n = \frac{1}{2^{\left(\frac{n}{2}\right)} \cdot \Gamma\left(\frac{n}{2}\right)},$$ here is the $\Gamma\left(\frac{n}{2}\right)$ **gamma function**.

The distribution function $F(z)$ of the χ^2-distribution is then the integral over the density function (i.e. the probability density from Eq. (21.1)) with the range of values from 0 to the value z we are interested in:

$$F(z) = \frac{1}{2^{\left(\frac{n}{2}\right)} \cdot \Gamma\left(\frac{n}{2}\right)} \cdot \int_0^z t^{\left(\frac{n-2}{2}\right)} \cdot e^{-\frac{t}{2}} dt, \quad z \geq 0$$

The typical shape of the χ^2-distribution is shown for different degrees of freedom in ◘ Fig. 21.1.

Mean μ and variance σ^2 of the chi-squared distribution thus depend solely on the degree of freedom $f = n$:
- The mean value μ has the value $\mu = n$;
- the variance is $\sigma^2 = 2n$.

21.1 · Chi-Square Distribution (χ^2-Distribution)

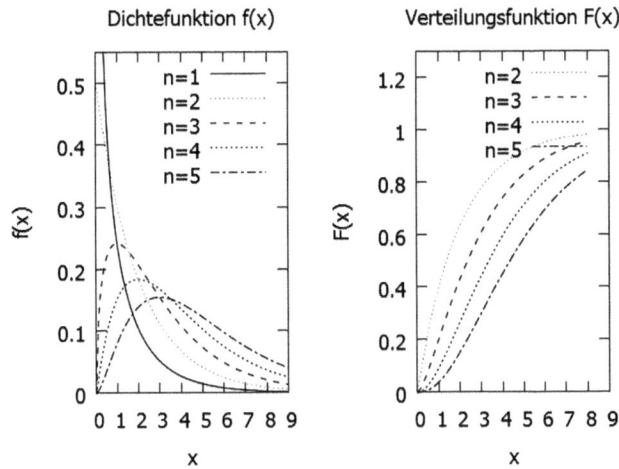

Fig. 21.1 Probability function of the χ^2-distribution for $n = 1, \ldots, 5$ and distribution function for $n = 2, \ldots, 5$

Also Here the Derivation Should Not Be Missing

For the mean value

$$E(X) = \mu = \int_0^\infty x \cdot f(x)\,dx$$

we obtain with the probability density of the chi-squared distribution:

$$E(X) = \mu = \frac{1}{2^{\left(\frac{n}{2}\right)} \Gamma\left(\frac{n}{2}\right)} \int_0^\infty x \cdot x^{\left(\frac{n-2}{2}\right)} e^{-\frac{x}{2}}\,dx = \frac{1}{2^{\left(\frac{n}{2}\right)} \Gamma\left(\frac{n}{2}\right)} \int_0^\infty x^{\left(\frac{n}{2}\right)} e^{-\frac{x}{2}}\,dx$$

$$E(X) = \mu = \frac{1}{2^{\left(\frac{n}{2}\right)} \Gamma\left(\frac{n}{2}\right)} \left[\underbrace{-2x^{\left(\frac{n}{2}\right)} e^{-\frac{x}{2}} \Big|_0^\infty}_{=0} + 2\frac{n}{2} \int_0^\infty x^{\left(\frac{n-2}{2}\right)} e^{-\frac{x}{2}}\,dx \right]$$

$$= n \cdot \underbrace{\frac{1}{2^{\left(\frac{n}{2}\right)} \Gamma\left(\frac{n}{2}\right)} \int_0^\infty x^{\left(\frac{n-2}{2}\right)} e^{-\frac{x}{2}}\,dx}_{=F(x \to \infty)=1}$$

The first summand in the square bracket becomes equal to 0 at both the upper and lower bounds. The remaining integral from partial integration is the distribution function $F(x)$, which tends to 1 for $x \to \infty$. Thus the expected value $E(X) = \mu = n$.

The variance $D^2(X) = \sigma^2$ is calculated here according to:

$$\sigma^2 = \int_0^\infty (x - \mu)^2 \cdot f(x)\,dx$$

Thus, using the probability density of the chi-squared distribution, we obtain:

$$D^2(X) = \sigma^2 = \frac{1}{2^{\left(\frac{n}{2}\right)} \Gamma\left(\frac{n}{2}\right)} \int_0^\infty (x - \mu)^2 \cdot x^{\left(\frac{n-2}{2}\right)} e^{-\frac{x}{2}}\,dx$$

Multiplying out gives:
$$D^2(X) = \sigma^2 = \frac{1}{2^{\left(\frac{n}{2}\right)}\Gamma\left(\frac{n}{2}\right)} \int_0^\infty \left(x^{\left(\frac{n+2}{2}\right)} - 2\mu x^{\left(\frac{n}{2}\right)} + \mu^2 x^{\left(\frac{n-2}{2}\right)} \right) e^{-\frac{x}{2}} dx$$

We write the last expression in terms of three integrals (instead of having to deal with a sum over three integrals, we sum three integrals instead):
$$D^2(X) = \sigma^2 = \frac{1}{2^{\left(\frac{n}{2}\right)}\Gamma\left(\frac{n}{2}\right)} \left(\int_0^\infty x^{\left(\frac{n+2}{2}\right)} e^{-\frac{x}{2}} dx - 2\mu \int_0^\infty x^{\left(\frac{n}{2}\right)} e^{-\frac{x}{2}} dx + \mu^2 \int_0^\infty x^{\left(\frac{n-2}{2}\right)} e^{-\frac{x}{2}} dx \right)$$

If we dissolve the parentheses (which is ultimately multiplying out again), we get:
$$D^2(X) = \sigma^2 = \frac{1}{2^{\left(\frac{n}{2}\right)}\Gamma\left(\frac{n}{2}\right)} \int_0^\infty x^{\left(\frac{n+2}{2}\right)} e^{-\frac{x}{2}} dx - \frac{2\mu}{2^{\left(\frac{n}{2}\right)}\Gamma\left(\frac{n}{2}\right)} \int_0^\infty x^{\left(\frac{n}{2}\right)} e^{-\frac{x}{2}} dx$$
$$+ \underbrace{\frac{\mu^2}{2^{\left(\frac{n}{2}\right)}\Gamma\left(\frac{n}{2}\right)} \int_0^\infty x^{\left(\frac{n-2}{2}\right)} e^{-\frac{x}{2}} dx}_{=\mu^2}$$

What does this (admittedly gigantic and somewhat confusing) formula mean now?
- The *last* integral gives the distribution function multiplied by the square of the expected value.
- The *mean* integral again gives the formula for the expected value, so that we get the expression—2 μ – μ in total at the point.
- Let us now consider the *first* integral:

$$D^2(X) = \sigma^2 = \frac{1}{2^{\left(\frac{n}{2}\right)}\Gamma\left(\frac{n}{2}\right)} \int_0^\infty x^{\left(\frac{n+2}{2}\right)} e^{-\frac{x}{2}} dx - 2\mu \cdot \mu + \mu^2 = \frac{1}{2^{\left(\frac{n}{2}\right)}\Gamma\left(\frac{n}{2}\right)} \int_0^\infty x^{\left(\frac{n+2}{2}\right)} e^{-\frac{x}{2}} dx - \mu^2$$

$$D^2(X) = \sigma^2 = \frac{1}{2^{\left(\frac{n}{2}\right)}\Gamma\left(\frac{n}{2}\right)} \left(\underbrace{\left.-2x^{\left(\frac{n+2}{2}\right)} e^{-\frac{x}{2}}\right|_0^\infty}_{=0} + 2\frac{n+2}{2} \int_0^\infty x^{\left(\frac{n}{2}\right)} e^{-\frac{x}{2}} dx \right) - \mu^2$$

The first summand in the large round bracket is equal to 0 at the lower bound, and *converges* to 0 at the upper bound. In the remaining integral from partial integration, we replace $x^{n.2}$ with the following expression:
$$x \cdot x^{\frac{n-2}{2}}$$

(This is another one of those mathematical tricks where a term is multiplied by 1 (here by x/x) So we get:
$$D^2(X) = \sigma^2 = \frac{1}{2^{\left(\frac{n}{2}\right)}\Gamma\left(\frac{n}{2}\right)} (n+2) \int_0^\infty x \cdot x^{\left(\frac{n-2}{2}\right)} e^{-\frac{x}{2}} dx - \mu^2$$

$$= (n+2) \cdot \underbrace{\frac{1}{2^{\left(\frac{n}{2}\right)}\Gamma\left(\frac{n}{2}\right)} \int_0^\infty x \cdot x^{\left(\frac{n-2}{2}\right)} e^{-\frac{x}{2}} dx}_{=E(x)} - \mu^2$$

$$D^2(X) = \sigma^2 = (n+2) \cdot n - \mu^2 = n^2 + 2 \cdot n - n^2 = 2 \cdot n$$

> For a large number of degrees of freedom $f = n > 100$, the chi-squared distribution can be replaced by a Gauss's normal distribution with parameters $\mu = n$ and $\sigma^2 = 2n$.

The tabulated values of the distribution function (◘ Table 21.1) play an important role in the context of test and trial procedures.

Illustrative examples, which are computationally not very complex (and need only little space—this part is already long enough anyway!), can hardly be found on this topic. Instead, we take the liberty of referring you to an illustration (with a detailed calculation path) of the (highly recommendable!) online presence ▶ matheguru.com.

▶ http://matheguru.com/stochastik/248-chi-quadrat-test.html

21.2 Student t-Distribution

The *last* distribution we will look at is the Student t-distribution (yes, that is the t from ▶ Sect. 18.3). This is used when the sample size is relatively small, i.e. you only have a few values available, but you still want to get reasonably reliable statistics. It is therefore the basis of the estimation methods we will deal with in ▶ Sect. 21.3 (of course, Harris also has a lot to say about this).

The random variable underlying this distribution is

$$T = \frac{X}{\sqrt{Y/n}}$$

Harris, Section 4.3: Comparison of means with Student's t-test

Thereby are:
- X is a standard normally distributed random variable,
- Y is a χ^2-distributed random variable.

$$f(t) = \frac{\Gamma\left(\frac{n+1}{2}\right)}{\sqrt{n\pi}\cdot\Gamma\left(\frac{n}{2}\right)} \cdot \frac{1}{\left(1+\frac{t^2}{n}\right)^{\left(\frac{n+1}{2}\right)}} \text{ with } -\infty < t < \infty$$

The prefactor is again chosen so that the area under the density function is normalised to the value 1:

$$\int_{-\infty}^{\infty} f(t)\,dt = \frac{\Gamma\left(\frac{n+1}{2}\right)}{\sqrt{n\pi}\cdot\Gamma\left(\frac{n}{2}\right)} \cdot \int_{-\infty}^{\infty} \frac{1}{\left(1+\frac{t^2}{n}\right)^{\left(\frac{n+1}{2}\right)}}\,dt = 1$$

> **For Those Who Enjoy Science History Anecdotes**
> This probability distribution was developed by the British statistician William S. Gosset. The reason why it did not enter the literature as the "Gosset distribution" is that at that time its author worked in the quality control department of a brewery (the name of which shall remain mercifully unmentioned here—although it still produces delicious types of beer…) and this company did not permit scientific publications based on the company's own products because it was generally feared that company secrets might be betrayed. For this reason, Gosset decided to use a pseudonym and published the work under the name "Student".

For different degrees of freedom, the probability density and distribution functions are shown in ◘ Fig. 21.2.

Chapter 21 · Parameter Estimates

Table 21.1 Chi-square distribution

Confidence level Degree of freedom	0.001	0.005	0.01	0.025	0.05	0.1	0.9	0.95	0.975	0.99	0.995	0.999
1	0.0000	0.0000	0.0002	0.0010	0.0039	0.0158	2.7055	3.8415	5.0239	6.6349	7.8794	10.8276
2	0.0020	0.0100	0.0201	0.0506	0.1026	0.2107	4.6052	5.9915	7.3778	9.2103	10.5966	13.8155
3	0.0243	0.0717	0.1148	0.2158	0.3518	0.5844	6.2514	7.8147	9.3484	11.3449	12.8382	16.2662
4	0.0908	0.2070	0.2971	0.4844	0.7107	1.0636	7.7794	9.4877	11.1433	13.2767	14.8603	18.4668
5	0.2102	0.4117	0.5543	0.8312	1.1455	1.6103	9.2364	11.0705	12.8325	15.0863	16.7496	20.5150
6	0.3811	0.6757	0.8721	1.2373	1.6354	2.2041	10.6446	12.5916	14.4494	16.8119	18.5476	22.4577
7	0.5985	0.9893	1.2390	1.6899	2.1673	2.8331	12.0170	14.0671	16.0128	18.4753	20.2777	24.3219
8	0.8571	1.3444	1.6465	2.1797	2.7326	3.4895	13.3616	15.5073	17.5345	20.0902	21.9550	26.1245
9	1.1519	1.7349	2.0879	2.7004	3.3251	4.1682	14.6837	16.9190	19.0228	21.6660	23.5894	27.8772
10	1.4787	2.1559	2.5582	3.2470	3.9403	4.8652	15.9872	18.3070	20.4832	23.2093	25.1882	29.5883
11	1.8339	2.6032	3.0535	3.8157	4.5748	5.5778	17.2750	19.6751	21.9200	24.7250	26.7568	31.2641
12	2.2142	3.0738	3.5706	4.4038	5.2260	6.3038	18.5493	21.0261	23.3367	26.2170	28.2995	32.9095
13	2.6172	3.5650	4.1069	5.0088	5.8919	7.0415	19.8119	22.3620	24.7356	27.6882	29.8195	34.5282
14	3.0407	4.0747	4.6604	5.6287	6.5706	7.7895	21.0641	23.6848	26.1189	29.1412	31.3193	36.1233
15	3.4827	4.6009	5.2293	6.2621	7.2609	8.5468	22.3071	24.9958	27.4884	30.5779	32.8013	37.6973
16	3.9416	5.1422	5.8122	6.9077	7.9616	9.3122	23.5418	26.2962	28.8454	31.9999	34.2672	39.2524
17	4.4161	5.6972	6.4078	7.5642	8.6718	10.0852	24.7690	27.5871	30.1910	33.4087	35.7185	40.7902

21.2 · Student t-Distribution

df												
18	4.9048	6.2648	7.0149	8.2307	9.3905	10.8649	25.9894	28.8693	31.5264	34.8053	37.1565	42.3124
19	5.4068	6.8440	7.6327	8.9065	10.1170	11.6509	27.2036	30.1435	32.8523	36.1909	38.5823	43.8202
20	5.9210	7.4338	8.2604	9.5908	10.8508	12.4426	28.4120	31.4104	34.1696	37.5662	39.9968	45.3147
21	6.4467	8.0337	8.8972	10.2829	11.5913	13.2396	29.6151	32.6706	35.4789	38.9322	41.4011	46.7970
22	6.9830	8.6427	9.5425	10.9823	12.3380	14.0415	30.8133	33.9244	36.7807	40.2894	42.7957	48.2679
23	7.5292	9.2604	10.1957	11.6886	13.0905	14.8480	32.0069	35.1725	38.0756	41.6384	44.1813	49.7282
24	8.0849	9.8862	10.8564	12.4012	13.8484	15.6587	33.1962	36.4150	39.3641	42.9798	45.5585	51.1786
25	8.6493	10.5197	11.5240	13.1197	14.6114	16.4734	34.3816	37.6525	40.6465	44.3141	46.9279	52.6197
26	9.2221	11.1602	12.1981	13.8439	15.3792	17.2919	35.5632	38.8851	41.9232	45.6417	48.2899	54.0520
27	9.8028	11.8076	12.8785	14.5734	16.1514	18.1139	36.7412	40.1133	43.1945	46.9629	49.6449	55.4760
28	10.3909	12.4613	13.5647	15.3079	16.9279	18.9392	37.9159	41.3371	44.4608	48.2782	50.9934	56.8923
29	10.9861	13.1211	14.2565	16.0471	17.7084	19.7677	39.0875	42.5570	45.7223	49.5879	52.3356	58.3012
30	11.5880	13.7867	14.9535	16.7908	18.4927	20.5992	40.2560	43.7730	46.9792	50.8922	53.6720	59.7031
40	17.9164	20.7065	22.1643	24.4330	26.5093	29.0505	51.8051	55.7585	59.3417	63.6907	66.7660	73.4020
50	24.6739	27.9907	29.7067	32.3574	34.7643	37.6886	63.1671	67.5048	71.4202	76.1539	79.4900	86.6608
60	31.7383	35.5345	37.4849	40.4817	43.1880	46.4589	74.3970	79.0819	83.2977	88.3794	91.9517	99.6072
70	39.0364	43.2752	45.4417	48.7576	51.7393	55.3289	85.5270	90.5312	95.0232	100.4252	104.2149	112.3169
80	46.5199	51.1719	53.5401	57.1532	60.3915	64.2778	96.5782	101.8795	106.6286	112.3288	116.3211	124.8392
90	54.1552	59.1963	61.7541	65.6466	69.1260	73.2911	107.5650	113.1453	118.1359	124.1163	128.2989	137.2084
100	61.9179	67.3276	70.0649	74.2219	77.9295	82.3581	118.4980	124.3421	129.5612	135.8067	140.1695	149.4493

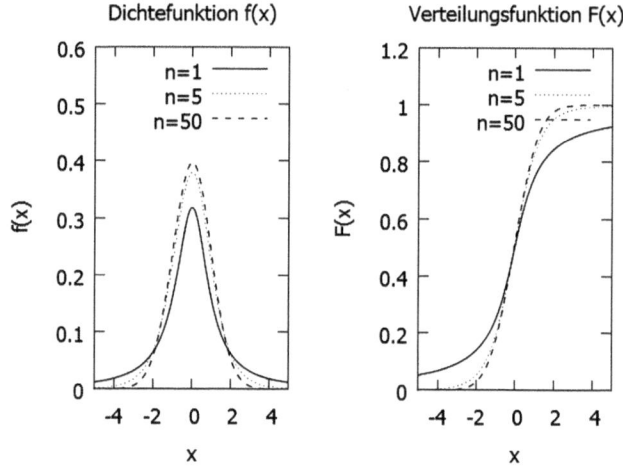

◘ Fig. 21.2 Student t-distribution for the degrees of freedom $n = 1$, $n = 5$ and $n = 50$

The expected value of the Student t-distribution is $E(X) = \mu = 0$ for $n > 1$, as one might expect for an axisymmetric probability density. Just be clear about what *axis symmetry* means:

$f(x) = f(-x)$, so the function values are mirrored at the ordinate.

As a consequence, the *center of gravity of* the function (the mean value, as we know it from ▶ Sect. 18.1) lies exactly at $\mu = 0$.

Then for the variance $D^2(X) = \sigma^2$ of the Student t-distribution holds:

$$D^2(X) = \sigma^2 = \frac{n}{n-2} \text{ für } n > 2. \tag{21.2}$$

> **Important**
> An important point that simplifies the handling of the Student t-distribution arises when the number of samples becomes sufficiently large:
> For n → ∞ the Student t-distribution changes to the standard normal distribution. So from $N > 100$ one may choose the "simplified" variant.

> **For Those Who Want to Know a Little More**
> For the sake of completeness, there should be a derivation here as well. But this requires a longer partial integration (or a series expansion) and would fill about three pages. At the end, the result would be Eq. (21.2), which is what you actually want to work with. That is why we have dispensed with it at this point.

▶ http://matheguru.com/stochastik/t-verteilung-students-t-verteilung.html

Instead of a calculation example, here is the link to an interactive illustration of the Student-t-distribution.

21.3 Estimation Methods

Having learned about the most common distributions used in the natural sciences, we now turn to the problem of obtaining information about the underlying (possibly unknown) distribution from measurement data and estimating the values for its parameters.

If we assume that the measured value distribution follows a known distribution, the parameters for this distribution must nevertheless be determined

21.3 · Estimation Methods

from the measured data. In such a case, we speak of a **point estimate**. Here, the values for the parameters must be estimated and confidence intervals must be specified in which the unknown parameters will lie with a given probability. The unknown parameters to be estimated are
(a) Mean values,
(b) Variances *and*
(c) Proportion values (probability in a binomial distribution).

To estimate the characteristic values or parameters, we use estimation or sampling functions that allow us to approximate the parameters we are looking for from a sample.

Consider a random variable X with a distribution function $F(X)$ determined by the parameters μ and σ^2. The estimator for the unknown mean $E(X) = \mu \approx \hat{\mu} = \bar{x}$ (where is $\hat{\mu}$ the estimated value) is the arithmetic mean for a sample n of size n with a random sample $x_1, x_2, \ldots, x_n$ and is given by:

$$\hat{\mu} = \bar{x} = \frac{1}{n}\sum_{i=1}^{n} x_i.$$

We denote the estimate for $\hat{\mu}$ the unknown mean μ to be estimated as a realisation of the function or also the estimator by: $\bar{X}$

$$\bar{X} = \frac{1}{n}\sum_{i=1}^{n} X_i$$

Here $X_1, X_2, \ldots, X_n$ are stochastically independent random variables that have the same distribution as the random variable X (namely μ and σ^2). Thus, on the one hand, we distinguish between the random variable of $\bar{X}$ the sampling function, which depends on n independent random variables $X_1, X_2, \ldots, X_n$, all of which have the distribution function $F(X)$. This estimator $\bar{X}$ is in turn a random variable of X and an estimator of the expected value μ we are looking for.

On the one hand, we thus have a mathematical sample with random variables $X_1, X_2, \ldots, X_n$ for the random variable X with the following expected value:

$$\mu = \bar{X} = \frac{1}{n}\sum_{i=1}^{n} X_i \left(\text{with the estimation – or sampling function } \bar{X}\right)$$

On the other hand, there is a concrete sample with the realisations $x_1, x_2, \ldots, x_n$ and the arithmetic mean:

$$\hat{\mu} = \bar{x} = \frac{1}{n}\sum_{i=1}^{n} x_i$$

We will use the sampling function $\bar{X}$ to explain what properties a sampling function should satisfy. The sampling function $\bar{X}$ has the expected value μ; since this is the desired value, we refer to our estimation function as $\bar{X}$ *expectation-true*. For the variance of the mean, as we saw with the standard deviation in ▶ Sect. 18.2:

$$D^2\left(\bar{X}\right) = \frac{\sigma^2}{n}$$

Thus, as the sample size n increases, the variance for the expected value decreases, i.e. the random variable $\bar{X}$ scatters less and less around the unknown mean μ. This property of the estimator or sampling function, that the variance becomes smaller as the sample size increases, is what we call **consistent**.

However, other estimator functions for the expected value μ are also conceivable, which are equally as $\bar{X}$, therefore we call the sampling function $\bar{X}$ efficient; one also speaks of the *most efficient sampling function*.

These three characteristics,
(a) faithful to expectations,
(b) consistent and
(c) efficient/effective (st),

are criteria that a good estimator or sampling function must meet.

Let us now turn to the parameter σ^2 for the deviation of the random variables $X_1, X_2, ..., X_n$ from the mean $\bar{X}$. As an estimator for the variance of σ^2 the distribution, we take the quantity σ^2 known from descriptive statistics and make an approximation:

$$\sigma^2 \approx \hat{\sigma}^2 = s^2 = \frac{1}{n-1}\sum_{i=1}^{n}(x-\bar{x}_i)^2$$

The corresponding sampling function for the variance of random variables $X_1, X_2, ..., X_n$ is given by:

$$S^2 = \frac{1}{n-1}\sum_{i=1}^{n}(X_i - \bar{X})^2 \tag{21.3}$$

This sampling function is expectation-true.

Thinking a Little Further

At this point we can also discuss why it is precisely the quantity S^2 calculated according to Eq. (21.3), and not—as one might suspect—the function

$$\hat{S}^2 = \frac{1}{n}\sum_{i=1}^{n}(X_i - \bar{X})^2 \tag{21.4}$$

is used, which can be taken as an expected value for

$$E\left((X_i - \bar{X})^2\right)$$

could be considered. For further discussion, we consider the sum occurring in both S^2 and in $\hat{S}^2$:

$$\sum_{i=1}^{n}(X_i - \bar{X})^2$$

We can write this sum multiplied out as:

$$\sum_{i=1}^{n}(X_i - \bar{X})^2 = \sum_{i=1}^{n}(X_i^2 - 2 \cdot X_i \cdot \bar{X} + \bar{X}^2)$$

With further algebraic transformations we find:

$$\sum_{i=1}^{n}(X_i^2 - 2 \cdot X_i \cdot \bar{X} + \bar{X}^2) = \sum_{i=1}^{n}X_i^2 - 2 \cdot \bar{X}\sum_{i=1}^{n}X_i + n \cdot \bar{X}^2 = \sum_{i=1}^{n}X_i^2 - n \cdot \bar{X}^2$$

Here we have made use of the sampling function for the mean.
Still valid:

$$\sum_{i=1}^{n}X_i^2 - n \cdot \bar{X}^2 = \sum_{i=1}^{n}X_i^2 - \frac{1}{n}\left(\sum_{i=1}^{n}X_i\right) \cdot \left(\sum_{j=1}^{n}X_j\right)$$

We divide the last double sum into two shares and get:

$$\sum_{i=1}^{n}X_i^2 - \frac{1}{n}\left(\sum_{i=1}^{n}X_i\right) \cdot \left(\sum_{j=1}^{n}X_j\right) = \sum_{i=1}^{n}X_i^2 - \frac{1}{n}\sum_{i=1}^{n}X_i^2 - \frac{1}{n}\sum_{i,j \neq i}^{n}X_i \cdot X_j$$

$$= \frac{n-1}{n}\sum_{i=1}^{n}X_i^2 - \frac{n \cdot (n-1)}{n}\bar{X}^2$$

21.3 · Estimation Methods

In the occurring double sum

$$\frac{1}{n}\sum_{i,j\neq i}^{n} X_i X_j = \frac{1}{n}\sum_{i=1}^{n} X_i \sum_{j\neq i, j=1}^{n} X_j$$

we make use of the definition of mean and substitute:

$$\sum_{i=1}^{n} X_i = n\bar{X}$$

When calculating the second sum

$$\sum_{j\neq i, j=1}^{n} X_j$$

it should be noted that we do *not* sum *over the total number of* random variables, but *one* value—namely X_i is left out of the sum, so we do not approximate this sum by the term $n\bar{X}$, but reduce the term by exactly *one* and set: $\bar{X}$

$$\sum_{j\neq i, j=1}^{n} X_j = (n-1)\bar{X}$$

For the double total, we get:

$$\frac{1}{n}\sum_{i,j\neq i}^{n} X_i X_j = \frac{1}{n}\sum_{i=1}^{n} X_i \sum_{j\neq i, j=1}^{n} X_j = \frac{1}{n}n\bar{X}(n-1)\bar{X} = \frac{n(n-1)}{n}\bar{X}^2.$$

So, in summary:

$$\sum_{i=1}^{n}(X_i - \bar{X})^2 = (n-1)\cdot\left[\frac{1}{n}\sum_{i=1}^{n} X_i^2 - \bar{X}^2\right] = (n-1)\cdot\sigma^2$$

However, as we have already seen in ▶ Eq. (18.11), the term in square brackets is the variance σ^2. The variance is therefore:

$$\sigma^2 = \frac{1}{n-1}\sum_{i=1}^{n}(X_i - \bar{X})^2$$

Thus, the estimator S^2 (from Eq. 21.3) is expectation-true for the variance, whereas the function from Eq. (21.4) is *not*.

However, for very large sample sizes, the estimator S2 approximates $\hat{S}^2$.

As an estimate of the unknown variance σ^2 we can use the following expression with the concrete sample $x_1, x_2, \ldots, x_n$:

$$\hat{\sigma}^2 = s^2 = \frac{1}{n-1}\sum_{i=1}^{n}(x_i - \bar{x})^2 \tag{21.5}$$

Now all that remains is to find an estimate for the proportion value p as a characteristic value for the binomial distribution. The relative frequency

$$\hat{p} = \frac{k}{n}$$

tells us something about how often an event A occurs in a sample of size n (namely, k times). This gives us an estimator for the proportion value with

$$\hat{P} = \frac{X}{n}$$

for the proportion value for the random variable X when a Bernoulli experiment is repeated n times.

> ❶ Please do not confuse this $\hat{P}$ with *Probability-P* from ▶ Sect. 20.3. This $\hat{P}$ here has a different meaning.

21.4 Maximum Likelihood Method

A method with which estimation functions for unknown parameters of probability functions or density functions can be obtained is the *maximum likelihood method* (i.e. "method for determining the maximum probability"). The derivation of this *likelihood function is based on the fact* that the probability L (for *likelihood*) for a concrete sample of size n with stochastically independent realisations $x_1, x_2, ..., x_n$ and corresponding (single) probabilities $f(x_1), f(x_2), ..., f(x_n)$ is given by the *product of* these (single) probabilities $L = f(x_1) \cdot f(x_2) \cdot \cdots \cdot f(x_n)$ (For those who like to have it especially mathematical: $L = \Pi f(x_i)$)

We refer to this function L as the *likelihood function*. The likelihood function (for *discrete* distributions) or density (for *continuous* distributions) $f(x)$ then depends on further parameters such as mean, variance, etc. We will denote these in the following with the general parameter ϑ, so that we can write:

$$L = F(x_1, \vartheta) \cdot f(x_2, \vartheta) \cdot \cdots \cdot f(x_n, \vartheta)$$

The occurrence of the sample $x_1, x_2, ..., x_n$ is particularly likely exactly when the function $L(\vartheta)$ becomes maximum. The determination of the maximum provides an estimate for $\hat{\vartheta}$ the unknown parameter ϑ *of* the distribution. We obtain a determination equation $\hat{\vartheta}$ for using equation:

$$\frac{\partial L(\vartheta)}{\partial \vartheta} = 0$$

The sample values $x_1, x_2, ..., x_n$ are now assumed to be constant (parameters) and go into the determination equation for:

$$\hat{\vartheta} = g(x_1, x_2, ..., x_n)$$

We call the associated estimator for the unknown parameter ϑ *of* the distribution $\Theta = g(X_1, X_2, ..., X_n)$.

> **Important**
> If the distribution has several unknown parameters, we obtain a system of equations for these parameters using the respective partial derivatives.
> The calculations are simplified by using the logarithm of the function instead of the likelihood function defined above.

▶ **Example**

Let's look at some examples of the different distributions covered in this part:

Binomial distribution (▶ Sect. 20.3)

In a binomial distribution, the parameter is the parameter p for the probability that an event A occurs. If this Bernoulli experiment is repeated n times, the probability that event A is realised k times is given by:

$$P(X = k) = \binom{n}{k} p^k \cdot q^{n-k}$$

The parameter p to be estimated is the so-called **proportion value**. If the experiment is performed n times, the following occurs
1. the event A exactly k times
2. the event $\bar{A}$ (with single probability $q = 1 - p$) then correspondingly $(n - k)$-times

Therefore, the term describing this sample realisation, which we need for the likelihood function $L(\vartheta) = L(p)$, is: $L(p) = p^k \cdot q^{n-k}$. By logarithmising we obtain:

$$\ln(L(p)) = k \cdot \ln(p) + (n-k) \cdot \ln(q)$$

21.4 · Maximum Likelihood Method

This function is now to be maximised, for this we calculate the partial derivative according to the proportion value p and then obtain a determination equation for p:

$$\frac{\partial \ln(L(p))}{\partial p} = \frac{k}{p} - \frac{n-k}{1-p} = 0 \text{ (necessary condition for the presence of a maximum)}$$

This equation is satisfied for the estimated value $\hat{p}$ mit $\hat{p} \neq 0, \hat{p} \neq 1$ if holds:

$$\frac{k}{\hat{p}} - \frac{n-k}{1-\hat{p}} = 0 \Leftrightarrow \frac{k}{\hat{p}} = \frac{n-k}{1-\hat{p}} \Leftrightarrow k - k\hat{p} = n\hat{p} - k\hat{p} \Leftrightarrow k = n\hat{p}$$

$$\hat{p} = \frac{k}{n}$$

This result means that the estimated value $\hat{p}$ corresponds to the relative frequency with which event A occurs. As an estimator or sampling function for an n-stage Bernoulli experiment with a random variable X, we use the function $\hat{p}$ with:

$$\hat{P} = \frac{X}{n}$$

Poisson distribution (▶ Sect. 20.5)

The parameter for a Poisson distribution is the parameter μ and the corresponding probability function is

$$f(x) = \frac{1}{x!} \mu^x e^{-\mu}$$

with realisations $x = 0, 1, 2,\ldots$ For the unknown parameter (in this case the mean) μ, we obtain the likelihood function with sample values $x_1, x_2, \ldots, x_n$:

$$L(\mu) = \frac{1}{x_1!} \cdot \mu^{x_1} e^{-\mu} \cdot \frac{1}{x_2!} \cdot \mu^{x_2} e^{-\mu} \cdot \ldots \cdot \frac{1}{x_n!} \cdot \mu^{x_n} e^{-\mu} = \frac{\mu^{x_1 + x_2 + \ldots + x_n}}{x_1! \cdot x_2! \cdot \ldots \cdot x_n!} \cdot e^{-n\mu}$$

Logarithmising the likelihood function then gives:

$$\ln(L(\mu)) = \ln\left(\frac{\mu^{x_1 + x_2 + \ldots + x_n}}{x_1! \cdot x_2! \cdot \ldots \cdot x_n!} \cdot e^{-n\mu}\right) = (x_1 + x_2 + \ldots + x_n)\ln(\mu) - \ln(x_1! \cdot x_2! \cdot \ldots \cdot x_n!) - n\mu$$

To maximise this function, we need the partial derivative with respect to μ and then obtain a governing equation for the unknown parameter μ:

$$\frac{\partial \ln(L(\mu))}{\partial \mu} = \frac{x_1 + x_2 + \ldots + x_n}{\mu} - n = 0 \text{ (necessary condition for a maximum)}$$

We obtain the sought estimation $\hat{\mu}$ with $\hat{\mu} \neq 0$ from:

$$\frac{x_1 + x_2 + \ldots + x_n}{\hat{\mu}} - n = 0 \Leftrightarrow \frac{x_1 + x_2 + \ldots + x_n}{\hat{\mu}} = n \Leftrightarrow \hat{\mu} = \frac{x_1 + x_2 + \ldots + x_n}{n}$$

The estimated value therefore corresponds to the mean value $\hat{\mu}\overline{x}$:

$$\overline{x} = \frac{x_1 + x_2 + \ldots + x_n}{n}$$

Therefore, the Poisson distribution estimator for the random variable X is:

$$\mu = \overline{X} = \frac{X_1 + X_2 + \cdots + X_n}{n} = \frac{1}{n}\sum_{i=1}^{n} X_i$$

Gaussian normal distribution (▶ Sect. 20.8)

As a final example of the calculation of estimators for unknown parameters of a distribution, we consider the Gaussian normal distribution, the particular importance of which has already been pointed out (several times). This distribution is determined by the parameters μ and σ^2. From the density function

$$f(x) = \frac{1}{\sqrt{2\pi}\sigma} e^{-\frac{1}{2}\left(\frac{x-\mu}{\sigma}\right)^2}$$

we determine the likelihood function $L(\mu, \sigma)$ with sample values $x_1, x_2, \ldots, x_n$ and obtain:

$$L(\mu,\sigma) = \frac{1}{\sqrt{2\pi}\sigma} e^{-\frac{1}{2}\left(\frac{x_1-\mu}{\sigma}\right)^2} \cdot \frac{1}{\sqrt{2\pi}\sigma} e^{-\frac{1}{2}\left(\frac{x_2-\mu}{\sigma}\right)^2} \cdot \ldots \cdot \frac{1}{\sqrt{2\pi}\sigma} e^{-\frac{1}{2}\left(\frac{x_n-\mu}{\sigma}\right)^2}$$

Multiplied out, we then get:

$$L(\mu,\sigma) = \frac{1}{\left(\sqrt{2\pi}\sigma\right)^n} e^{-\frac{(x_1-\mu)^2+(x_2-\mu)^2+\ldots+(x_n-\mu)^2}{2\sigma^2}}$$

Logarithmising the likelihood function yields:

$$\ln(L(\mu,\sigma)) = -n\ln(\sqrt{2\pi}) - n\ln(\sigma) - \frac{(x_1-\mu)^2+(x_2-\mu)^2+\ldots+(x_n-\mu)^2}{2\sigma^2}$$

We obtain the desired maximum from the zeros of the partial derivatives according to the parameters μ and σ:

$$\frac{\partial \ln(L(\mu,\sigma))}{\partial \mu} = \frac{(x_1-\mu)+(x_2-\mu)+\ldots+(x_n-\mu)}{\sigma^2} = \frac{1}{\sigma^2}\sum_{i=1}^{n}(x_i-\mu) = 0$$

And

$$\frac{\partial \ln(L(\mu,\sigma))}{\partial \sigma} = -\frac{n}{\sigma} + \frac{(x_1-\mu)^2+(x_2-\mu)^2+\ldots+(x_n-\mu)^2}{\sigma^3} = 0$$

Finally, the governing equations for the estimated values $\hat{\mu}$ and $\hat{\sigma}$:

$$\frac{1}{\hat{\sigma}^2}\sum_{i=1}^{n}(x_i-\hat{\mu}) = 0 \text{ and } -\frac{n}{\hat{\sigma}} + \frac{(x_1-\hat{\mu})^2+(x_2-\hat{\mu})^2+\ldots+(x_n-\hat{\mu})^2}{\hat{\sigma}^3} = 0$$

The first equation yields with some algebraic transformations:

$$\frac{1}{\hat{\sigma}^2}\sum_{i=1}^{n}(x_i-\hat{\mu}) = 0 \Leftrightarrow \sum_{i=1}^{n}(x_i-\hat{\mu}) = 0 \Leftrightarrow \sum_{i=1}^{n}x_i = \sum_{i=1}^{n}\hat{\mu} \Leftrightarrow \sum_{i=1}^{n}x_i = n\cdot\hat{\mu}$$

Thus:

$$\hat{\mu} = \frac{1}{n}\sum_{i=1}^{n}x_i = \overline{x}$$

Thus, the estimated value $\hat{\mu}$ is the mean of the sample $\overline{x}$.

From the second equation for $\hat{\sigma}$ with $\hat{\sigma} \neq 0$ we get:

$$-\frac{n}{\hat{\sigma}} + \frac{(x_1-\hat{\mu})^2+(x_2-\hat{\mu})^2+\ldots+(x_n-\hat{\mu})^2}{\hat{\sigma}^3}$$

$$= 0 \Leftrightarrow \frac{(x_1-\hat{\mu})^2+(x_2-\hat{\mu})^2+\ldots+(x_n-\hat{\mu})^2}{\hat{\sigma}^3} = \frac{n}{\hat{\sigma}}$$

$$\hat{\sigma}^2 = \frac{(x_1-\hat{\mu})^2+(x_2-\hat{\mu})^2+\ldots+(x_n-\hat{\mu})^2}{n} = \frac{1}{n}\sum_{i=1}^{n}(x_i-\overline{x})^2$$

The estimators that now result from the maximum likelihood method for the random variable X are:

$$\mu = \overline{X} = \frac{1}{n}\sum_{i=1}^{n}X_i \text{ and } \hat{S}^2 = \frac{1}{n}\sum_{i=1}^{n}(X_i-\overline{X})^2$$

Here, the estimator of variance $\hat{S}^2$ (Eq. 21.4) is not expectation-true (as we saw in the in-depth box of ▶ Sect. 21.3), but converges to variance σ^2 (S^2, from Eq. 21.3) for large sample sizes. ◄

21.5 Confidence Intervals for the Unknown Parameters ϑ of a Distribution

In the previous example, we saw that for the Gaussian normal distribution, the estimator $\hat{S}^2$ for the variance from Eq. (21.4) is not expectation-true, but tends toward the variance σ^2 (from Eq. 21.3) for large sample sizes. We must therefore ask ourselves how accurate and certain the point estimates made are. To do this, we will determine confidence intervals within which the computed parameters ϑ lie with a (pre-specified) large probability γ. This usually very large probability γ (typically $\gamma = 0.95$ or $\gamma = 0.99$) is also referred to in this context as the **confidence level**, the *confidence level*, or the *statistical confidence*. (You are already familiar with this term from ▶ Sect. 18.3).

The bounds of the confidence intervals are calculated from *estimators* or *sampling functions that* depend on the values of the concrete sample $x_1, x_2, \ldots, x_n$, the type of distribution and the specified probability γ. For example, if we choose the confidence level with $\gamma = 0.95$, it means that our calculated estimate for the parameter ϑ (mean, variance, or proportion value) is within the bounds of the confidence interval with a probability of 95%. Conversely, we have an error probability of $\alpha = 1 -> \gamma$; thus, with this probability, the calculated estimate does *not* lie within the confidence interval.

> If we run a sample of size n a total of 100 times and set the confidence level very high ($\gamma = 0.95$), the unknown parameter ϑ lies in about 95 confidence intervals calculated from our samples. However, in about 5 samples we make a wrong decision because their confidence intervals just do *not* contain the parameter ϑ.

Now that we have described the influence of the confidence level γ and the resulting error probability, we come to the calculation of the limits of the confidence interval. For this we need corresponding sampling or estimation functions, which then provide the associated interval bounds for the concrete sample. These estimators will in turn depend on the distribution of the random variable X and the parameter ϑ. Let's get started:

Confidence interval for an unknown expected value μ of a normal distribution with known variance σ^2

For the estimation of the expected value, we used the estimator

$$\bar{X} = \frac{1}{n}\sum_{i=1}^{n} X_i$$

The following apply: Mean $E(\bar{X}) = \mu$ and variance $D^2(\bar{X}) = \frac{\sigma^2}{n}$

Through the transformation

$$Z = \frac{\bar{X} - \mu}{\sigma / \sqrt{n}}$$

we obtain a standard normal distribution for the random variable Z with mean $E(Z) = 0$ and variance $D^2(Z) = 1$. For this standard normal distribution, we first specify a confidence level $\gamma = 1 - \alpha$. The random variable Z is then supposed to lie with probability γ in a symmetric interval $-c \leq Z \leq c$, i.e. $P(-c \leq Z \leq c) = \gamma = 1 - \alpha$ (◨ Fig. 21.3).

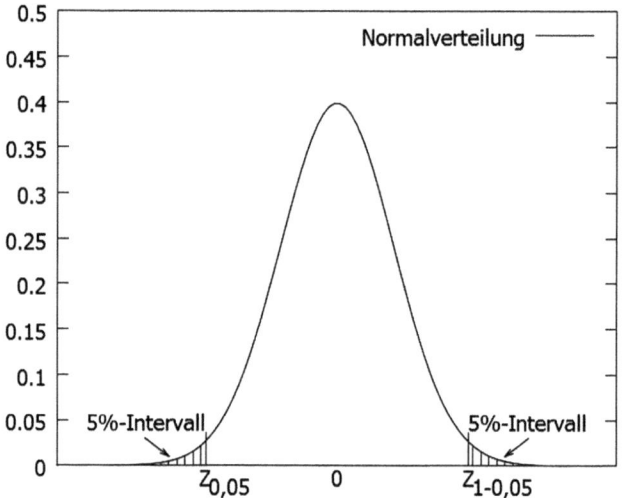

◨ **Fig. 21.3** The 5 and 95% quantiles of the standard normal distribution

We can read the location for the upper bound c from ▶ Table 20.1 at the location of $1 - (\alpha/2)$, and the lower bound of the confidence interval is then $-c$. Thus we know that the standardised variable Z lies in the interval $-c \leq Z \leq c$. For our normally distributed random variable $\bar{X}$ then holds:

$$-c \leq Z \leq c \Leftrightarrow -c \leq \frac{\bar{X} - \mu}{\sigma/\sqrt{n}} \leq c$$

By Way of Illustration
So here we take a transformation of the Gaussian distribution, first shifting the distribution curve so that the mean = 0 (which is our $-\mu$), and then "slimming" the function by dividing it by its own width ($\sigma/\sqrt{n}$) to get width 1:

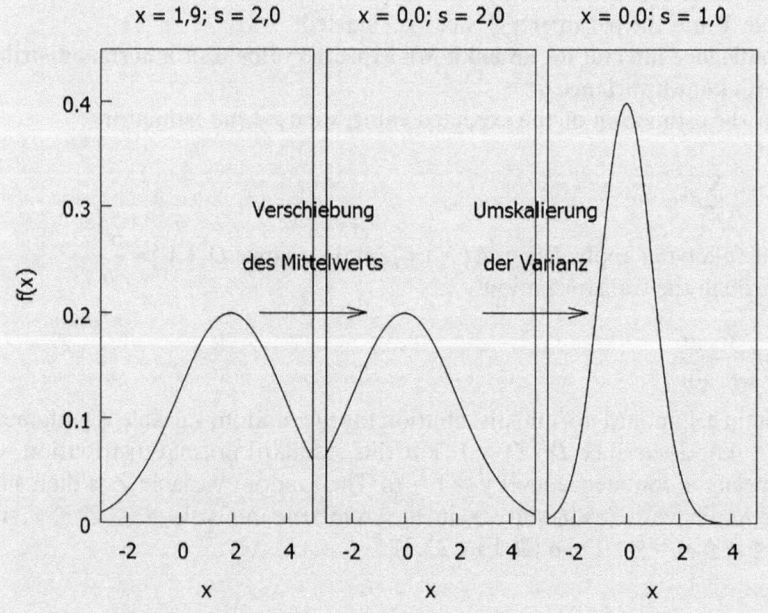

Transformation of a Gaussian distribution with $\mu = 1.9$ and $\sigma = 2$ to a standard normal distribution

21.5 · Confidence Intervals for the Unknown Parameters ϑ of a Distribution

We can then use this to determine a confidence interval for the unknown expected value μ:

$$-c \leq \frac{\bar{X} - \mu}{\sigma/\sqrt{n}} \leq c \Leftrightarrow -c\frac{\sigma}{\sqrt{n}} \leq \bar{X} - \mu \leq c\frac{\sigma}{\sqrt{n}}$$

$$\Leftrightarrow c\frac{\sigma}{\sqrt{n}} \geq \mu - \bar{X} \geq -c\frac{\sigma}{\sqrt{n}} \Leftrightarrow -c\frac{\sigma}{\sqrt{n}} \leq \mu - \bar{X} \leq c\frac{\sigma}{\sqrt{n}}$$

In a nutshell:

$$-c\frac{\sigma}{\sqrt{n}} + \bar{X} \leq \mu \leq c\frac{\sigma}{\sqrt{n}} + \bar{X}$$

Thus, we have obtained estimators for the bounds of the confidence interval. For the lower bound C_u of the interval we have the sampling function:

$$C_u = -c\frac{\sigma}{\sqrt{n}} + \bar{X}$$

We then calculate the upper bound C_o using the above estimator: the left part gives the lower bound, the right part gives the upper bound:

$$C_o = c\frac{\sigma}{\sqrt{n}} + \bar{X}.$$

For a concrete sample of size n with values $x_1, x_2, ..., x_n$ and an estimate $\bar{x}$ for the mean with known variance σ^2, we obtain as confidence interval for the unknown expected value μ:

$$-c\frac{\sigma}{\sqrt{n}} + \bar{x} \leq \mu \leq c\frac{\sigma}{\sqrt{n}} + \bar{x} \tag{21.6}$$

▶ **Example**

In the synthesis of a high-performance polymer (not to be specified here), a precisely defined amount of a chemical (also kept secret) must be added, otherwise the polymer loses its structural properties. Since the manufacturer refuses to disclose statements about the volume of the chemical used, but he tells us that the variance of the chemical volume must be $\sigma^2 = 0.20$ in order to obtain a reliable batch, we use a method, also not to be specified here, to check which amount of the mentioned chemical was *actually used*. For a sample size of $n = 50$ samples, the determination of the mean $\bar{x} = 102.40$ mL. We set 95% as the confidence level. What is the confidence interval for the unknown expected value μ?

The upper bound c is obtained from the distribution function of the standard normal distribution with $\phi(c) = 1 - \alpha/2$, in our case $\alpha = 5\% = 0.05$. In ▶ Table 20.1 we read at the point $\phi(c) = 0.975$ that $c = 1.96$.

Since we now know the value for c in addition to the mean $\bar{x}$, of the standard deviation σ and the sample size n, we obtain for the limits according to Eq. (21.6):

$$-1{,}96\sqrt{\frac{0{,}2}{50}} + 102{,}40 \leq \mu \leq 1{,}96\sqrt{\frac{0{,}2}{50}} + 102{,}40 \text{ also } 102{,}276 \leq \mu \leq 102{,}524.$$

Thus, the confidence interval is [102.28; 102.52]. ◀

Confidence interval for an unknown expected value μ of a normal distribution with unknown variance σ^2

We proceed analogously to the above case and specify a confidence level $\gamma = 1 - \alpha$ for our sample of size n. However, in this case, in addition to the estimator $\bar{X}$ for the expected value, we must now also use the sampling func-

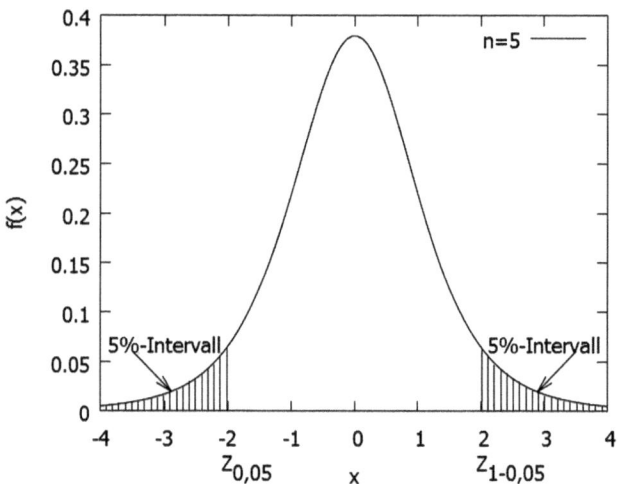

Fig. 21.4 Acceptance range of a Student t-distribution with a degree of freedom $n = 5$ and a two-sided hypothesis test

- Harris, Section 4.1: Gaussian distribution
- Harris, Section 4.3: Comparison of means with Student's t-test
- Harris, Section 4.5: t-tests with spreadsheets
- Harris, Section 4.7: The method of least squares
- Harris, Section 4.9: Least squares worksheet

tion S^2 for the variance. The transformed random variable we are working with is:

$$T = \frac{\bar{X} - \mu}{S/\sqrt{n}}$$

On the use of spreadsheet programs for various aspects of statistics, see also

This random variable belongs to the Student t distribution (known from ▶ Sect. 21.2) with degree of freedom $f = n - 1$ (◘ Fig. 21.4).

Together with the chosen confidence level, we can determine the upper bound c using ▶ Table 18.1:

$$c = t\left(1 - \frac{\alpha}{2}, f\right)$$

This t is, of course, the same t from ▶ Sect. 18.3 that we encountered again in the Student t distribution (▶ Sect. 21.2). The resulting number is then immediately reused and gives us the factor by which the standard deviation is multiplied and thus describes the absolute range around the mean in which the expected value is to be found. So then it holds:

$$-c \leq T \leq c \Leftrightarrow -c \leq \frac{\bar{X} - \mu}{S/\sqrt{n}} \leq c$$

For the concrete sample at hand with the mean $\bar{x}$ and variance s^2 (which can be calculated directly from the sample values $x_1, x_2, \ldots, x_n$), we can then specify the confidence interval for the expected value μ:

$$-c \leq \frac{\bar{x} - \mu}{s/\sqrt{n}} \leq c \Leftrightarrow -c\frac{s}{\sqrt{n}} \leq \bar{x} - \mu \leq c\frac{s}{\sqrt{n}}$$

$$\Leftrightarrow c\frac{s}{\sqrt{n}} \geq \mu - \bar{x} \geq -c\frac{s}{\sqrt{n}} \Leftrightarrow -c\frac{s}{\sqrt{n}} \leq \mu - \bar{x} \leq c\frac{s}{\sqrt{n}}$$

$$\bar{x} - c\frac{s}{\sqrt{n}} \leq \mu \leq \bar{x} + c\frac{s}{\sqrt{n}} \tag{21.7}$$

21.5 · Confidence Intervals for the Unknown Parameters ϑ of a Distribution

> Two things are particularly worth mentioning here:
> — For n → ∞, as mentioned at the end of ▶ Sect. 21.2, the Student t distribution becomes the standard normal distribution.
> — Typically, for large sample sizes ($n > 30$), one can replace the unknown standard deviation σ with the standard deviation of the sample s and then use the procedure to determine the confidence interval of an unknown mean of a normal distribution with variance σ^2.

▶ Example

The sample of the diameter of 20 dowels gives a mean value $\bar{x} = 7.8$ mm with a standard deviation of $s = 0.5$ mm. Note that we have now calculated both the mean and the variance/standard deviation from 20 measured values! We set the confidence level for the confidence interval (quite arbitrarily) to 99%. (It could have been 95%, but we're going with 99% now).

The probability of error should therefore be $\alpha = 1 - 0.99 = 0.01$. The confidence interval is now determined with Eq. (21.7). This time we obtain the missing quantity c from the Student t-distribution, since the size is too small to calculate with the normal distribution. We get the value for c from ▶ Table 18.1 for $t_{m, 1-\alpha/2}$, m is the degree of freedom of the t-distribution with m = n − 1. We get $t_{19;0,995} = 2.539.483 = c$. Now we substitute everything into Eq. (21.7), and we get:

$$7.8 - 2.539483 \frac{0,5}{\sqrt{20}} \leq \mu \leq 7.8 + 2.539483 \frac{0.5}{\sqrt{20}} \text{ also } 7.52 \leq \mu \leq 8.08.$$

The confidence interval (without taking into account the units of these quantities) is thus [7.5; 8.1]. ◀

- **Confidence interval for the unknown variance σ^2 of a normal distribution**

In quality assurance, the size of the variance plays an important role, since it gives a measure of the deviations of the random variables from their mean value. As a rule, the variance σ^2 will not be known for a production taking place in reality. Thus, we need to compute an estimate of the variance from the sample values and determine its confidence interval for a given confidence level $\gamma = 1 - \alpha$. In our considerations, we assume a normally distributed random variable X with a sample $x_1, x_2, \ldots, x_n$ whose variance σ^2 we approximate with the sampling function S^2. Furthermore, we form a random variable Z with the unknown variance, for which then holds:

$$Z = (n-1)\frac{S^2}{\sigma^2}$$

> Please note: We are not talking about the standard deviation s ("small s") here, but about the *sampling function* S ("large S"). Do not confuse!

This quantity Z is χ^2-distributed with a degree of freedom $f = n - 1$, so we can determine an interval that satisfies the condition $P(-c \leq Z \leq c) = \gamma = 1 - \alpha$ for the confidence level.

From the quantiles of the χ^2-distribution, we can use the equations.

$$F(c_1) = \frac{1}{2}(1-\gamma) = \frac{\alpha}{2}$$

the lower limit c_1 and with the equation

$$F(c_2) = \frac{1}{2}(1+\gamma) = 1 - \frac{\alpha}{2}$$

determine the upper limit c_2 of the confidence interval.

Thus we obtain $c_1 \leq Z \leq c_2$ resp.

$$c_1 \leq (n-1)\frac{S^2}{\sigma^2} \leq c_2$$

Rearranging the inequality according to the unknown variance σ^2 yields:

$$(n-1)\frac{S^2}{c_1} \geq \sigma^2 \geq (n-1)\frac{S^2}{c_2} \text{ bzw. } (n-1)\frac{S^2}{c_2} \leq \sigma^2 \leq (n-1)\frac{S^2}{c_1} \qquad (21.8)$$

Within this interval, the true variance can then be found with probability $\gamma = 1 - \alpha$.

> ▶ **Example**
>
> Staying with the example of dowels, we turn to the question of a confidence interval for the variance σ^2. Since the sample contains n = 20 dowels and the standard deviation is s = 0.5, we take this as an estimate of the variance with $S^2 = 0.25$. We set $\alpha = 0.01$ as the probability of error. In Eq. (21.8), we still need the values of c_1 and c_2 so that we can determine the confidence interval. For this purpose, we again need the quantiles of the χ^2-distribution from ◘ Table 21.1: $c_1 = \chi^2_{n-1;\alpha/2}$ and $c_2 = \chi^2_{n-1;1-\alpha/2}$ for the concrete sample with n = 20 we get:
>
> $c_1 = \chi^2_{19;0.005} = 6.8440$ and $c_2 = \chi^2_{19;0.995} = 38.5823$. Putting this into Eq. (21.8), we get for the bounds on the variance:
>
> $$19 \cdot \frac{0.25}{38.5823} \leq \sigma^2 \leq 19 \cdot \frac{0.25}{6.8440} \text{ also } 0.1231 \leq \sigma^2 \leq 0.6940.$$
>
> With probability 99%, the true variance σ^2 (again without units) lies in the interval [0.12; 0.69.]. ◀

❯ Again, there are two facts that make dealing with bills like this easier:
  — From the confidence interval for the variance σ^2 it is very easy to derive a corresponding interval for the standard deviation σ by simple root extraction.
  — As a rule, instead of the confidence level γ, the error probability α is specified in concrete applications. The corresponding statistical certainty is then $\gamma = 1 - \alpha$.

■ **Confidence interval for the unknown proportion value p of a binomial distribution**

As a prerequisite for determining the confidence interval for the (unknown) proportion value p at which an event A occurs, we need a large sample so that we can take the quantiles of the standard normal distribution to help us. The size results from the relation:

$$n \cdot p \cdot (1-p) \geq 9$$

> **As a Reminder**
> In ▶ Sect. 20.3 we learned the rule of thumb that the binomial distribution may be replaced by the *normal distribution if* $N - p - (1 - p) \geq 9$.

Where n is the sample size and $\hat{p}$ the estimate for the unknown proportion value p for the occurrence of event A. The estimated value is then obtained from the relative frequency with which event A occurred in the sample. Thus, if A is encountered k times, we obtain as the estimated value

$$\hat{p} = \frac{k}{n}$$

21.5 · Confidence Intervals for the Unknown Parameters θ of a Distribution

for the proportion value p. To determine a confidence interval, we specify a confidence level γ. For the binomial distribution we obtain with the expected value $\mu = n - p$ and the standard deviation

$$\sigma = \sqrt{np(1-p)}$$

out of transformation

$$Z = \frac{n\hat{p} - np}{\sqrt{np(1-p)}}$$

a (approximately standard normally distributed) standardised random variable Z for the sampling function

$$\hat{p} = \frac{X}{n}$$

The limits of the confidence interval are given by the probability P:

$$P\left(-c \leq \frac{n\hat{p} - np}{\sqrt{np(1-p)}} \leq c\right) = \gamma = 1 - \alpha$$

P indicates the probability of $[-c, c]$ finding the random variable in Z the interval. For the unknown parameter p of the binomial distribution we obtain with the relation

$$-c \leq \frac{n\hat{p} - np}{\sqrt{np(1-p)}} \leq c$$

and some transformations

$$-c\frac{\sqrt{np(1-p)}}{n} \leq \hat{p} - p \leq c\frac{\sqrt{np(1-p)}}{n} \Leftrightarrow c\frac{\sqrt{np(1-p)}}{n} \geq p - \hat{p} \geq -c\frac{\sqrt{np(1-p)}}{n}$$

finally the sought information about the interval to be considered:

$$\hat{p} - c\frac{\sqrt{np(1-p)}}{n} \leq p \leq \hat{p} + c\frac{\sqrt{np(1-p)}}{n}$$

In this relation, the interval limits for the proportion value p contain both the parameter p itself and the sampling function $\hat{p}$. These two quantities are now approximated by the estimated value $\hat{p}$ determined from the sample, and in this way we obtain the confidence interval for the proportion value p with:

$$\hat{p} - c\frac{\sqrt{n\hat{p}(1-\hat{p})}}{n} \leq p \leq \hat{p} + c\frac{\sqrt{n\hat{p}(1-\hat{p})}}{n} \tag{21.9}$$

Thus, the proportion value p lies in this interval with probability γ.

> ### ▶ Example
> A machine is used 350 days a year. The probability of failure is $q = 0.20$, and the machine can therefore be used with a probability $p = 0.80$. What is the confidence interval for the probability of use p with a confidence level $\varepsilon = 0.95$, i.e. a probability of error of $\alpha = 1 - \varepsilon = 0.05$?
>
> Since the sample is $n \cdot p \cdot q = 350 \cdot 0.80 \cdot 0.20 = 56 > 9$, we may use the normal distribution to determine c—needed in Eq. (21.9). We obtain the value for c from $\phi(c) = 1 - \alpha/2 = 1 - 0.025 = 0.975$. In ▶ Table 20.1, we find $\phi(1.96) = 0.975$, i.e., $c = 1.96$. Now we have all the quantities to determine the confidence interval for p:

$$0.80 - 1.96 \cdot \frac{\sqrt{350 \cdot 0.80 \cdot 0.20}}{350} \leq p \leq 0.80 + 1.96 \cdot \frac{\sqrt{350 \cdot 0.80 \cdot 0.20}}{350} \text{ also } 0.758 \leq p \leq 0.842$$

The betting probability ranges from $0.76 \leq p \leq 0.84$ with a probability of error of $\alpha = 0.05$. ◀

- **Confidence interval for an unknown expected value μ of *any* distribution**

If there is no information about the distribution of a random variable X, its mean is again determined using the estimator function. It reads:

$$\bar{X} = \frac{1}{n} \sum_{i=1}^{n} X_i$$

Here we need the random variables X_i, their distribution and the variance D^2 for which holds:

$$D^2(\bar{X}) = \frac{\sigma^2}{n}$$

> **Important**
> For very large sample sizes (with $n > 30$), the **central limit theorem** applies. This states nothing other than that *whenever* enough samples are drawn, i.e. n is sufficiently large, the sum of arbitrarily distributed (!) random variables is *normally distributed*.
> This really applies to *all* distributions, whether binomial, logarithmic, or otherwise.

It can then be concluded that the estimator can be considered to be $\bar{X}$ *approximately normally distributed*. We can then resort to the procedures described in this section to calculate the confidence interval.

> ▶ **Example**
> Let's get back to the diameter determination of dowels. This time, the lack (?) of accuracy did not leave us alone, and to be on the safe side, we significantly increased the sample size and determined the diameter of 50 dowels. We obtained $\bar{x} = 8.16$ mm as the mean value and $s = 0.64$ mm as the standard deviation. With a confidence level of 95%, we now want to determine the expected value. We revert to Eq. (21.7) and use the normal distribution to compute the value of c. We have encountered $\phi(c) = 1 - \alpha/2 = 0.975$ a few times now, and we know that the value of c is 1.96 again. Now we substitute everything into Eq. (21.7):
>
> $$8.16 - 1.96 \cdot \frac{0.64}{\sqrt{50}} \leq \mu \leq 8.16 + 1.96 \cdot \frac{0.64}{\sqrt{50}} \text{ also } 7.983 \leq \mu \leq 8.337.$$
>
> The confidence interval for the true dowel diameter ranges from 7.98 to 8.34 mm with an error probability of 5%. ◀

> If the distribution is *unknown*, a sample size *n as large as possible is* desirable, because the larger *n* is, the smaller the differences in the methods are, whether the quantiles of the standard normal distribution (presence of a known variance) or the quantiles of the Student t-distribution (unknown variance) are used to calculate the limits.

21.6 Parameter Tests

Parameter tests are generally used to test statements (e.g. information from a manufacturer on the quality of products) or assumptions that are formulated as hypotheses. However, since such a hypothesis does not necessarily *have to be* true, it is contrasted with an alternative hypothesis. Such hypotheses cannot therefore be identified as valid or false, i.e. proven or disproven, with 100% certainty within the framework of statistical methods.

> We can only classify hypotheses on statistical random variables as true or false with a probability that is (as) high as possible; in any case, there is a so-called probability of error in the parameter tests—the (quantifiable) probability that we have made a *wrong decision.*

■ **Statistical hypotheses and parameter tests**

First, let's take a closer look at some terms that are crucial here: What exactly do we mean by:
- Hypothesis,
- Alternative hypothesis,
- Probability of error *and*
- Parameter test?

The formulation of a statistical hypothesis involves a statement about a random variable in a statistical aggregate. This can be, for example, assumptions about the mean of a normally distributed variable or conjectures about the equality of the unknown means in two normal distributions (This topic is again addressed by Harris *and* Skoog).

We call the expression of any hypothesis the **null hypothesis** (*H*0); it is exactly this hypothesis that is to be tested by means of a parameter test. The counter formulation of the made assumption—about a probability distribution or one of its parameters ϑ—is then the *alternative hypothesis*, which is called H_1. The significance of such a test is determined by the significance level or **significance number.** The lower the error probability α (determined before the test) with $0 < \alpha < 1$, the lower will be the risk of a wrong decision based on the performed parameter test.

But before we now explain what a false decision is, we must first make a decision: The decision whether we want to accept or reject the null hypothesis is made by means of a test variable *T*, which is chosen according to the problem and for which the interval limits c_1 and c_2 are subsequently calculated. The interval limits, in turn, depend on the significance level or the probability of error and the probability distribution (corresponding to the problem). The area enclosed by the interval bounds is called the **acceptance area** or *non-critical* area or *non-rejection area*. We refer to the set outside the interval bounds as the *critical range* or *rejection range*. Finally, a test value $\hat{t}$ is computed based on a sample for the test variable *T*, which serves as an estimate $\hat{\vartheta}$ for the parameter ϑ. The two cases that can now occur allow us to decide whether we can accept (or, more clearly, *not* reject) the hypothesis, or whether we reject it instead.
- In case the test value $\hat{t}$ is within the non-critical range, the null hypothesis is not rejected.
- In the other case, namely that the value from the sample $\hat{t}$ is in the critical range (i.e. outside the acceptance range), the null hypothesis is rejected.

Harris, Section 4.3: Comparison of means with Student's *t*-test
Skoog, Annex A.3: Statistical tests (hypothesis tests)

And having made such a decision, we can of course be wrong.

> **Important**
> We distinguish between two types of wrong decisions:
> 1. If the null hypothesis is *rejected* even though it is *true*, there is an **error of the first kind.**
> 2. If, on the other hand, the null hypothesis is *not rejected* even though it is *not true*, we speak of a **second kind of error**.

Typically, one should choose the null hypothesis such that, if the decision is wrong, the more unpleasant consequences lead to a mistake of the *first* kind.

> ▶ **Example**
> First, a simple, albeit abstruse, example: let's talk about the colour of beach balls. You see a blue beach ball in a store while on vacation at the beach. You might now make the null hypothesis: "All beach balls are blue." As soon as you see even a single beach ball of a different colour, you know that your null hypothesis was not true: so you must reject the hypothesis. Thus we have decided *against* our null hypothesis and made no mistake.
>
> But if you don't see another colour anywhere, and you're still desperate to make a decision (for whatever reason), there are two options:
> - You *stick to* your null hypothesis, but then later learn that there are other colours after *all*, then you have made a second kind of mistake.
> - You yourself doubt that there should only be blue beachballs and reject the null hypothesis. In this case, you have only made a mistake of the first kind (and only then!) when it turns out reliably (and provably) that *only* blue waterballs are produced worldwide and all previous ones in other colours have been destroyed. What damage has been done by rejecting the surprisingly correct null hypothesis? None, actually. So it's no big deal.

▶ https://de.wikipedia.org/wiki/Russells_Teekanne

Now such a scenario is unlikely, but possible. However, it is very tricky to prove a negative (in this case, the non-existence of non-blue beachballs). For fun, you can try to prove that there is *not a* teapot somewhere between the Earth and Mars, orbiting the Sun like a planet, but so tiny that none of our telescopes will ever detect it!

But sometimes it's a little more difficult—which you can observe again and again in pharmaceutical research. Suppose you were dealing with a drug for which no side effects have been observed so far. Which null hypothesis makes more sense/is more advisable here?
- Null hypothesis 1: "This drug might have side effects after all that we just don't know about yet, so we'd better do more testing before it hits the market."
- Null hypothesis 2: "The drug has not caused any side effects so far, so it won't have any. We can put it on the market."

Let's look at both null hypotheses in turn.
- If null hypothesis 1 turns out to be correct, it was obviously good to test the drug extensively first: problems with side effects were avoided. This is good in principle, but for the patients who would have benefited from the "actual effect" of the drug (and who might not have cared about the side effects), it is of course unfortunate that the drug was not available (perhaps the side effects would have been much less bad than the untreated course of the disease).

However, if null hypothesis 1 is *false*, we would have made a mistake of the *first* kind. In this case, a good drug would have come onto the market too late or even not at all for some patients due to the hesitancy of the producers.
- With null hypothesis 2, on the other hand, there is the ideal case: the drug works and has no side effects.

But there is also the *worst-case scenario*: the drug has (perhaps even serious) side effects.

Let us think of the example of Thalidomide®: This *second* type of error caused severe damage to many thousands of foetuses that were still unborn at the time the drug was administered. Especially in view of the fact that this drug was supposed to be "only" a sleeping pill and sedative (also for pregnant women), it would have been advisable to resort to null hypothesis 1 in case of doubt.

Therefore, as mentioned above, in case of doubt, one should always combine the possible worse consequences with an error of the *first* kind. ◄

■ Planning and execution of parameter tests

When planning parameter tests, we usually assume that the probability distribution of the variable X is known. Typically, we assume a normal distribution whose parameters μ and σ are not known. First, we formulate the null hypothesis H_0 and the corresponding alternative hypothesis H_1:
- Null hypothesis H_0: $\vartheta = \vartheta_0$
- Alternative hypothesis H_1: $\vartheta \neq \vartheta_0$

With the null hypothesis we formulate our assumption that the unknown parameter ϑ has the value ϑ_0. The alternative of this assumption is that the parameter ϑ takes a value other than ϑ_0. Since in the alternative formulation the values of ϑ can be larger or smaller than ϑ_0, we refer to this case as a *two-sided* parameter test.

In the next step, we specify a significance level α ($0 < \alpha < 1$). By doing so, we specify the probability with which the null hypothesis is rejected even though it is correct (error of the first kind); we refer to this probability as the probability of error. As values for α we choose $\alpha = 0.05 = 5\%$ or $\alpha = 0.01 = 1\%$ (small values).

The parameter test is performed with a test variable T that depends on n random variables $X_1, X_2, \ldots, X_n$, all of which have the same probability distribution as X. This test variable is a concrete sampling function with $T = g(X_1, X_2, \ldots, X_n)$.

The selected significance level α now yields the critical limits c_u and c_o. These limit the interval in which the test variable T lies with a probability $\gamma = 1-\alpha$. The range $c_u \leq T \leq c_o$ is the *non-critical* range, the range outside this interval is called the *critical* range.

In the penultimate step of the parameter test, we calculate the value $\hat{t}$ for the test variable T from

$$\hat{t} = g(x_1, x_2, \ldots, x_n)$$

with concrete sample values $x_1, x_2, \ldots, x_n$ for the random variables $x_1, x_2, \ldots, x_n$.

In the last step, we check whether the value $\hat{t}$ for the test variable T falls into the non-critical range, i.e., whether holds:

$$c_u \leq \hat{t} \leq c_o$$

- If this holds, the null hypothesis H_0 is accepted (or not rejected) at a significance level α.
- If the test value $\hat{t}$ falls within the critical range, then the null hypothesis H_0 is not tenable and is rejected in favour of the alternative hypothesis H_1.

> ► **Example**
> We assume a normally distributed aggregate whose expected value is $\mu_0 = 12.25$. *No information is available about the standard deviation σ*. A sample of n = 15 now yielded a mean $\bar{x} = 11.8$ with a standard deviation s = 0.3. Is this deviation significant at a significance level with $\alpha = 0.01$ or just random? Such a test is called a *t*-test.
> - Setting up the null hypothesis: we are optimistic and believe that the expected value is equal to 12.25. $H_0: E(X) = \mu_0$.
> - Significance level: As specified, $\alpha = 0.01$.
>
> $$\text{Test variable}: T = \frac{\bar{x} - \mu_0}{s/\sqrt{n}} = \frac{\bar{x} - \mu_0}{s} \cdot \sqrt{n}$$
>
> With our metrics, we get $T = -5.809$.
> - Critical region: The quantiles of the t-distribution—see ► Table 18.1—give the value $T_{14;0.995} = 2.976843$ for a degree of freedom $m = n - 1 = 14$ for a significance level $1-\alpha/2 = 0.995$.
> - Test decision: Since $|T| > t_{m,1-\alpha/2}$ with $|-5.809| > 2.976843$ holds, the null hypothesis cannot be retained. Thus, we can assume that the mean is significantly $\bar{x}$ different from the expected value μ_0. ◄

- **Possible sources of error during parameter tests**

Again, we need to distinguish between first and second type errors in the manner described above. Once again:
- An error of the first kind occurs when a (actually) correct null hypothesis is rejected.
- An error of the second kind is made when the (actually) wrong null hypothesis is accepted as correct after all.

Let us first consider the type of a first kind error. When carrying out parameter tests, we set the significance level α at the beginning, which also specifies the risk of a first kind error. If the test variable falls into the critical range, i.e. lies outside the acceptance range—and the probability of this is just α, then the null hypothesis is rejected, but the certainty that this decision is also *correct* lies only at $1 - \alpha$. With a residual probability of even α, we have made a wrong decision because we have rejected the null hypothesis.

Because we can be *wrong about the* decision with a probability α, α is also popularly called the *probability of error*.

Let us now turn to the second kind of error. We keep the null hypothesis, although it is actually *false*. It is easy to imagine the consequences of such an error! Let us return to the medical agents: If their side effects are estimated to be harmless, the risk lies completely with the patient if it turns out that this is *not the* case.

We therefore also refer to a second type of error as *consumer risk*.

That's why you should always choose the null hypothesis such that the more unpleasant consequences lead to an error of the first kind (yes, you're reading that for the third time right now, but we *really* care about that).

One reason for the existence of an error of the second kind is that the distribution that is actually present is different from the assumed distribution (e.g. shifted).

◘ Figure 21.5 shows the first and second type errors of an assumed distribution—shifted against the actual distribution.

- **Parameter tests for the equality of parameters (μ or σ)**

At the end of this chapter, we present test procedures in which the equality of mean values or standard deviations of *two* series of measurements is to be examined, or in which measured values are to be compared with quantities

21.6 · Parameter Tests

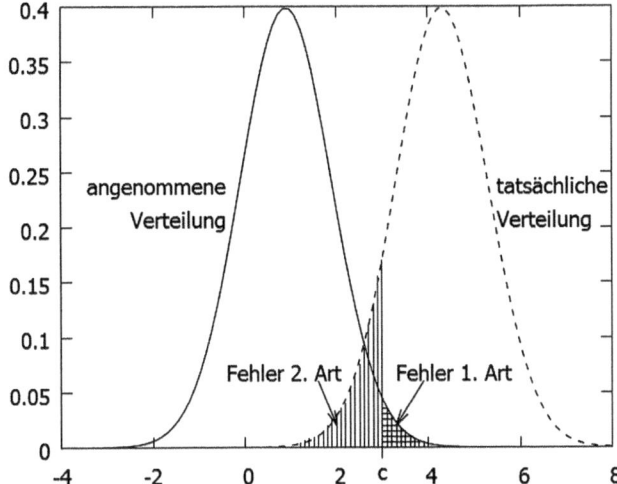

Fig. 21.5 Error 1 and 2 type

known from the literature. (Once again, a look at Harris may be of advantage here.)

When testing for equality of means $\bar{x}_1$ and $\bar{x}_2$ two series of measurements, it is important to consider whether the underlying standard deviations σ_1 and σ_2 of the series of measurements are the *same* or different because a larger/smaller scatter is expected in one of the measurement methods used.

Let us first consider the case where the two measurement series belong in principle to the same totality and have the same standard deviation. As an example, let us consider our two dowel measurement series.

Harris, Section 4.3: Comparison of means with Student's *t*-test

> ▶ **Example**
> We want to investigate whether the different diameters of the dowels obtained in the two series of measurements are only different by chance or whether they differ *significantly*. In the following, we work through our five-point program for parameter tests, in which we first set up a null hypothesis, determine a significance level, calculate the test variable, determine the critical range and *only then* make a decision. Let us again summarise the known data of the measurement series: Measurement series 1 has a circumference $n_1 = 20$, a mean $\bar{x}_1 = 7.8$ mm and a standard deviation $s_1 = 0.5$ mm, measurement series 2 has a circumference $n_2 = 50$, a mean $\bar{x}_2 = 8.16$ mm and a standard deviation $s_2 = 0.64$.
> - Null hypothesis: we are again optimistic and assume that the mean values differ only by *chance*. $H_0: E(X_1) = E(X_2)$
> - We choose $\varepsilon = 0.95$ as the significance level, i.e., a probability of error $\alpha = 0.05$.
> - Since we assume that the standard deviation is the same in both cases, we calculate the joint standard deviation s with:
>
> $$s = \sqrt{\frac{\sum_{i=1}^{n_1}(\bar{x}_1 - x_i)^2 + \sum_{j=1}^{n_2}(\bar{x}_2 - x_j)^2}{n_1 + n_2 - 2}} = \sqrt{\frac{s_1^2(n_1-1) + s_2^2(n_2-1)}{n_1 + n_2 - 2}}$$
>
> We receive: $s = \sqrt{\dfrac{0.5^2 \cdot (20-1) + 0.64^2 \cdot (50-1)}{20 + 50 - 2}} = 0.6042$
>
> We obtain the test variable T with:
>
> $$T = \frac{|\bar{x}_1 - \bar{x}_2|}{s}\sqrt{\frac{n_1 n_2}{n_1 + n_2}}$$

The value for the test variable is T = 2.252.

- This value of T must now be compared with the value from the Student t-istribution for m = $n_1 + n_2 - 2$ = 68 degrees of freedom from the level $1 - \alpha/2$ = 0.975, i.e. with $t_{68;0.975}$. Unfortunately, ▶ Table 18.1 does not provide us with the concrete value for m = 68 degrees of freedom, but it must lie between the values of m = 60 and m = 70. For m = 60 we read off the value $t_{60;0.975}$ = 1.670649, for m = 70 the value $t_{70;0.975}$ = 1.666914. *Our test value T is thus in each case clearly above the t-value.*
- Since the test value T is *above the* value of t, we *cannot* accept the hypothesis—the means are equal. The probability that the mean values are equal after *all*—and that we therefore make an error of the first kind—is less than 5%. ◀

As another case that can occur when comparing means, we examine the same samples using two measurement methods—of different precision. The test that we use in such a context to examine the mean values is called *paired t-test*. The data set is best presented in tabular form, so that we can already calculate the difference between the measured values of the first and second method at this point.

▶ Example

The alcohol content is determined for 10 blood samples using two methods:

Measurements of the alcohol content of 10 blood samples with 2 different methods

- Null hypothesis: we assume that the mean of the second series of measurements is significantly smaller than the mean $\bar{x}_1$. H_0: $E(X_2) < E(X_1)$. This is then a *one-sided* test, i.e. the critical range for the test value is at higher values, since we assume a deviation towards lower values. The *alternative hypothesis* is accordingly: $E(X_2) > E(X_1)$. Now we test for H_0:
- We choose ε = 0.95 as the significance level, i.e., a probability of error α = 0.05.
- The test variable T in this example is:

$$T = \frac{|\bar{d}|}{s/\sqrt{n}} = \frac{|\bar{d}|}{s} \cdot \sqrt{n}$$

Using the above values, we obtain a test variable T = 0.41677.

- This value of T must now be compared with the value from the Student t-distribution for m = n − 1 = 9 degrees of freedom from the level $1 - \alpha$ = 0.95, i.e. with $t_{9;0.95}$. In ▶ Table 18.1 we read the value $t_{9;0.95}$ = 1.833113. Our test value T is *below the* t-value.
- Since the test value T is less than the value of t, we can *keep the* hypothesis—method 2 gives smaller readings (Could still be wrong, of course. That's how it is with decisions).

Sample	Measured value x_1 Method 1 (expressed in g/kg)	Measured value x_2 Method 2 (g/kg)	Difference d = $x_1 - x_2$ (g/kg)
1	0.86	0.92	−0.06
2	0.23	0.15	0.08
3	0.54	0.55	−0.01
4	0.04	006	−0.02
5	0.20	0.,17	0.03
6	1.03	1.10	−0.07

21.6 · Parameter Tests

Sample	Measured value x_1 Method 1 (expressed in g/kg)	Measured value x_2 Method 2 (g/kg)	Difference $d = x_1 - x_2$ (g/kg)
7	0.99	1.10	−0.11
8	0.05	0.06	−0.01
9	0.76	0.69	0.07
10	0.12	0.10	0.02
Mean values	0.482	0.490	−0.008
Standard deviations	0.3997	0.4362	0.0607

Let us now compare the mean value of a series of measurements with a value known from the literature (or otherwise reliably).

▶ **Example**

We turn again to our dowels and compare the mean value of our first series of measurements—as a reminder: n = 20, $\bar{x}_1$ = 7.8 mm and s_1 = 0.5 mm—with the manufacturer's specifications. According to these, the diameter of the dowels should be μ_0 = 8.2 mm. At a significance level of 95%, we now want to test whether our measurement result deviates from the official data only *by chance* or whether there is a *significant* deviation. So let us go through the scheme for the parameter tests:

— Null hypothesis: we assume that the mean value deviates from the manufacturer's specifications only *by chance*. H_0: $E(X_1) = \mu_0$.
This is again a *two-sided* test, i.e. the critical range for the test value can be at higher or also lower values, since deviations can occur in both directions. As significance level we choose $\varepsilon = 0.95$, i.e. an error probability of $\alpha = 0.05$.

— The test variable T in this example is:

$$T = \frac{|\bar{x} - \mu_0|}{s/\sqrt{n}} = \frac{|\bar{x} - \mu_0|}{s} \cdot \sqrt{n}$$

Using the above values, we obtain a test variable T = 3.5777.

— This value of T must now be compared with the value from the Student t-distribution for $m = n - 1 = 19$ degrees of freedom from the level $1 - \alpha/2 = 0.975$, i.e. with $t_{19;0.975}$. In ▶ Table 18.1 we read the value $t_{19;0.975}$ = 1.729133. Our test value T is clearly above the *t*-value.

— Since the test value T is *above the* value of *t*, we *cannot* maintain the hypothesis—the measurement deviates from the manufacturer's specifications only by chance. The probability that the mean value *does* correspond to the manufacturer's specifications—i.e. that we are making an error of the first kind—is less than 5%. ◀

The last parameter test we would like to present here is the so-called *F-test*, which is used to compare standard deviations (Harris also refers to the F-test again).

In this test, the ratio of the standard deviations s_1 and s_2 of two series of measurements is formed in such a way that the larger value of the standard deviations is in the numerator and the smaller in the denominator.

The quantity that is calculated is called F, and the following applies: $F = s_1^2/s_2^2$ (here the standard deviation s_1 is therefore greater than s_2). This calculated value is then compared with the values to be read in ◻ Table 21.2. If the calculated value for F is greater than the tabulated value, the difference in the standard deviations can be considered significant.

Harris, Section 4.4: Comparison of standard deviations with the F-test

Table 21.2 F-values for a significance level of 95%. The notation is $F(0.95; m_1, m_2)$ with the degrees of freedom m_1 for the first measurement series and m_2 for the second data set

	2	3	4	5	6	7	8	9	10	11	12	13	14	15	16	17	18	19	20	21	22	23	24	25	26	27	28	29	30	∞
2	19.00	19.16	19.25	19.30	19.33	19.35	19.37	19.38	19.40	19.40	19.41	19.42	19.42	19.43	19.43	19.44	19.44	19.44	19.45	19.45	19.45	19.45	19.45	19.46	19.46	19.46	19.46	19.46	19.46	19.50
3	955	9.28	9.12	9.01	8.94	8.89	8.85	8.81	8.79	8.76	8.74	8.73	8.71	8.70	8.69	8.68	8.67	8.67	8.66	8.65	8.65	8.64	8.64	8.63	8.63	8.63	8.62	8.62	8.62	8.53
4	6.94	6.59	6.39	6.26	6.16	6.09	6.04	6.00	5.96	5.94	5.91	5.89	5.87	5.86	5.84	5.83	5.82	5.81	5.80	5.79	5.79	5.78	5.77	5.77	5.76	5.76	5.75	5.75	5.75	5.63
5	5.79	5.41	5.19	5.05	4.95	4.88	4.82	4.77	4.74	4.70	4.68	4.66	4.64	4.62	4.60	4.59	4.58	4.57	4.56	4.55	4.54	4.53	4.53	4.52	4.52	4.51	4.50	4.50	4.50	4.36
6	5.14	4.76	4.53	4.39	4.28	4.21	4.15	4.10	4.06	4.03	4.00	3.98	3.96	3.94	3.92	3.91	3.90	3.88	3.87	3.86	3.86	3.85	3.84	3.83	3.83	3.82	3.82	3.81	3.81	3.67
7	4.74	4.35	4.12	3.97	3.87	3.79	3.73	3.68	3.64	3.60	3.57	3.55	3.53	3.51	3.49	3.48	3.47	3.46	3.44	3.43	3.43	3.42	3.41	3.40	3.40	3.39	3.39	3.38	3.38	3.23
8	4.46	4.07	3.84	3.69	3.58	3.50	3.44	3.39	3.35	3.31	3.28	3.26	3.24	3.22	3.20	3.19	3.17	3.16	3.15	3.14	3.13	3.12	3.12	3.11	3.10	3.10	3.09	3.08	3.08	2.93
9	4.26	3.86	3.63	3.48	3.37	3.29	3.23	3.18	3.14	3.10	3.07	3.05	3.03	3.01	2.99	2.97	2.96	2.95	2.94	2.93	2.92	2.91	2.90	2.89	2.89	2.88	2.87	2.87	2.86	2.71
10	4.10	3.71	3.48	3.33	3.22	3.14	3.07	3.02	2.98	2.94	2.91	2.89	2.86	2.85	2.83	2.81	2.80	2.79	2.77	2.76	2.75	2.75	2.74	2.73	2.72	2.72	2.71	2.70	2.70	2.54
11	3.98	3.59	3.36	3.20	3.09	3.01	2.95	2.90	2.85	2.82	2.79	2.76	2.74	2.72	2.70	2.69	2.67	2.66	2.65	2.64	2.63	2.62	2.61	2.60	2.59	2.59	2.58	2.58	2.57	2.40
12	3.89	3.49	3.26	3.11	3.00	2.91	2.85	2.80	2.75	2.72	2.69	2.66	2.64	2.62	2.60	2.58	2.57	2.56	2.54	2.53	2.52	2.51	2.51	2.50	2.49	2.48	2.48	2.47	2.47	2.30
13	3.81	3.41	3.18	3.03	2.92	2.83	2.77	2.71	2.67	2.63	2.60	2.58	2.55	2.53	2.51	2.50	2.48	2.47	2.46	2.45	2.44	2.43	2.42	2.41	2.41	2.40	2.39	2.39	2.38	2.21
14	3.74	3.34	3.11	2.96	2.85	2.76	2.70	2.65	2.60	2.57	2.53	2.51	2.48	2.46	2.44	2.43	2.41	2.40	2.39	2.38	2.37	2.36	2.35	2.34	2.33	2.33	2.32	2.31	2.31	2.13
15	3.68	3.29	3.06	2.90	2.79	2.71	2.64	2.59	2.54	2.51	2.48	2.45	2.42	2.40	2.38	2.37	2.35	2.34	2.33	2.32	2.31	2.30	2.29	2.28	2.27	2.27	2.26	2.25	2.25	2.07
16	3.63	3.24	3.01	2.85	2.74	2.66	2.59	2.54	2.49	2.46	2.42	2.40	2.37	2.35	2.33	2.32	2.30	2.29	2.28	2.26	2.25	2.24	2.24	2.23	2.22	2.21	2.21	2.20	2.19	2.01
17	3.59	3.20	2.96	2.81	2.70	2.61	2.55	2.49	2.45	2.41	2.38	2.35	2.33	2.31	2.29	2.27	2.26	2.24	2.23	2.22	2.21	2.20	2.19	2.18	2.17	2.17	2.16	2.15	2.15	1.96
18	3.55	3.16	2.93	2.77	2.66	2.58	2.51	2.46	2.41	2.37	2.34	2.31	2.29	2.27	2.25	2.23	2.22	2.20	2.19	2.18	2.17	2.16	2.15	2.14	2.13	2.13	2.12	2.11	2.11	1.92
19	3.52	3.13	2.90	2.74	2.63	2.54	2.48	2.42	2.38	2.34	2.31	2.28	2.26	2.23	2.21	2.20	2.18	2.17	2.16	2.14	2.13	2.12	2.11	2.11	2.10	2.09	2.08	2.08	2.07	1.88
20	3.49	3.10	2.87	2.71	2.60	2.51	2.45	2.39	2.35	2.31	2.28	2.25	2.22	2.20	2.18	2.17	2.15	2.14	2.12	2.11	2.10	2.09	2.08	2.07	2.07	2.06	2.05	2.05	2.04	1.84
21	3.47	3.07	2.84	2.68	2.57	2.49	2.42	2.37	2.32	2.28	2.25	2.22	2.20	2.18	2.16	2.14	2.12	2.11	2.10	2.08	2.07	2.06	2.05	2.05	2.04	2.03	2.02	2.02	2.01	1.81
22	3.44	3.05	2.82	2.66	2.55	2.46	2.40	2.34	2.30	2.26	2.23	2.20	2.17	2.15	2.13	2.11	2.10	2.08	2.07	2.06	2.05	2.04	2.03	2.02	2.01	2.00	2.00	1.99	1.98	1.78
23	3.42	3.03	2.80	2.64	2.53	2.44	237.	2.32	2.27	2.24	2.20	2.18	2.15	2.13	2.11	2.09	2.08	2.06	2.05	2.04	2.02	2.01	2.01	2.00	1.99	1.98	1.97	1.97	1.96	1.76
24	3.40	3.01	2.78	2.62	2.51	2.42	2.36	2.30	2.25	2.22	2.18	2.15	2.13	2.11	2.09	2.07	2.05	2.04	2.03	2.01	2.00	1.99	1.98	1.97	1.97	1.96	1.95	1.95	1.94	1.73
25	3.39	2.99	2.76	2.60	2.49	2.40	2.34	2.28	2.24	2.20	2.16	2.14	2.11	2.09	2.07	2.05	2.04	2.02	2.01	2.00	1.98	1.97	1.96	1.96	1.95	1.94	1.93	1.93	1.92	1.71
26	3.37	2.98	2.74	2.59	2.47	2.39	2.32	2.27	2.22	2.18	2.15	2.12	2.09	2.07	2.05	2.03	2.02	2.00	1.99	1.98	1.97	1.96	1.95	1.94	1.93	1.92	1.91	1.91	1.90	1.69
27	3.35	2.96	2.73	2.57	2.46	2.37	2.31	2.25	2.20	2.17	2.13	2.10	2.08	2.06	2.04	2.02	2.00	1.99	1.97	1.96	1.95	1.94	1.93	1.92	1.91	1.90	1.90	1.89	1.88	1.67
28	3.34	2.95	2.71	2.56	2.45	2.36	2.29	2.24	2.19	2.15	2.12	2.09	2.06	2.04	2.02	2.00	1.99	1.97	1.96	1.95	1.93	1.92	1.91	1.91	1.90	1.89	1.88	1.88	1.87	1.65
29	3.33	2.93	2.70	2.55	2.43	2.35	2.28	2.22	2.18	2.14	2.10	2.08	2.05	2.03	201.	1.99	1.97	1.96	1.94	1.93	1.92	1.91	1.90	1.89	1.88	1.88	1.87	1.86	1.85	1.64
30	3.32	2.92	2.69	2.53	2.42	2.33	2.27	2.21	2.16	2.13	2.09	2.06	2.04	2.01	1.99	1.98	1.96	1.95	1.93	1.92	1.91	1.90	1.89	1.88	1.87	1.86	1.85	1.85	1.84	1.62
∞	3.00	2.60	2.37	2.21	2.10	2.01	1.94	1.88	1.83	1.79	1.75	1.72	1.69	1.67	1.64	1.62	1.60	1.59	1.57	1.56	1.54	1.53	1.52	1.51	1.50	1.49	1.48	1.47	1.46	1.00

21.6 · Parameter Tests

> ► **Example**
> Let us look at the standard deviations for the different methods of alcohol control from our penultimate example. The first method gives a standard deviation of $s_1 = 0.3997$ and for the second method it is $s_2 = 0.4362$ for a sample size of $n = 10$ in each case, which is then 9 degrees of freedom. Since the latter method has the larger value, the F-value is calculated from:
> $F = s_2^2/s_1^2 = 0.4362^2/0.3997^2 = 1.1910$. Since the tabulated F value is $F(0.95;9;9) = 3.18$, the calculated value is *less than* the tabulated value and we can consider the standard deviations as *not* significantly different. ◄

❓ Questions

7. It is known about a high-performance alloy (which shall not be named in detail here) that its cobalt content should ideally be 52.35%. When five samples of a batch are examined, the following values are obtained (via XRF—see Part V of "Analytical Chemistry I"): 52.15; 52.78; 52.93; 52.26 and 52.41%. Does this batch with a confidence interval of 95% meet the desired requirements?

8. The titer was determined for several stock solutions of presumably the same concentration; for a series of measurements with $n = 15$, the variance is $s_2 = 0.2$ for these samples. Calculate the confidence interval for the variance with a probability of 95%.

9. Let us return to task 2: We had looked at the standard deviation in a multiple titration of sodium hydroxide solution against hydrochloric acid. We had found slight fluctuations (expressed as mean value and standard deviation). If another titration were now carried out: Within what interval would the expected measured value (consumption of hydrochloric acid) lie with a probability of 95%? Within which interval, if the probability is to be 99%?

Once again, the entire Chapter 4 of Harris is highly recommended: There you will find numerous example calculations.

Harris, Chapter 4: Statistics

Validation of Methods

Contents

22.1 Standard Addition – 310

22.2 Internal Standard and External Standard – 312

Harris, Section 4.8: Calibration curves

In this chapter, we will apply the considerations and (computational) procedures made previously to quantitative methods in chemistry. In Part I of "Analytical Chemistry I" you have already become familiar with important topics such as quality assurance and calibration. And it is precisely these topics that we are now revisiting from the point of view of statistics. The concrete goal we are pursuing is: Determination of the (unknown—otherwise we would not need to determine it) concentration $c(x)$ of an analyte whose presence we have already previously determined by a purely qualitative method.

22.1 Standard Addition

Following the procedure described in Part I of "Analytical Chemistry I", a calibration series is established by adding known amounts of the analyte, i.e. using the method or measuring instrument (e.g. a spectrophotometer) to be validated, the concentrations of the established series are determined several times (at least three times). Before further evaluation, we recommend a graphical representation of the values. In particular, outliers can be identified relatively easily in this way. Let us look at the measured values given in ◘ Table 22.1. (In case they look familiar to you: Yes, these are the measured values from Part I of "Analytical Chemistry I", only extended by two series of measurements).

If we represent these data in a diagram, as shown in ◘ Fig. 22.1, it turns out that almost all points lie approximately on a straight line. However, one point (the measured value $y = 11.9$ for $x = 5$) stands out. We can classify this value as an outlier—you will see how this works in ▶ Chap. 23—and must remove it from the data set for the further calculations.

A further correction, which becomes necessary, has its cause in possible influences of other substances/impurities, which the sample to be examined—without the analyte—can contain. The measurement of a sample without analyte is called *blank sample* (you already know this from Part I of "Analytical Chemistry I"), and we subtract the corresponding mean value from our measured values—the outlier is now already no longer included. In our example the mean value of the blank sample is 0.1. With this value the corrected ◘ Table 22.2 is obtained.

In ◘ Fig. 22.2, the mean values of y are plotted against the variable x; a linear plot is readily apparent. We calculate the compensation line as seen in ▶ Sect. 19.1, and then plot this together with the mean values (as shown in ◘ Fig. 22.2).

The calculated compensation line $y = 1.97 \cdot x + 0.0875$ is used to determine the unknown concentration/analyte quantity x of a sample with at a measured signal y by interpolation.

◘ Table 22.1 Arbitrary measured values for creating a calibration curve

x-value	1	2	3	4	5	6	7	8
y-value	2.0	4.5	6.0	7.5	10.0	12.4	14.0	15.5
y-value	2.1	4.4	6.3	7.6	11.9	12.4	14.3	15.7
y-value	2.2	4.5	5.9	7.4	9.9	12.5	14.5	15.8

22.1 · Standard Addition

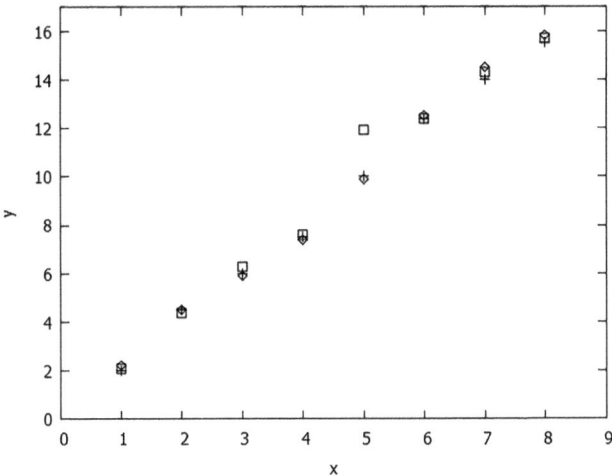

Fig. 22.1 Measured values

Table 22.2 Corrected measured values without outliers, evaluation with mean value and standard deviation

x-value	1	2	3	4	5	6	7	8
y-value	1.9	4.4	5.9	7.4	9.9	12.3	13.9	15.4
y-value	2.0	4.3	6.2	7.5	–	12.3	14.2	15.6
y-value	2.1	4.4	5.8	7.3	9.8	12.4	14.4	15.7
Mean value	2.00	4.37	5.97	7.40	9.85	12.33	14.17	15.57

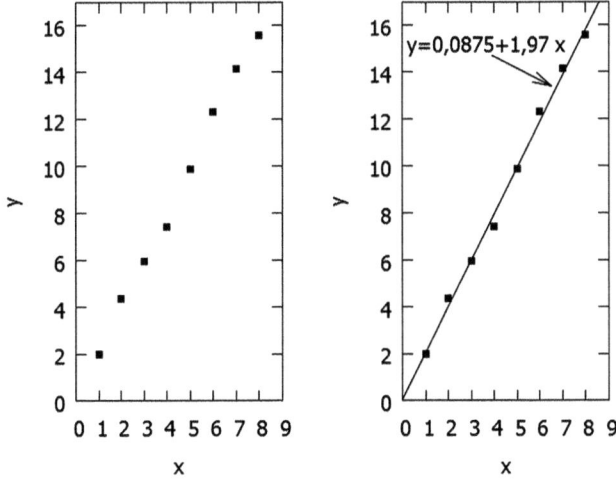

Fig. 22.2 Mean values of the calibration series and compensation line

> **▶ Example**
>
> If the measurement signal of the unknown sample is at $y_p = 8.5$, the corresponding concentration of the analyte can be read off from the graph by measuring the height/ordinate $y = 8.5$ on the compensation line and reading off the corresponding concentration by projection onto the abscissa/x-axis, which is at $x_p = 4.27$, as can be seen from the graph.
>
>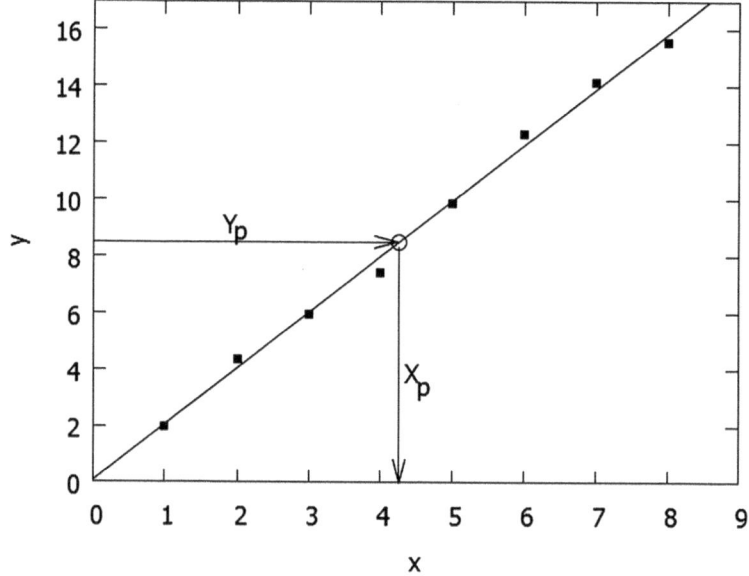
>
> Interpolation of a measured value on the basis of the regression line
>
> If you prefer to calculate, you can of course do so. The measured value y_p is inserted into the equation for the compensation line and the equation is converted to x_p: $8.5 = 1.97 \cdot x_p + 0.0875$. From this follows:
>
> $$x_p = (8.5 - 0.0875)/1.97 = 4.27(03).$$ ◀

Two other calibration methods are used: the *internal/inner* standard and the *external/outer* standard.

22.2 Internal Standard and External Standard

Although these methods differ chemically, they will be treated together, since the mathematical procedure of these methods is analogous to that described in ▶ Sect. 22.1. First, however, let us briefly return to the different chemical variants:

- **Addition of an analyte standard**

Here, the analyte to be quantified is added to the analyte solution itself in a precisely known quantity, thus significantly amplifying the measurement signal obtained. This procedure is useful when the concentration of the solution to be analyzed is linearly related to the measured value (i.e., the signal). (Think, for example, of Lambert-Beer's law.) Practically, the addition step can be carried out very well reproducibly because all samples treated with the addition are diluted in the *same way*.

The addition of such a standard allows, for example, matrix effects to be corrected (or eliminated); accordingly, this method can also be used where the matrix otherwise causes difficulties or varies greatly (such as with biological samples). The evaluation is then performed with a calibration function related to the standard in question.

22.2 Internal Standard and External Standard

Unfortunately, at least two analyses must always be carried out for each sample to be examined (once with, once without the addition), which considerably increases the workload. In addition, the relationship between concentration and signal is not known to remain truly linear over all conceivable concentration ranges (we have known this at least since Part IV of "Analytical Chemistry I"; again, Mr. Lambert and Mr. Beer must be consulted). The fact that the volume changes due to the standard addition must be taken into account in *all* associated calculations (and is often forgotten …).

- **Use of an internal standard**

A compound is required which is suitable as an internal standard and which does not react with the analyte or any matrix components, and which also distributes itself evenly in the sample (again, biological samples can cause us difficulties). If these conditions are fulfilled, the internal standard also leads to very well reproducible measurement results. (You may already be familiar with such internal standards from chromatography since Part III of "Analytical Chemistry I"). Systematic errors (analyte losses or concentrations due to evaporating solvent) can be easily compensated here, because a systematic error affects all samples equally. Another great advantage of this technique is that many samples can be measured in this way (automatically, if necessary) and they all fall back on the same calibration.

Of course, this technique also has its disadvantages: Only a few matrix effects can be corrected, and if systematic errors are ultimately due to the internal standard used, this can easily be overlooked. In the case of solids or some biosamples, the problem arises that the internal standard may not be distributed homogeneously throughout the sample to the desired extent, and last but not least, one is occasionally faced with the problem of finding a substance suitable as an internal standard in the first place.

- **Use of an external standard**

An external standard must give measured values as similar as possible to those of the analyte in question. If this is the case, the reproducibility is pleasingly high and the number of systematic sources of error is largely minimised. Particularly for routine operation, in which a large number of samples are to be analysed (preferably automated again, as we mentioned in Part IV), this technique is extremely well suited—once an appropriate external standard has been found.

However, similar to the internal standards, the problem arises that systematic errors can easily be overlooked, and if the nature of the samples differs fundamentally from time to time, especially with regard to their matrix (again: think of biological samples), any matrix effects can be extremely difficult to compensate for.

A nice summary of the various advantages and disadvantages of the different techniques of working with a standard is provided by the (highly recommended) website chemgapedia.

▶ http://www.chemgapedia.de/vsengine/vlu/vsc/de/ch/3/anc/croma/kalibrierung.vlu.html

Let us now return to the general procedure for quantitative determination:

In the case of the internal standard, a substance that is chemically *similar* but *not identical to the* analyte is added to the sample to be investigated and the corresponding calibration solutions in a known quantity (in contrast to standard addition, in which *the analyte itself is* added in a known quantity). Then the chemical equipment, e.g. for chromatography, is applied to the sample and the prepared solutions. The measured data sets are then treated with the procedures described in ▶ Sect. 22.1 (graphing, testing for outliers, correcting for the blank sample if necessary, calculating the balance line and finally determining the amount of analyte in the sample).

For the external standard, reference samples with different, known concentrations of the analyte are prepared. These standard samples are measured using the assay method to be validated. Here it is essential that the concentration expected in the sample lies in the range taken by the standard solutions/calibration solutions. All measured values—e.g. concentration values—from the external standards and the sample are then subjected to a graphical inspection and, examined for outliers ... and if everything is then linear, we calculate the compensation line and determine the amount of analyte in the actual sample.

▶ **Example**

The concentration of β-carotene of a sample is to be determined. The intensity value measured by a spectroscopic method is $I_p = 59.8$. Six solutions with 10–60% carotene content were used as reference samples, resulting in the measured values given in the table.

Standard solutions with known amounts of β-carotene

Sample	1	2	3	4	5	6
c/%	10	20	30	40	50	60
I	26.1	39.4	54.5	64.3	76.9	87.0

The graphical representation shows the dependence of the intensity of a spectroscopic method (for example photometry) of the calibration solutions on the β-carotene concentration.

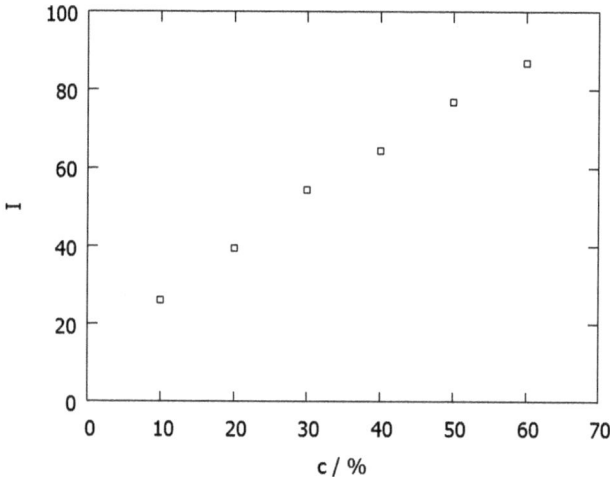

Intensity of six standard solutions against the concentration of β-carotene
Visual inspection of the graph shows a good linear curve without outliers. Now the statistical evaluation with determination of the compensation lines is carried out—see ▶ Sect. 19.1.
Evaluation of the measured values

22.2 Internal Standard and External Standard

i	1	2	3	4	5	6	Totals	Mean values
$c_i/\%$	10	20	30	40	50	60	210	35
I_i	26.1	39.4	54.5	64.3	76.9	87.0	348.2	58,033
c_i^2	100	400	900	1600	2500	3600	9100	1,516,667
$c_i \cdot I_i$	261	788	1635	2572	3845	5220	14,321	2,386,833

Using the formulas from ▶ Eq. (19.7), we calculate the slope dimension m and the intercept b of the compensation line:

$$m = \frac{\sum_i (x_i \cdot y_i) - N\overline{xy}}{\sum_i x_i^2 - N(\overline{x})^2} = \frac{14321 - 6 \cdot 35 \cdot 58.0\overline{3}}{9100 - 6 \cdot 35^2} \approx 1.2194$$

$$b = \overline{y} - m \cdot \overline{x} = 58.0\overline{3} - 1.2194 \cdot 35 \approx 15.3533$$

This is the value we get if we use the *exact* value for the slope measure m, not the approximate one, i.e. 1.2194. Don't be surprised: If you work with the *approximate* values in the line above (namely the 1.2194), you get a *slightly different* value: 15.3543. (But the number 1.2194... just went further!).

The straight line equation for the intensity is: $I = 1.2194 \cdot c + 15.3533$ and is plotted with the data in the figure.

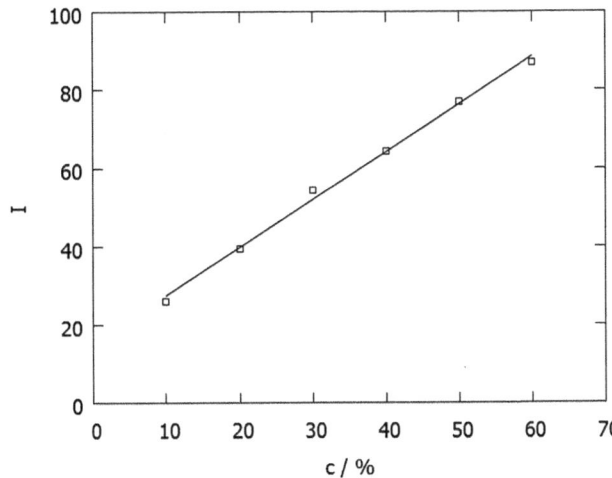

Balancing line with the data on β-carotene concentration

Finally, the *actual* measuring point is entered in the straight line, so that the value of the concentration can be read off. Alternatively, the value for the sample intensity is inserted into the straight line equation and this is converted according to the (sought) concentration. From the sample value $I_p = 59.8$ follows:

$$59.8 = 1.2194 \cdot c_p + 15.3543$$

If we solve the equation for concentration, we get:

$$c_p = (59.8 - 15.3543)/1.2194 = 36.449$$

The concentration of the sample is therefore $c_p = 36.4\%$.

Here is a note on the accuracy of the calculation of the parameters m and b of a compensation line. Due to the distribution of the measured values, the individual measurement points are subject to fluctuations, which we have

Harris, Section 4.7: The method of least squares
Skoog, Appendix A.4: The method of least squares

already learned about as variances (see ▶ Sect. 18.2). (Harris and Skoog also deal with this subject.) However, this variance does not only affect the dependent variable y (e.g. the measurand intensity I), but also the independent variable x (e.g. the given concentration c) and, of course, the expression $x \cdot y$ formed from measurands also has a scatter, which is called *covariance*. For the definition of the measures of dispersion we resort to the variances according to ▶ Eq. (18.9):

$$S_x^2 = (N-1) \cdot s_x^2 = \sum_{i=1}^{N}(x_i - \bar{x})^2 \text{ for the measurands } x_i \text{ and} \tag{22.1}$$

$$S_y^2 = (N-1) \cdot s_y^2 = \sum_{i=1}^{N}(y_i - \bar{y})^2 \text{ for the measurands } y_i. \tag{22.2}$$

We calculate the covariance with:

$$S_{xy} = \sum_{i=1}^{N}(x_i - \bar{x})(y_i - \bar{y}) \tag{22.3}$$

Using Eqs. (22.1) and (22.3), the slope measure m of the linear equation can be compactly represented: $m = S_{xy} / S_x^2$. How good the assumed linearity actually is is determined by the coefficient of determination B:

$$B = \frac{\sum_{i=1}^{N}(\hat{y}_i - \bar{y})^2}{\sum_{i=1}^{N}(y_i - \bar{y})^2} = \frac{S_{\hat{y}}^2}{S_y^2} \tag{22.4}$$

▶ http://xkcd.com/1725/

> In some textbooks, instead of the coefficient of determination B, the correlation coefficient R is preferred. In this case, $B = R^2$. It is important to note that R can also assume negative values (possible values are $-1 \leq R \leq 1$), while B is (logically) always positive.

In this equation, the variables $\hat{y}_i$ are the values from the straight line equation—so to speak the function values theoretically belonging to the location x_i. If these are identical to the measured values y_i, then the variances are also identical and the coefficient of determination B is 1. This in turn means that our measured values provide a perfect straight line. The smaller B is, the worse the representation of the data by means of a balancing line. B is thus a measure of the goodness of fit of data to a straight line. We denote the deviation of the variables y_i from the regression line by s_r:

$$s_r = \sqrt{\frac{\sum_{i=1}^{N}(y_i - (m \cdot x_i + b))^2}{N-2}} \tag{22.5}$$

Harris, Section 4.7: The method of least squares
Skoog, Appendix A 4.2: Determining the line with the smallest square deviations (compensation line)

Using this deviation, we calculate the scatter for the slope m and the intercept b. (Again, we are in complete agreement with Harris and Skoog here).

We obtain the scatter s_m of the slope m and s_b of the ordinate section b by:

$$s_m^2 = \frac{s_r^2}{S_x^2} \text{ and } s_b^2 = \frac{s_r^2 \sum x_i^2}{n \sum x_i^2 - (\sum x_i)^2} = \frac{s_r^2 \sum x_i^2}{n S_x^2}$$

Let's apply these ideas to our intensity measurement example below.

22.2 Internal Standard and External Standard

▶ **Example**

We extend the above table with the measured intensities and enter the theoretical intensities I_{theor} resulting from the compensation curve and the squared differences between the measured value and the theoretical value in the following table; for the measured intensities and concentrations we calculate the squared deviations from the corresponding mean values in each case.

Determination of the deviations of the parameters for the compensation line

i	1	2	3	4	5	6	Totals	Mean values
$c_i/\%$	10	20	30	40	50	60	210	35
I_i	26.1	39.4	54.5	64.3	76.9	87.0	348.2	58,033
c_i^2	100	400	900	1600	2500	3600	9100	1516,667
$c_i \cdot I_i$	261	788	1635	2572	3845	5220	14,321	2386,833
I_{theor}	27,548	39,742	51,936	64,130	76,324	88,518		
$(I_i-I_{theor})^2$	20967	0,11696	6,5741	0,0289	0,33178	2,3043	11,4527	
$(I_{theor}-\bar{I})^2$	929,335	334,561	37,173	37,173	334,561	929,335	2602,138	
$(I_i-\bar{I})^2$	1019,72	347,189	12,482	39,275	355,964	839,087	2613,717	
$(c_i-\bar{c})^2$	625	225	25	25	225	625	1750	

From the sample size $n = 6$ and the calculated sums we can now determine the standard deviation from the regression line:

$$s_r^2 = \frac{\sum_{i=1}^{N}(y_i - y_{theor})^2}{N-2} = \frac{11.4527}{6-2} = 2.8632$$

In the next step, we use this quantity to calculate the deviations for the parameters m and b:

$$s_m^2 = \frac{s_r^2}{S_x^2} = \frac{2.8632}{1750} = 0.001636 \rightarrow s_m = 0.04047$$

and

$$s_b^2 = \frac{s_r^2 \cdot \sum x_i^2}{S_x^2 \cdot n} = \frac{2.8632 \cdot 9100}{1750 \cdot 6} = 2.8414 \rightarrow s_b = 1.6857$$

The balance line can now be specified as follows:

$$I(c) = (1.22 \pm 0.04) \cdot c + (15.35 \pm 1.69)$$

The coefficient of determination B is:

$$B = \frac{\sum_{i=1}^{N}(\hat{y}_i - \bar{y})^2}{S_y^2} = \frac{2602.138}{2613.717} = 0.9956$$

Thus, we have a very good linear relationship between the measured intensities and the corresponding concentrations.

> Since the use of only two points logically *always* results in a straight line, significantly more (at least 5) points must always be measured to determine a calibration line. (However, there is a rule that is especially popular among physicists and physicians: "If you want a straight line, take two measurements. If you want an *origin line*, take *one* measurement." This rule should be taken with a grain of salt).

Questions

10. For the measurement of the concentration-dependent absorption of a lead-containing sample, the following calibration series is used as a basis:

Measured values of the calibration line for determining the lead content of an unknown sample

i	1	2	3	4	5
x_i/mg	0.400	1.200	2.000	2.800	3.600
A	0.020	0.043	0.071	0.093	0.116

Make a graph and determine the balance line.

11. The measured absorbance of a sample containing lead is $A = 0.052$. Using the linear equation from Task 10, determine the Pb concentration of this sample

Outlier Tests

Contents

23.1   **Dixon Q-Test** – 320

23.2   **4σ-Environment** – 322

23.3   **Grubbs Test** – 322

23.4   **Summary: This Time in Keywords** – 323
23.4.1   Experimental Errors – 323
23.4.2   Error Propagation – 324
23.4.3   Measured Value Distribution – 324
23.4.4   Parameter Estimates – 326
23.4.5   Validation of Methods – 326

23.5   **Outlier Tests** – 326

         Further Reading – 332

Outlier tests must be carried out before each calculation of statistical quantities in order to check the quality of the measurements. A data point that qualifies as an outlier must not be taken into account in further calculations under any circumstances, as otherwise you will get incorrect results. Such tests may only be performed *once* in a data set—this is partly because outliers are indeed rare events and partly because taking out values would then change the distribution a lot. Imagine that you spot two suspicious measurement points A and B. If you actually eliminate A in a first step, then the characteristic values (mean and standard deviation) of the distribution change in such a way that candidate B fits all at once. If, on the other hand, you start with candidate B, then it is possible that this point is an outlier on the basis of the distribution, and accordingly the point is excluded from further considerations. But what happens with point A, possibly *this* point now fits the remaining distribution. You can see the dilemma here when you examine suspicious measurement points *one after the other* for their outlier properties.

> Therefore, *all* possible outliers must be eliminated in *one* step and not *sequentially* by applying an outlier test multiple times.

We will look at three outlier tests below:
- Dixon Q-test
- 4σ-environment *and*
- Grubbs test

23.1 Dixon Q-Test

A very simple test is the so-called Q-test, in which the measurement data are first sorted by size. Then the quotient is calculated:

$$Q = \left| \frac{x - x_n}{x_{max} - x_{min}} \right| \tag{23.1}$$

x: the data point to be examined (the first or the last value)
x_n: Neighbouring point to x
x_{max}: Maximum value in the data set
x_{min}: Minimum value in the data set

The values for $Q(x)$ obtained in this way are compared with table values—see ◻ Table 23.1—which provide a measure of whether and with what probability the data point under investigation is an outlier.

If the value of *Q is* greater than the critical value, *x is* an outlier to be rejected with probability $1 - \alpha$.

23.1 · Dixon Q-Test

Table 23.1 Comparison values for Q, N: number of measurement points, α: significance level

N	α = 0,001	α = 0,002	α = 0,005	α = 0,01	α = 0,02	α = 0,05	α = 0,1	α = 0,2
3	0,999	0,998	0,994	0,988	0,976	0,941	0,886	0,782
4	0,964	0,949	0,921	0,889	0,847	0,766	0,679	0,561
5	0,895	0,869	0,824	0,782	0,729	0,643	0,559	0,452
6	0,822	0,792	0,744	0,698	0,646	0,563	0,484	0,387
7	0,763	0,731	0,681	0,636	0,587	0,507	0,433	0,344
8	0,716	0,682	0,633	0,591	0,542	0,467	0,398	0,314
9	0,675	0,644	0,596	0,555	0,508	0,436	0,370	0,291
10	0,647	0,614	0,568	0,527	0,482	0,412	0,349	0,274
15	0,544	0,515	0,473	0,438	0,398	0,338	0,284	0,220
20	0,491	0,464	0,426	0,393	0,356	0,300	0,251	0,193
25	0,455	0,430	0,395	0,364	0,329	0,277	0,230	0,176
30	0,430	0,407	0,371	0,342	0,310	0,260	0,216	0,165

> **► Example**
> Measured data:
>
> 0,189; 0,167; 0,187; 0,183; 0,186; 0,182; 0,181; 0,184; 0,181; 0,177
>
> Sorted by size you get:
>
> 0,167; 0,177; 0,181; 0,181; 0,182; 0,183; 0,184; 0,186; 0,187; 0,189
>
> Potential outlier is 0.167, since the distance from the first to the second value is greater than the distance between the second to last value.
>
> Let us now calculate the value of Q using Eq. (23.1):
>
> $$Q = \left|\frac{0.167 - 0.177}{0.167 - 0.189}\right| = 0.455$$
>
> The number of measurement points is $N = 10$, so we look in ◘ Table 23.1 for $N = 10$ to see if the calculated value of Q appears.
>
> According to the table, the Q value is above the value of 0.412 corresponding to the significance level $\alpha = 0.05$ and below 0.482 with the significance level of $\alpha = 0.02$. Consequently, the value of 0.167 must be rejected with a probability greater than 95% and less than 98%. ◄

23.2 4σ-Environment

Another criterion with which an outlier can be identified is based on the distance of the suspicious measured variable x from the mean value $\bar{x}_t$—which in this case, however, is determined *without* including the critical value (hence $\bar{x}_t$ [for test], not $\bar{x}$). The standard deviation σ_t (the same applies here, it is the standard deviation *without* the critical value, i.e. not σ) is also calculated for this test *without* including the potential outlier. Now, if the conspicuous value x lies outside a $k\sigma_t$-environment of $\bar{x}_t$—it should be noted that typically $k = 4$ is chosen—then we qualify the value x as an outlier and exclude it from further calculations—this is how bullying works in statistics. However, please remember that 99.73% of all measured values lie in the 3σ-environment of the Gaussian distribution alone—as we saw in ► Sect. 18.3.

> **► Example**
> Let's apply this method to the readings from our example above: 0.189; 0.167; 0.187; 0.183; 0.186; 0.182; 0.181; 0.184; 0.181; 0.177.
>
> If we calculate mean $\bar{x}_t$ and level deviation σ_t—mind you *without* the influence of the critical value 0.167—then we get: $\bar{x}_t = 0.1833$ and $\sigma_t = 0.0036$. If we now calculate the fourfold environment, the interval covers the range from 0.1688 to 0.1979. Since the value 0.167 lies outside the interval, it is identified as an outlier and discarded. ◄

23.3 Grubbs Test

Harris, Section 4.6: Grubbs test for an outlier

In this test, the distance of the value under test from the mean $\bar{x}$—here the mean is taken over *all*—including the critical values (!)—and then put in relation to the standard deviation σ. We call the number thus obtained G, and it holds:

$$G = \frac{|x - \bar{x}|}{\sigma} \qquad (23.2)$$

This value is now to be compared with tabulated values: If the value of G exceeds the tabulated value $G_{0.95}$ or $G_{0.99}$, then with a probability of 95 or 99%

Table 23.2 Comparison values $G_{0.95}$ and $G_{0.99}$ for different sample sizes N

N	$G_{0.95}$	$G_{0.99}$	N	$G_{0.95}$	$G_{0.99}$	N	$G_{0.95}$	$G_{0.99}$
3	1,1531	1,1546	15	2,4090	2,7049	80	3,1319	3,5208
4	1,4625	1,4925	16	2,4433	2,7470	90	3,1733	3,5632
5	1,6714	1,7489	17	2,4748	2,7854	100	3,2095	3,6002
6	1,8221	1,9442	18	2,5040	2,8208	120	3,2706	3,6619
7	1,9381	2,0973	19	2,5312	2,8535	140	3,3208	3,7121
8	2,0317	2,2208	20	2,5566	2,8838	160	3,3633	3,7542
9	2,1096	2,3231	25	2,6629	3,0086	180	3,4001	3,7904
10	2,1761	2,4097	30	2,7451	3,1029	200	3,4324	3,8220
11	2,2339	2,4843	40	2,8675	3,2395	300	3,5525	3,9385
12	2,2850	2,5494	50	2,9570	3,3366	400	3,6339	4,0166
13	2,3305	2,6070	60	3,0269	3,4111	500	3,6952	4,0749
14	2,3717	2,6585	70	3,0839	3,4710	600	3,7442	4,1214

there is an outlier. For different circumferences N the comparison values are given in Table 23.2.

Let's apply this method to our example as well.

► **Example**

If we calculate the mean and standard deviation over our measured values 0.189; 0.167; 0.187; 0.183; 0.186; 0.182; 0.181; 0.184; 0.181; 0.177 then we get: $\bar{x} = 0.1817$ and $\sigma = 0.006\,201$. The value of G is now calculated using Eq. (23.2), and we get:

$$G = |0.167 - 0.1817| \, 0.006\,201 = 2.3706.$$

For the test, we need the comparison values when the sample size is $N = 10$, the tabulated comparison values are $G_{0.95} = 2.1761$ and $G_{0.99} = 2.4097$. Since the size G is between these values, the value of 0.167 must be rejected as an outlier with a probability greater than 95% and less than 99%. ◄

❓ **Questions**

12. One task for all three outlier tests: determine whether there is an outlier among each of the following values: 23.42; 23.66; 24.10; 22.99; 24.99; 23.01.

23.4 Summary: This Time in Keywords

23.4.1 Experimental Errors

A distinction is made:
- Systematic errors, e.g. as a result of incorrect calibration of the measuring device used. These constantly cause measured values that are too high or too low.
- Statistical errors/random errors are errors that result from a scattering of the individual measured values.

23.4.2 Error Propagation

Three key points are particularly important:
- Mean value ($\bar{x}$): arithmetic mean of all measured values
- Standard deviation (s): Measure of the scatter of the measured values
- Confidence interval: Intervals within which the mean value lies with a certain (high) probability.

23.4.3 Measured Value Distribution

Important formulas:
  Overview of the measured value distributions

23.4 · Summary: This Time in Keywords

Designation	Individual probability p_i	Distribution function $F(x)$	Expected value	Variance
Discrete uniform distribution	$1/N$ for $i = 1, \ldots, N$	$\sum_{i, x_i \leq x} \dfrac{1}{N}$	$\dfrac{N+1}{2}$	$\dfrac{N^2 - 1}{12}$
Two-point distribution	$1-p$ for $x = 0$ p for $x = 1$ 0 else	0 for $x < 0$ $1 - p$ for $0 \leq x < 1$ 1 for $x \geq 1$	p	$p(1-p)$
Binomial distribution	$\binom{n}{k} p^k (1-p)^{n-k}$ for $k = 1, 2, \ldots, n$	$\sum_{k=1}^{x-1} \binom{n}{k} p^k (1-p)^{n-k}$	$n \cdot p$	$n \cdot p \cdot (1-p)$
Hypergeometric distribution	$\dfrac{\binom{M}{k}\binom{N-M}{n-k}}{\binom{N}{n}}$ with $0 \leq k \leq n \leq N$, $k \leq M \leq N$ and $n - k \leq N - M$	$\sum_{k=1}^{x-1} \dfrac{\binom{M}{k}\binom{N-M}{n-k}}{\binom{N}{n}}$	$\dfrac{n \cdot M}{N}$	$\dfrac{nM(N-M)(N-n)}{N^2(N-1)}$
Poisson distribution	$\dfrac{1}{k!}(Np)^k e^{-Np}$ $k = 0, 1, 2, \ldots$	$e^{-Np} \sum_{k \leq x} \dfrac{(Np)^k}{k!}$	$n \cdot p$	$n \cdot p$

Designation	Density function $f(x)$	Distribution function $F(x)$	Expected value	Variance
Continuous uniform distribution	$\dfrac{1}{x_2 - x_1}$ for $x_1 \leq x \leq x_2$ 0 else	$\dfrac{x - x_1}{x_2 - x_1}$ for $x_1 \leq x \leq x_2$ 0 for $x < x_1$ 1 for $x > x_2$	$\dfrac{x_1 + x_2}{2}$	$\dfrac{(x_2 - x_1)^2}{12}$
Exponential distribution	0 for $x < 0$ $\lambda e^{-\lambda x}$ for $x \geq 0$	0 for $x < 0$ $1 - e^{-\lambda x}$ for $x \geq 0$	$\dfrac{1}{\lambda}$	$\dfrac{1}{\lambda^2}$
Gaussian distribution	$\dfrac{1}{\sqrt{2\pi}\sigma} e^{-\frac{1}{2}\left(\frac{x-\mu}{\sigma}\right)^2}$ for $-\infty < x < \infty$	$\dfrac{1}{\sqrt{2\pi}\sigma} \int_{-\infty}^{x} e^{-\frac{1}{2}\left(\frac{t-\mu}{\sigma}\right)^2} dt$ for $-\infty < x < \infty$	μ	σ^2
Logarithmic normal distribution	$\dfrac{1}{\sqrt{2\pi}\sigma \cdot x} e^{-\frac{1}{2}\left(\frac{\ln(x) - \mu}{\sigma}\right)^2}$ for $x > 0$ 0 for $x \leq 0$	$\dfrac{1}{\sqrt{2\pi}\sigma} \int_{0}^{x} \dfrac{1}{t} e^{-\frac{1}{2}\left(\frac{\ln(t) - \mu}{\sigma}\right)^2} dt$ for $x > 0$ 0 for $x \leq 0$	$e^{\mu + \frac{\sigma^2}{2}}$	$e^{2\mu + \sigma^2}\left(e^{\sigma^2} - 1\right)$

23.4.4 Parameter Estimates

Here, depending on the situation, different approaches come into play:
(a) Chi-squared distribution: distribution of a random variable Z that is a sum over squared random variables X_i—which are themselves standard-normally distributed. Application in the estimation of confidence intervals for variances
(b) Student t-distribution: test distribution used for small sample sizes, transitions to the standard normal distribution for large samples.
(c) Estimation methods: Determination of estimated values for means and standard deviations from samples
(d) Maximum likelihood method: Method used to obtain estimates for unknown parameters of probability functions or density functions,

Other important terms:
1. Confidence intervals: intervals within which the estimates/characteristic values of distributions with a specified level (confidence level) lie.
2. Parameter tests: Testing of statements or assumptions that are formulated as hypotheses. However, since such a hypothesis does not necessarily have to *be* true, it is contrasted with an alternative hypothesis. General procedure of parameter tests:
   (a) Formulation of the null hypothesis
   (b) Determining the significance level
   (c) Calculation of the test variable
   (d) Determination of the critical range
   (e) Test decision
3. Error of the first type: a (actually) correct null hypothesis is rejected
4. Error of the second type: a (actually) false null hypothesis is accepted as correct.

23.4.5 Validation of Methods

Three important methods and approaches:
- Standard addition/standard addition: Creation of calibration series for the verification of a method and for the quantitative determination of analytes, addition of known amounts of the analyte to the sample, creation of equilibrium lines, determination of the sample
- Internal standard: controlled addition of a substance chemically *similar to* the analyte, creation of calibration lines/compensation lines, determination of the sample
- External standard: Preparation of reference samples with different known concentrations of the analyte. These standard samples are measured with the test method to be validated; calculation of the compensation lines, determination of the sample

23.5 Outlier Tests

Again, there are different approaches:
- Dixon Q-test:

$$\text{Test size } Q = \left| \frac{x - x_n}{x_{max} - x_{min}} \right|$$

23.5 Outlier Tests

x is the potential outlier, x_n is the closest value to x, $x_{max} - x_{min}$ is the range of the data set; comparison with tabulated values is used to decide whether x is an outlier or not.

- 4σ-environment: if the potential outlier x_k is within the 4σ-environment from the mean ($x_t - 4st < x_k < x_t + 4st$), then x_k is *not* an outlier, otherwise the value is qualified as an outlier, x_t and s_t are mean and standard deviation without the potential outlier.
- Grubbs test:

  Test size $G = \dfrac{|x - \bar{x}|}{\sigma}$

  This method is based on the distance of the potential outlier x from the mean in relation to the standard deviation. By comparing this value with tabulated values, it is decided whether x is an outlier or not.

■■ Answers

1. $\bar{x} = 23.56$, $s^2 = 0.2029$, $s = 0.4504$, $\sigma_x = 0.1702$ (here we first calculate with more figures to avoid rounding errors), final result $x = 23.6 \pm 0.2$ (here again only *one* decimal place because of the significant figures).

2. According to ► Eq. (18.1), this gives a mean value $\bar{x} = 24.42$ mL. Since n(HCl) = n(NaOH) and (as we know) $c = n/V$, we arrive at n(NaOH) = 24.42 mmol in V(NaOH) = 20.00 mL, so c(NaOH) = 1.221 mol/L. For the measured values (i.e.: V(HCl)) we then get a standard deviation $s = 0.128\ 6468$ according to ► Eq. (18.8). This means that the actual amount of substance of HCl used could also be larger or smaller by this value. So we *could* also use this amount of hydrochloric acid to calculate the "variation" of the amount of sodium hydroxide, but this would not make much sense, because we have assumed here that the volume of sodium hydroxide to be titrated is really *exactly* 20.00 mL. In a real experiment, however, deviations would also occur here (perhaps we titrate only 19.998 mL, or there are actually 20.01 mL in the vessel). Here, therefore, in order to make more precise statements, we would have to enter the beautiful field of error propagation, which we will not go into until ► Chap. 19. However, you will encounter this task again.

3. Here R is the universal gas constant, which you already know from *general chemistry*. If we now want to determine the absolute and relative error for the activation energy, we resort to ► Eq. (19.1) to determine the error propagation and obtain the error ΔE_a:

$$|\Delta E_a| = \left|\frac{\partial E_a}{\partial T_1} \Delta T_1\right| + \left|\frac{\partial E_a}{\partial T_2} \Delta T_2\right| + \left|\frac{\partial E_a}{\partial k_1} \Delta k_1\right| + \left|\frac{\partial E_a}{\partial k_2} \Delta k_2\right|$$

Now the formula for the activation energy is derived once to T_1, once to T_2, then to k_1 and to k_2—this way we get the four contributions we need:

$$\Delta E_a = \left|R\left(\frac{T_2}{T_1 - T_2} - \frac{T_1 T_2}{(T_1 - T_2)^2}\right)\ln\left(\frac{k_1}{k_2}\right)\Delta T_1\right| + \left|R\left(\frac{T_1}{T_1 - T_2} + \frac{T_1 T_2}{(T_1 - T_2)^2}\right)\ln\left(\frac{k_1}{k_2}\right)\Delta T_2\right|$$
$$+ \left|R\frac{T_1 T_2}{T_1 - T_2} \cdot \frac{1}{k_1} \Delta k_1\right| + \left|R\frac{T_1 T_2}{T_1 - T_2} \cdot \frac{(-1)}{k_2} \Delta k_2\right|$$

If we reshape the equation a little, we get for the *absolute* error:

$$|\Delta E_a| = \left|\left(R\frac{T_2^2 \Delta T_1 + T_1^2 \Delta T_2}{(T_1-T_2)^2}\right)\ln\left(\frac{k_1}{k_2}\right)\right| + \left|R\frac{T_1 T_2}{T_1 - T_2}\cdot\left(\frac{\Delta k_1}{k_1} + \frac{\Delta k_2}{k_2}\right)\right|$$

For the *relative* error, we still divide this equation by E_a:

$$\left|\frac{\Delta E_a}{E_a}\right| = \frac{\left|\left(R\frac{T_2^2 \Delta T_1 + T_1^2 \Delta T_2}{(T_1-T_2)^2}\right)\ln\left(\frac{k_1}{k_2}\right)\right| + \left|R\frac{T_1 T_2}{T_1 - T_2}\cdot\left(\frac{\Delta k_1}{k_1} + \frac{\Delta k_2}{k_2}\right)\right|}{\left|R\frac{T_1 T_2}{T_1 - T_2}\ln\left(\frac{k_1}{k_2}\right)\right|}$$

In a nutshell:

$$\left|\frac{\Delta E_a}{E_a}\right| = \left|\frac{T_2^2 \Delta T_1 + T_1^2 \Delta T_2}{T_1 T_2 (T_1 - T_2)}\right| + \left|\frac{\frac{\Delta k_1}{k_1} + \frac{\Delta k_2}{k_2}}{\ln\left(\frac{k_1}{k_2}\right)}\right|$$

And now with the numerical values:
— The *relative* error is:

$$\left|\frac{\Delta E_a}{E_a}\right| = \left|\frac{300^2 \cdot 0.3 + 290^2 \cdot 0.3}{290 \cdot 300(290-300)}\right| + \left|\frac{0.05 + 0.05}{\ln\left(\frac{1.1\cdot 10^{-3}}{2.3\cdot 10^{-3}}\right)}\right|$$

$$= |-0.06003| + |-0.1356| = 0.1956 = 19.56\%$$

The relative error $|\Delta E_a/E_a|$ is therefore 19.6%.
The *absolute* mistake is:

$$|\Delta E_a| = \left|8{,}314\frac{300^2\cdot 0.3 + 290^2\cdot 0.3}{(290-300)^2}\ln\left(\frac{1.1\cdot 10^{-3}}{2.3\cdot 10^{-3}}\right)\text{Jmol}^{-1}\right|$$

$$+ \left|8{,}314\frac{290\cdot 300}{290-300}\cdot(0.05+0.05)\text{Jmol}^{-1}\right| |\Delta E_a| = |-3202.95 \text{ Jmol}^{-1}|$$

$$+ |-7233.18 \text{ Jmol}^{-1}| = 10436.13 \text{ Jmol}^{-1}$$

— So the *absolute* error is $|\Delta E_a| = 10{,}4k$ J mol^{-1}. (Do not forget to round sensibly!)

4. By logarithmising the temperature formula we get: $\ln(T) = \ln(k) + (1-\kappa)\ln(V)$
The required transformation is: $y = \ln(T)$, $x = \ln(V)$, the slope measure is $m = 1 - \kappa$ and the ordinate intercept is $b = \ln(k)$. We use the procedure analogous to the table in the box "Linear regression with different measured values":

I	1	2	3	4	5	Totals	Mean values
V/L	1	2	3	4	5		
T/K	401	302	259	229	211		
$x_i = \ln(V_i)$	0	0,693	1,099	1,386	1,609	4,787	0,9574
$y_i = \ln(T_i)$	5,994	5,710	5,557	5,434	5,352	28,047	5,6094

23.5 Outlier Tests

I	1	2	3	4	5	Totals	Mean values
x_i^2	0	0,4802	1,2078	1,9210	2,5889	6,1979	1,2396
$x_i \cdot y_i$	0	3,9570	6,1071	7,5315	8,6114	26,2070	5,2414

With the sums and averages calculated, we have everything at hand and determine the slope measure m and the intercept b using the formula ▶ Eq. (19.7):

$$m = \frac{\sum_i (x_i \cdot y_i) - N \cdot \bar{x} \cdot \bar{y}}{\sum_i x_i^2 - N(\bar{x})^2} = \frac{26.2070 - 5 \cdot 0.9574 \cdot 5.6094}{6.1979 - 5 \cdot (0.9574)^2} = -0.3995$$

and

$$b = \bar{y} - m \cdot \bar{x} = 5.6094 - (-0.3995) \cdot 0.9574 = 5.9919$$

Thus, the equation of the straight line is: $y = -0.3995 \cdot x + 5.9919$. With the assignments from the transformation, $\ln(k) = b = 5.9919$, so $k = 400.1724$, and with $1 - \kappa = m = -0.3995$, $\kappa = 1.3995$.

If we use this information for the temperature equation, we get the relation $T = 400.1724 \cdot V^{-0.3995}$.

5. With ▶ Eq. (20.8) we obtain

— for 5 × heads, 5 × tails:
$$P(X = 5) = \binom{10}{5} 0.5^5 \cdot 0.5^5 = 0.2461 = 24.61\%$$

— for 6 × heads, 4 × tails:
$$P(X = 6) = \binom{10}{6} 0.5^6 \cdot 0.5^4 = 0.2051 = 20.51\%$$

— for 10 × heads, 0 × tails:
$$P(X = 10) = \binom{10}{10} 0.5^{10} \cdot 0.5^0 = 0.000977 = 0.0977\%$$

6. The probability of functioning is:

$$p = \frac{(443 - 23) \cdot 1 + 23 \cdot 0}{443} = 0.9481$$

The probability of failure is then $q = 1 - p = 0.0519$. Thus, the system functions with a probability of 94.8%.

7. We solve this problem with the *two-sided* Student t-test, in which we compare our expected value with a literature value. In this test, we still need to estimate the mean and standard deviation of the batch beforehand.

The arithmetic mean $\bar{c}_{Ni} = 52.506\%$ and the standard deviation is $s_{Ni} = 0.335\%$ with a sample size of $n = 5$.

(a) Null hypothesis: We assume that the mean value deviates from the manufacturer's specifications only by chance. $H_0: E(X_1) = \mu_0$. This is again a two-sided test, i.e. the critical range for the test value is at higher or lower values, since deviations can occur in both directions.

(b) Specified as a significance level $\varepsilon = 0.95$, i.e., a probability of error $\alpha = 0.05$.

(c) The test variable T in this example is:

$$T = \frac{|\bar{x} - \mu_0|}{s / \sqrt{n}} = \frac{|\bar{x} - \mu_0|}{s} \cdot \sqrt{n}$$

Using the above values, we obtain a test variable $T = 1.0385$.

(d) This value of T must now be compared with the value from Student t-distribution for $m = n - 1 = 4$ degrees of freedom from the level $1 - \alpha/2 = 0.975$, i.e. with $t_{4;0.975}$. In ▶ Table 18.1 we read the value $t_{4;0.975} = 2.131847$. Our test value T is clearly below the t-value.

(e) Since the test value T is lower than the value of t, we can keep the hypothesis—the measurement deviates from the manufacturer's specifications only by chance.

8. Here the chi-square distribution is used. For ▶ Eq. (21.8) we still need the values c_1 and c_2 so that we can determine the confidence interval, for this we need the quantiles of the χ^2-distribution from ▶ Table 21.1: $c_1 = \chi^2_{n-1;\alpha/2}$ and $c_2 = \chi^2_{n-1;1-\alpha/2}$ for the concrete sample with $n = 15$ result in: $c_1 = \chi^2_{14;0.025} = 5.6287$ and $c_2 = \chi^2_{14;0.975} = 26.1189$. Putting this into ▶ Eq. (21.8), we get for the limits of the variance:

$$14 \cdot \frac{0.2}{26.1189} \leq \sigma^2 \leq 14 \cdot \frac{0.2}{5.6287} \text{ thus } 0.1072 \leq \sigma^2 \leq 0.4975.$$

9. As a reminder, the characteristic values: sample size $n = 5$, mean value $\bar{x} = 24.42$ mL and standard deviation $s = 0.128\,6468$ mL. Because of the small sample we calculate with the t-distribution and fall back on ▶ Eq. (21.7)

$$\bar{x} - c\frac{s}{\sqrt{n}} \leq \mu \leq \bar{x} + c\frac{s}{\sqrt{n}}$$

We obtain the value for c from ▶ Table 18.1 $c = t4_{;0.975} = 2.131847$ and the interval is:

$$24.42 - 2.131\,847 \cdot \frac{0.128\,6468}{\sqrt{5}} \leq \mu \leq 24.42 + 2.131\,847 \cdot \frac{0.128\,6468}{\sqrt{5}}$$
$$\rightarrow 24.297 \leq \mu \leq 24.543$$

For a 99% confidence interval, the factor is $c = t4_{;0.995} = 3.746947$, and the corresponding interval is:

$$24.42 - 3.746\,947 \cdot \frac{0.128\,6468}{\sqrt{5}} \leq \mu \leq 24.42 + 3.746\,947 \cdot \frac{0.128\,6468}{\sqrt{5}}$$
$$\rightarrow 24.204 \leq \mu \leq 24.636$$

10. Again, we revert to the tabular form for calculating the balance line. Calculation of the sums and mean values for the parameter determination of the compensation line

I	1	2	3	4	5	Totals	Mean values
x_i/mg	0,400	1,200	2,000	2,800	3,600	10,000	2,000
A	0,020	0,043	0,071	0,093	0,116	0,3430	0,0686
x_i^2	0,160	1,440	4,000	7,840	12,960	26,400	
$x_i \cdot A$	0,0080	0,0516	0,1420	0,2744	0,4176	0,8936	

The graph of the measuring points is shown in the figure below. We obtain the slope measure m with:

$$m = \frac{\sum_i (x_i \cdot y_i) - N \cdot \bar{x} \cdot \bar{y}}{\sum_i x_i^2 - N(\bar{x})^2} \text{ used : } m = \frac{0.8936 - 5 \cdot 2.000 \cdot 0.0686}{26.400 - 5 \cdot (2.000)^2} = 0.03244$$

23.5 Outlier Tests

We obtain the intercept b from: $b = \bar{y} - m \cdot \bar{x} = 0.00372$. The straight line equation is then:
$A = 0.03244 \cdot x + 0.00372$.
Here is the graph with data and compensation line:

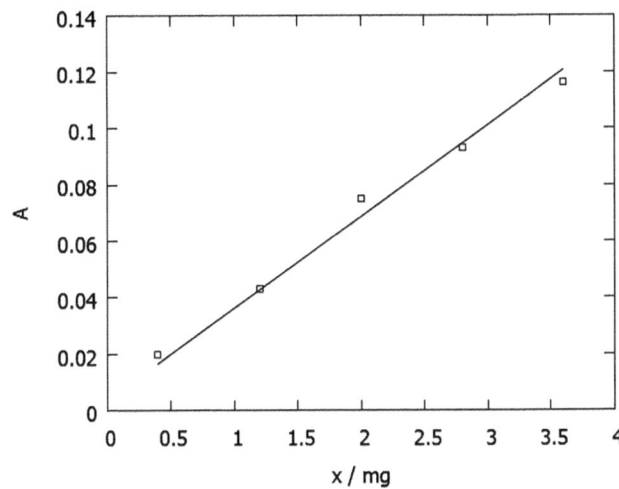

Pb absorption data and compensation line

11. In the equation from problem 9 ($A = 0.03244 \cdot x + 0.00372$) we substitute the value for A and solve for the concentration x *we are* looking for:

$$0.052 = 0.03244 \cdot x + 0.00372 \rightarrow x = (0.052 - 0.00372)/0.03244 = 1.4883$$

In the sample, the concentration of Pb is $x = 1.488$ mg.

12. For the *Dixon test*, we sort the data: 22.99; 23.01; 23.42; 23.66; 24.10; 24.99. The potential outlier is 24.99 because the distance to the second highest value is 0.89, which is larger than the distance between the two smallest values (0.02). The test statistic is Q is given by Eq. (23.1):

$$Q = \left|\frac{24.10 - 24.99}{22.99 - 24.99}\right| = 0.445$$

In ◘ Table 23.1, we see if the calculated value of Q for $N = 6$ shows up. The value for Q *is* above the value 0.387 corresponding to the significance level $\alpha = 0.2$ and below 0.484 with the significance level of $\alpha = 0.1$. Consequently, the measured value 24.99 must be rejected with a probability greater than 80% and less than 90%. However, since we are more than 10% likely to be wrong (and that is not a particularly small probability of error!), the value should be retained.

For the outlier test using the *4σ environment*, we need to determine the mean and standard deviation *without* the critical value. The mean value for the data 22.99; 23.01; 23.42; 23.66; 24.10 is $\bar{x} = 23.44$ and the standard deviation is $s = 0.4668$. If we add four times the standard deviation to the mean value we get the comparison value $x_m = 25.30$, thus the measured value is within the environment and must *not* be treated as an outlier

Using the *Grubbs test*, we need the mean and standard deviation of *all the* data points: $\bar{x} = 23.70$ and the standard deviation is $s = 0.7595$, the test statistic is now calculated using ◘ Table 23.2, and we get:

$$G = |24.99 - 23.70|/0.7595 = 1.6985.$$

For the test itself, we now still need the comparison values with a sample size $N = 6$; the tabulated comparison values are $G_{0.95} = 1.8221$ and $G_{0.99} = 1.9442$. Since the size G is below these values, the value 24.99 should be kept.

Further Reading

Bartsch H-J (2007) Taschenbuch Mathematischer Formeln. Carl Hanser, München

Binder H-J, Buhrow J, Just G, Meisel A, Mühlig H, Oelschlägel D (1993) Mathematik für Chemiker. Deutscher Verlag für Grundstoffindustrie, Leipzig

Brunner G, Brück R (2013) Mathematik für Chemiker. Springer, Berlin

Cann AJ (2007) Maths from Scratch for Biologists. Wiley & Sons, München

Dubben H-H, Beck-Bornholdt H-P (2003) Der Schein der Weisen—Irrtümer und Fehlurteile im täglichen Denken. Rowohlt, Hamburg

Dubben H-H, Beck-Bornholdt H-P (2011) Mit an Wahrscheinlichkeit grenzender Sicherheit—Logisches Denken und Zufall. Rowohlt, Hamburg

Dubben H-H, Beck-Bornholdt H-P (2013) Der Hund, der Eier legt—Erkennen von Fehlinformationen durch Querdenken. Rowohlt, Hamburg

Gieck K, Gieck R (1995) Technische Formelsammlung. Gieck, Germering

Harris DC (2014) Lehrbuch der Quantitativen Analyse. Springer, Heidelberg

Horstmann D (2008) Mathematik für Biologen. Spektrum, Heidelberg

Monk P (2006) Maths for Chemistry. Oxford University Press, New York

Papula L (2009) Mathematik für Naturwissenschaftler und Ingenieure, Band 3. Vieweg + Teubner, Wiesbaden

Skoog DA, Holler FJ, Crouch SR (2013) Instrumentelle Analytik. Springer, Heidelberg

Walz G, Zeilfelder F, Rießinger T (2005) Brückenkurs Mathematik. Elsevier/Spektrum, München

You might have noticed that in this part of the book the fundamentals of statistics were presented very mathematically—and that is a good thing, since you are not only supposed to be able to *use* the mathematical tools required here, but also should have understood it (at least the fundamentals of it—you will find a summary of much of it also in Harris.) Just in case this was a bit *too much of maths* for you, we recommend the books by Dubben and Beck-Bornholdt: The authors basically managed to put a square peg into a round hole by explainig the fundamentals of statistics both correct and demonstratively clear, using ... rather amusing examples now and then (as e.g. the title "Der Hund, der Eier legt" shows, translating as "The Oviparous Dog"). And while you are at it, you will also find out interesting details, which problems scientists and/or medical practitioners occasionally have with one or the other aspect of statistics.

Supplementary Information

Glossary – 334

Index – 353

© Springer-Verlag GmbH Germany, part of Springer Nature 2025
U. Ritgen, *Analytical Chemistry II*, https://doi.org/10.1007/978-3-662-68710-9

Glossary

Abscissa The correct technical term for the x-axis in an x/y diagram.

Absorbance (E) (also: Absorption A) Quantity describing the relationship between the analyte solution and the blank sample with regard to their interaction with the incident light. (The more strongly the analyte absorbs one or the other wavelength, the greater is the absorbance at this wavelength.)

Absorption Taking up of energy (in spectroscopy usually) in the form of electromagnetic radiation/photons, which changes the energy content of the absorbing particle (atom, molecule, ion, etc.).

Acceptance range (or better: non-rejection range) The acceptance range comprises the numbers for a test variable of a hypothesis at which the correctness of this hypothesis is to be assumed.

Acid error In the case of strongly acidic solutions, i.e. those with an extremely high content of hydroxonium ions, the measurement of the pH value with a glass electrode leads to a value that is a little (but noticeably and significantly) too high. A full explanation for this phenomenon has yet to be found.

Acronym Abbreviation for a term consisting of several words, usually with the initial letters of all (or most) of the corresponding words strung together; the resulting sequence of letters is then pronounced (and often treated) as an "independent word". The acronym most commonly used in analytics is probably "laser": *Light Amplification by Stimulated Emission of Radiation*. Numerous more complex NMR techniques are named with an acronym, occasionally using intermediate or final letters to construct said acronym (such as in COrrelationSpectroscopY, COSY).

Activity (a_x) The activity of a dissolved ion X; linked to the concentration via the activity coefficient. Please do not be confused: In some textbooks, the formula symbol for the activity is a capital A, in others a small a is used (in analogy to the concentration of a substance, to which the formula symbol c belongs). A uniform worldwide standard has not yet been established.

Activity coefficient (γ) Substance-specific factor (and dependent on the selected reaction conditions) that indicates the reduced reactivity of correspondingly interacting ions in the case of non-ideal solutions (in which the dissolved particles are not free of interaction).

Adduct A general term for compounds formed by the reaction of two suitable molecules; upon adduct formation, by-products are commonly not formed. Examples of adducts are the result of the interaction of electron donors and electron acceptors (Lewis acids and bases) or the products of Diels-Alder and other pericyclic reactions, in which no by-products are split off.

Adsorbate The material that is attached to the surface of the adsorbing material (the adsorbent) during an adsorption process.

Adsorption "Sticking" of a substance to the surface of another material; in contrast to a<u>b</u>sorption, the adsorbed substance remains on the surface and does not penetrate deeper into the adsorbate's structure.

Aglycone A non-sugar component that has taken the place of an H atom of a free glycoside; the part of a glycosylated substance that is not a carbohydrate.

Alkali error When the sodium content of an alkaline analyte solution is high, the glass electrode also responds to the Na^+ ions; accordingly, a pH value that is too low is measured. Despite the common name for this phenomenon, other alkali metal than sodium ions have a much less pronounced effect.

α-cleavage Cleavage occurring within a molecule with (at least) one heteroatom which has (at least) one free electron pair, in which the bond is broken which exists between the chain members in α- and β-position to the heteroatom in question. The free electron pair at the hetero atom stabilises the positive charge occurring in the course of mass spectrometry.

α-decay Radioactive decay process in which a radioactive isotope spontaneously releases a helium cation (He^{2+}) from the atomic nucleus; this leads to a decrease in the mass number of said isotope by 4 and a decrease in the atomic number by 2—real elemental transformation takes place.

Allyl cleavage Cleavage occurring within an unsaturated molecule, in which the bond "next but one" to the C-C double bond is broken, so that a (radical) allyl ion ($CH_2 = CH\text{-}CH_2^{+\bullet}$) is formed. (Of course, any H atom in this structure can also be replaced by any other substituent). By interaction of the positive charge with the π-portion of the C-C double bond, the charge of the allyl ion is delocalised and thus stabilised.

Glossary

Amalgam Any alloy of a metal with mercury (solution of a metal in mercury).

Amorphous Solid whose individual components are randomly connected/interlinked in three dimensions. An amorphous solid has no long-range order and only a partially pronounced short-range order.

Ampere (A) Unit of current intensity. $1\ A = 1\ C\ s^{-1}$.

amu See u

Analgesic Painkiller

Anisotropy The magnetic behaviour of many chemical bonds depends on the spatial orientation of the atoms involved, i.e. the magnetic shielding of the atoms involved in or immediately adjacent to a bond depends on their relative spatial orientation. Of particular importance here are multiple bonds and cyclically conjugated systems, especially aromatics.

Anode The electrode at which oxidation takes place; in electrogravimetry, the counter electrode.

anomeric centre The chirality centre that is formed during the formation of the cyclic hemi- or full-acetal/-ketal from the open-chain form of a saccharide; it is located in the immediate vicinity of the "ring-oxygen".

Anthropogenic Technical term for "elicited by humans", "caused by humans", "man-made".

Antibody A protein produced by the (animal or human) immune system in response to an antigen; it interacts with the unfamiliar substance (usually without forming new covalent bonds) and thus blocks it, preventing the antigen from exerting its toxic or otherwise undesirable effect.

Antigen (Bio)molecules (usually proteins, but can also be carbohydrates or other compounds) to which the immune system of an animal or human organism responds in order to contain or completely prevent the undesired effect of the unfamiliar molecule.

Antonym Antonyms are words with opposite meanings; chemical example: noble/non-noble (often called: base)

Aptamer Short DNA or RNA sequences that interact (in single or double-stranded state) specifically with an analyte.

ATP (adenosine triphosphate) The most important energy transporter of practically all cells.

Autosampler Device for automatic sample feeding.

Azo coupling Electrophilic aromatic substitution of a substituent (usually an H atom) by an (aryl) diazonium cation to form an azo bridge; synthetic route for obtaining azo dyes.

Azo dye Synthetic dyes with an azo bridge ($-N = N-$) as chromophore.

Bell curve See Gaussian distribution

Benzyl cleavage Fragmentation analogous to allyl cleavage in mass spectrometry; here, stabilisation of the (radical) cation occurs through interaction with the π-electrons of the aromatic system.

Bernoulli experiment If only two results/outcomes are possible in an experiment (e.g. when tossing a coin—and please don't make any quibbles about the fact that it could at least in principle land on the edge …), this is known in statistics as a Bernoulli experiment.

Best value The best value is the result that comes closest to the true value for the measurand.

β-cleavage Fragmentation analogous to α-cleavage, in which, however, the bond existing between the β- and γ-atoms to the heteroatom is broken. A special case of β-cleavage, in which an additional H-shift occurs, is referred to in mass spectrometry as a McLafferty rearrangement, and in organic synthesis (in which other reaction conditions prevail) as an En reaction.

β-decay Radioactive decay process in which a neutron of a radioactive nuclide transforms into a proton. In ordinary β⁻ decay, an electron (e⁻, in the formula symbolism of nuclear chemistry: β⁻) and an antineutrino ($\bar{v}$, uncharged) are released in the process. An alternative decay process leads to a positron (e⁺ or β⁺) and a neutrino (v, also uncharged).

Betaine Group name for zwitterionic compounds: Molecules that carry no net charge overall, but for which no Lewis formula can be established without charge separation.

Binomial distribution Discrete probability distribution of independent trials, each with only two possible outcomes.

Blank sample A term commonly used in the life sciences for a sample that contains neither the analyte nor a substance very similar to the analyte (such as an enzyme or similar) and therefore does not lead to the characteristic (colour) reaction, just as the negative control (with which the blank sample should not be confused!) does not. The blank sample ensures that the "additional reagents" required for the detection reaction alone do not lead to a false-positive result.

Broadband decoupling *See Decoupling.*

Calomel electrode Common reference electrode in potentiometry, based on the reduction of monovalent mercury to elemental mercury; if the potential of such an electrode is kept constant by a saturated potassium chloride solution, it is also referred to as a *standard calomel electrode* (SCE). Since in the presence of the KCl solution the activity of the counter ions of the monovalent mercury (Hg_2Cl_2) is not $a(Cl^-) = 1$, the following applies: $E(Hg^+/Hg) < E^0(Hg^+/Hg)$.

Calorimetry General: Method for determining heat quantities; of particular importance in analysis is (dynamic) differential calorimetry, *see there*.

Cambridge half-life In 1990, IUPAC recommended value for the half-life of the carbon isotope ^{14}C: $t_{1/2}(^{14}C) = 5715 \pm 30$ years.

Cathode The electrode at which reduction takes place; in electrogravimetry, the working electrode.

Central Limit Theorem Important statement from statistics: If enough samples are taken (than one is sufficiently large), the sum of arbitrarily distributed random variables is normally distributed.

Challenger disaster In 1986, the space shuttle "Challenger" broke apart shortly after launch; the entire crew died.

Chemical equivalence Chemical equivalence of different atomic nuclei (of the same atomic type) is said to exist when the nuclei in question are in the same environment, i.e. have the same type of bonding partner and cannot be distinguished from one another even with regard to any anisotropy effects. For some analytes, chemical equivalence is only given above a certain (usually quite low) temperature, because only then—for example with sufficient thermal excitation for free rotation around a single bond—can the individual nuclei actually no longer be distinguished.

Chemical ionisation (CI) Ionisation method commonly used in mass spectrometry in which the analytes interact with smaller molecules (short-chain hydrocarbons, water, ammonia, etc.) that have been ionised in advance and are converted to radical cations. Chemical ionisation is less aggressive than electron impact ionisation, therefore analytes ionised in this way experience significantly less fragmentation.

chemical shift (δ) A measure of how easy (or how difficult) it is to excite an atomic nucleus to resonance; dependent on various factors (shielding due to increased or decreased electron density, effect of anisotropy, etc.). The chemical shift is always given in relation to a reference substance (in ^1H- and ^{13}C-NMR: tetramethylsilane, TMS). The chemical shift depends on the magnetic field strength applied during the recording of the corresponding NMR spectrum and is commonly given in ppm.

Chemisorption Adsorption of molecules or other multi-atomic assemblies onto the surface of a substance, whereby the cohesion between adsorbate and adsorbent is brought about by bonds within the adsorbate being weakened or even completely dissolved, while new bonds are formed between the adsorbed substance and the adsorbent. Unlike physisorption, the adsorbed substance in the chemisorbed state therefore changes its chemical properties, e.g. becoming more reactive.

Chromophore Functional group that facilitates excitation of an electron system, so that the absorption of the compound is shifted into the VIS range.

Clark electrode Sensor system for amperometric quantification of dissolved oxygen.

COLOC (COrrelation through LOng-range Coupling) 2D NMR experiment in which short-distance and long-distance coupling are used to elucidate the structure. The most common is the CH-COLOC variant, in which $^2J_{CH}$ and occasionally even $^3J_{CH}$ couplings are considered in addition to the 1JCH couplings.

Complete survey In a complete survey, all objects of a population are included in a study. (Then is either n very large or n very small. Mostly the latter).

Completeness relation Describes the occurrence of a certain event. (For example, if there is an electron, it must also be somewhere ...)

Compound nucleus The (excited) atomic nucleus produced during neutron activation by the target isotope capturing a neutron and thus increasing its mass number by 1. Often, neutron capture is followed by β⁻-decay, so that a neutron of the affected nucleus converts to

Glossary

a proton (and an high-energy electron is emitted). In this process, the mass number does not change, but the atomic number does. (It increases by 1—yes, that is indeed elementary transformation.)

Concentration chain A galvanic cell constructed from two half-cells, each containing the same electrode and the same electrolytes, differing only in their concentration. The potential resulting from this concentration difference can be calculated using the Nernst equation.

Concentration potential The diffusion potential that is formed in electrogravimetry where, due to the interaction of the analyte solution with the electrode, a different (lower) analyte concentration prevails in its immediate vicinity than in the "rest" of the solution.

Concerted reaction A reaction in which several bonds are broken or newly formed at the same time; usually they occur via several simultaneous electron shifts. The decisive factor for concerted reactions of any kind is that no intermediate steps occur (or can be detected).

Conductive electrode Any electrode that only passes on a potential without itself being involved in the potential-forming reaction.

Confidence interval Interval from statistics that indicates the precision of the estimate of a parameter (for example, a mean value).

Confidence interval *See confidence interval*

Confidence level (confidence level, statistical certainty) (γ). *See confidence level*

Confidence level (γ) Such a level is set to determine the confidence interval within which the parameter (e.g. the true mean) can be found with just the specified statistical confidence. The higher this confidence level is chosen, the larger the confidence interval will be.

Connectivity The connectivity of a molecule describes which atoms interact with each other through direct, directed bonds; constitutional isomers differ in their connectivity. Example: ethanol (CH_3-CH_2-OH) and dimethyl ether (CH_3-O-CH_3).

Consistently stable firm, stable, here: as the size of a sample increases, the estimated mean and the true mean come closer and closer together.

Continuous For small changes in the function argument, the changes in the function values are correspondingly small.

Conventional ^{14}C-age Designation for age determinations using radiocarbon dating, which are based on a half-life $t_{1/2}(^{14}C) = 5730 \pm 40$ years; also referred to as **Libby half-life**. In the meantime, the Cambridge half-life is considered authoritative.

Conversion Radiationless transition of a system from one state to another. A distinction is made between:

External conversion (EC) Conversion of an excited system can occur by direct interaction of the analyte under consideration with a solvent molecule, which is itself energetically excited (although this is difficult to observe, let alone quantify). The more easily a solvent can be excited, the more likely an external conversion will occur; for this reason, the solvent used in a fluorescence experiment has a great influence on the intensity of the quantifiable fluorescence.

Internal conversion (IC) Radiationless transition of an electronically and/or vibratorily excited system from one state to another without changing the multiplicity state of the system; part of the previously absorbed energy is released to its environment (usually: the solvent).

Correlation spectroscopy (COSY, *COrrellation SpectroscopY*) 2D NMR experiments that allow statements to be made about which atomic nuclei couple with each other (and which do not) on the basis of the couplings that occur; a distinction is made between homonuclear variants, in which the same atomic species is considered in both dimensions (such as HH-COSY), and heteronuclear techniques, in which different atomic nuclei are relevant in the two dimensions (e.g. CH-COSY).

coulometric titration Alternative term for galvanostatic coulometry

Counter electrode The electrode of an electrolysis system at which the second half-cell reaction takes place, which does not cause a chemical change in the analyte; in electrogravimetry this is the anode, which accordingly provides the electrons required for the reduction at the cathode.

Coupling constants In contrast to the absolute position of the NMR signals of a spectrum (specified in MHz), the distances of the individual lines of a multiplet are not dependent on the selected measurement conditions: They are constant (substance-specific), therefore they can be specified with absolute values (in Hz).

cpm (*Counts Per Minute*) Unit used to indicate the measurement results of a photon or scintillation counter.

Cramer's rule With the help of Cramer's rule linear systems of equations can be solved, the rule is also known as the **determinant method**, since determinants must be calculated to solve the system of equations.

Cross-contamination In flow injection analysis: The running together of different "analyte plugs" when the respective samples are injected too close to each other. To be avoided.

Crystalline A solid is said to be crystalline if the individual constituents exhibit long-range order in addition to a short-range order. In the case of polymers, there is also the "special case" of semi-crystallinity: Here, in addition to disordered (amorphous) areas, there are also zones in the three-dimensional structure of the polymer chains in which some chains are ordered (anti-)parallel and are thus locally crystallised. (The extent of the (microscopic) crystallinity of a polymer has a decisive influence on its macroscopic properties).

Current The quantity of charge that flows through an electric circuit within a unit of time. The unit of the electrical current is the ampere (A), thus stating the *amperage*.

Current density Intensity of an electrical current per unit area at the electrode surface under consideration; the current density is commonly given in the unit A/m^2 or A/cm^2. (When using table values, please note which unit has been used in each case.)

Cyclic voltammetry *See Voltammetry, cyclic*

D The hydrogen isotope ^2H is one of the (extremely) few isotopes for which a separate element symbol has been established: D. (This honour has otherwise only been bestowed on the (radioactive) hydrogen isotope ^3H: Tritium has the element symbol T.)

Dalton (Da) common name for the unit of the atomic mass, u (*see there*); particularly popular in biochemistry.

Deactivation processes Radiationless transitions that take place in an electronically and/or vib./rot. excited system and do not lead to observable fluorescence. Deactivation processes decrease the quantum yield.

Decay radiation The radiation that occurs when an atomic nucleus activated by neutron capture becomes radioactive itself due to the changed mass number and emits, among other things, a γ-photon in the course of decay.

Decomposition voltage The voltage required to actually start electrolysis; the decomposition voltage is in principle higher than the pure redox potential of the analyte to be electrolysed (mainly because of the overvoltage).

Decoupling Coupling can be avoided in ^{13}C-NMR by selective irradiation of the coupling frequencies (or the frequency range in which the couplings are expected to lie). If one covers the entire range, so that both J_{CC} and J_{CH} are avoided, one speaks of **broadband decoupling**. Via **selective decoupling**, in which a different frequency range is irradiated, it can be achieved that only CC interactions are omitted, but any information that can be obtained by CH couplings is not lost.

Defect In solid-state chemistry, we speak of defects when lattice sites of a crystal structure remain unoccupied (vacant) due to a disorder in the lattice or due to admixtures of other substances.

Definitional domain Set of all objects (typically numbers) that lead to well-defined statements in more specified contexts.

Degrees of freedom In statistics, the degree of freedom corresponds to the number of independent observed values minus the number of estimable parameters; not to be confused with the degrees of freedom from IR spectroscopy.

δ (delta) In NMR, common sign for the chemical shift, usually given in ppm.

Density function *See probability density*

Depsipeptide A peptide-like molecule whose main chain also contains at least one ester group. Depsipeptides include many (but not all) macrolide antibiotics such as valinomycin.

De-Shielding Increased sensitivity of an atomic nucleus to the effect of the magnetic field caused by a reduction in the electron density and other effects. De-shielded atomic nuclei can be more easily excited to resonance (by less energetic excitation radiation or already in the somewhat weaker magnetic field). This leads to a low-field shift of the signal concerned. Opposite: shielding.

Determinant Method *See Cramer's rule*

Deuterated In a deuterated solvent, all hydrogen atoms are replaced by ^2H isotopes (i.e. deuterium, D); accordingly, one uses D_2O for water, $CDCl_3$ for chloroform, for dimethyl sulfoxide $(CD_3)_2SO$ is used, etc. Of course, these must first be synthesised (which can be quite laborious); they are available in specialised shops (but correspondingly expensive).

Glossary

Diaphragm Membrane permeable only to certain particles

Diels-Alder Reaction Pericyclic reaction in which new C-C bonds are formed by concerted displacement of electrons (or: electron density) and, if necessary, multiple bonds are converted into single bonds. You will encounter this and similar cycloadditions again in advanced organic chemistry.

Differential scanning calorimetry (DSC) Measurement method used to determine the amount of heat/energy required to cause phase transformations and the like.

Differential thermal analysis (DTA) Investigation of the behaviour of an analyte at increasing temperature in comparison with a reference substance; allows statements to be made about exothermic or endothermic processes that occur at increased temperature. Differential thermal analysis is mainly used for polymers and other non-metallic materials.

Diffusion potential Potential difference resulting from the different mobility of different charge carriers; occurs e.g. at the interface between an electrolyte solution and an electrode, but also at the interface between two different electrolyte solutions, i.e. for example at the two ends of a salt bridge, which provides for ion exchange between two half cells of a galvanic element.

Dimensional analysis To check whether the interaction of different units ultimately leads to a result with the correct unit or not. In the case of derived units (such as resistance, Ω), it is necessary to keep in mind what these units are "composed of" (1 Ω = 1 V/A, etc.).

Dimensionless A dimensionless quantity is any quantity without a unit; typical representatives of dimensionless quantities in chemistry are electronegativity or activity.

Direct Dissociation *See Dissociation*

Direct potentiometry Determination of the concentration of an analyte from the potential difference to a reference electrode.

Discrete Distinguishable, separable, countable. (Again an example that technical language does not necessarily have to be compatible with everyday language.)

Displacement *See chemical shift*

Dissociation, direct (DD) In fluorescence experiments: immediate bond cleavage caused by the irradiated excitation radiation; does not lead to fluorescence and tends to increase with increasing energy content of the excitation radiation used (but of course also depends on the stability of the analyte under consideration). Please do not confuse this with predissociation.

Doublet *See Multiplett*

DSC *See Differential Calorimetry*

DTA *See Differential thermal analysis*

EC *See Conversion*

EIA (Enzyme-linked immunosorbent assay) *See ELISA*

Electrochemical Series *See Galvanic Series*

Electrode In galvanic cells: the (metallic) conductor that is not in solution (i.e. not dissolved), but is in contact with the solution so that electrochemical reactions are possible.

Electrolysis Electrolysis occurs where the application of electric current forces a redox reaction that would "normally" occur in the opposite direction; the opposite of what occurs in a galvanic cell.

Electrolyte In galvanic cells: the substance in solution that is dissociated to form cations and anions and therefore conducts current; the solution in contact with the electrode so that an electrochemical reaction can occur.

Electron impact ionisation (EI) The most common technique in mass spectrometry for ionising analytes; these are bombarded with an electron beam and thus converted to the corresponding radical molecule cations via the loss of one of their own electrons. These then undergo further fragmentation with the splitting off of neutral radicals or neutral molecules. In most cases, only single ionisation takes place (with $z = 1$), but depending on the type of analyte, multiple ionisation is also possible.

Electron spray ionisation (ESI) A method commonly used in mass spectrometry for ionising analytes that are not transferred into the gas phase as pure substances but are introduced into an electric field in the form of a solution in the presence of a dry gas; used especially for large analytes that cannot be vaporised in pure form.

ELISA (*Enzyme Linked ImmunoSorbent Assay*) Used for the detection of analytes; the ELISA is based on the interaction of antigen and antibody and leads to an enzymatically catalysed colour change (or: formation).

Emission Loss of "excess" energy in the form of photons with (usually characteristic) energy content/wavelenght.

En reaction Pericyclic reaction in which an allyl hydrogen atom changes the bonding partner. The corresponding analogue with subsequent fragmentation in mass spectrometry is called McLafferty rearrangement.

Ensemble Total, population, basic set of all objects belonging together.

Error analysis Mathematical methods for calculating the errors and deviations from a quantity to be measured.

Error propagation law Gaussian method for calculating the error of a variable which cannot be measured directly and which depends on various other measured variables. The individual errors determine the error of the quantity to be calculated.

Errors A distinction is made between statistical and systematic errors. In the case of systematic errors, the deviations are unilaterally directed and are caused by causes that can in principle be determined. Under the same conditions, systematic errors cannot be detected in repeated measurements. **Random error fluctuates in** a series of measurements around the value to be measured. Even with an identical experimental setup, one never obtains identical measured values with repeated measurements: such deviations are random errors.

Estimators Estimators are used to obtain estimates from samples and thereby obtain information about unknown parameters of a population. They are the basis for calculating point estimates and determining confidence intervals.

Expectation dispersion An estimator is expectation dispersion if its expected value is equal to the true value of the parameter to be estimated.

Expected value The expected value is the numerical value of a random variable that it assumes on average. (The expected value does not necessarily have to be reached or may even be unattainable. If you roll a standard dice—the one with six sides—often enough, then add up all the dice results and divide by the number of rolls, you get an expected value of 3.5. Actively rolling the dice to obtain that value is ... tricky, to say the least.)

External Conversion *See Conversion*

Extrapolation The estimation of expected measured values beyond the (experimentally) assured range; however, with every extrapolation there is the danger that the (linear) relationship between concentration and resulting measured value no longer exists.

Faraday constant (F) The charge of one mole of electrons; since an electron carries an elementary (negative) charge, the value for the Faraday constant is $(1.602 \times 10^{-19} \text{ C}) - (6.022 \times 10^{23} \text{ mol}^{-1}) = 9.649 \times 10^4$ coulombs/mol; usually given as 96,485 C.

Faraday's laws Physical laws describing the relationship between current intensity, charge quantity, current flow, and analyte deposition.

Fast Atom Bombardment **(FAB)** Mass spectrometric technique for the investigation of analytes which undergo decomposition upon the attempt of vapouring them.

FIA *See Flow Injection Analysis*

Field desorption (FD) Variant of field ionisation in which analytes are desorbed from the surface of a metal wire to which they have previously been applied in the presence of an electric field.

Field ionisation (FI) Ionisation of analytes by applying a strong electric field (about 10^{10} V/m). Under these conditions, molecular ions are obtained which allow the m/z ratio of an analyte of previously unknown mass to be determined; fragmentation is largely eliminated.

First order/first type electrode Electrode whose potential depends directly on the concentration of the electrolyte solution surrounding it.

Flow injection analysis (FIA) Analysis technique in which the analyte solution is injected into a capillary in which there is a continuous flow of a flowing medium containing any necessary reagent required for detection; permits automated analysis of numerous samples within a comparatively short period of time. The variant known as **sequential injection analysis**, which uses considerably fewer chemicals and can also be automated, does not involve a continuous flow but a (computer-)controlled variable flow of the same fluid, in which the direction of flow can also be reversed by means of pumps.

Fluorescence quantum yield (Φ) Ratio of fluorescence photons to the number of total irradiated excitation photons; $\Phi \leq 1$.

Fluorescence spectrometry Variant of spectrophotometry in which the extent to which irradiated excitation photons cause the fluorescence of corresponding analytes is determined. In **time-resolved fluorescence spectrometry**, the phosphorescence properties (yes, really not fluorescence!) of lanthanide ions (usually in chelated form) are exploited.

Fluorescence If a substance is excited by electromagnetic radiation of wavelength λ_1 to emit electromagnetic radiation of wavelength λ_2, for which the following applies: wavelength $\lambda_2 > \lambda_1$, fluorescence is present. Fluorescence stops almost immediately after the excitation ceases. Not to be confuses with phosphorescence.

Fluorophore Functional group or molecular moiety that is readily excited to produce fluorescence. Complete molecules that exhibit these properties are also occasionally referred to as fluorophores—especially when they are increased in reactivity by minor chemical modifications (which do not affect the fluorescence properties) and then form adducts with analytically relevant (bio)molecules via covalent bonds. Thus, the term fluorophore is to be used analogously to the term chromophore (known from Part IV of "Analytical Chemistry I").

Galvanic element Two electrochemical half-cells separated from each other, in which, when they are conductively connected, a potential difference builds up so that an electrical current flows, whereby oxidation or reduction occurs in the respective half-cells.

Galvanic Series Listing of the electrochemical standard potentials of various redox pairs, related to the reference value of the normal hydrogen electrode.

Galvani voltage The voltage resulting from the internal potential difference between two phases in contact with each other, such as the (metallic) electrode of a half-cell and the electrolyte. The Galvani voltage of a half-cell cannot be measured directly.

γ-radiation Energy-rich electromagnetic radiation that is released in the course of a radioactive decay process; can be quantified.

Gamma function (Γ) Special function used in statistics. As a special property, the gamma function for a natural number (n + 1) gives the value n! ($\Gamma(n + 1) = n!$), so it has to do with combinatorics.

Gaussian algorithm Procedure for solving systems of linear equations.

Gaussian distribution One of the most important continuous distributions; it is symmetrical and characterised by expected value and variance. The shape of the density function resembles that of a bell, which is why it is often referred to as a bell curve.

GC-MS Combination of gas chromatography (GC) and mass spectrometry (MS): The individual components of a mixture present in the gas phase are each examined by mass spectrometry.

Germinal The term geminal substituents is used when they are bonded to the same carbon atom, for example in 1,1-dichloroethane (CH_3-$CHCl_2$). There are also compounds with geminal functional groups, although most geminal diols convert spontaneously to the corresponding carbonyl groups under water splitting, for example, H_3C-$CH(OH)_2$ → CH_3-CHO (ethanal, acetaldehyde) + H_2O; not to be confused with *vicinal* (e.g. vicinal diols are often stable).

Glass temperature (T_G)/glass transition Below a substance-specific temperature, an amorphous substance (usually a polymer) has glass-like properties, i.e. it is (more or less) transparent and very brittle (not very elastic). Above this glass temperature, i.e. after the glass transition, such a substance, depending on the chemical composition and degree of cross-linking of the individual polymer molecules, either melts completely or at least softens considerably (it then becomes *rubbery*).

Haber-Bosch synthesis The synthesis of ammonia from the elements under high pressure in the presence of a catalyst.

Half cells The parts of a galvanic element in which either reduction or oxidation takes place. For a complete galvanic element, both half-cells are always required, since oxidation can never take place without the associated reduction, and vice versa.

Half-life ($t_{1/2}$) The half-life of a radioactive isotope indicates the period of time in which half of the radioactive atoms present undergo decay. The half-life is completely independent of the amount of material present: Whether 1 mole is present or only 100 atoms: After the half-life has elapsed, in each case half of them will have decayed; accordingly, the radioactivity of a sample continues to decrease exponentially. So far, no way has been found to determine *which* atom enters into this nuclear chemical reaction and which does not: according to the current model of physics, this is in principle undetermined and thus proceeds purely statistically.

Half-step potential ($E_{1/2}$) The potential at the inflection point and midpoint of the current-rise curve in a polarogram; also the inflection point of the current-fall curve in cyclic voltammetry. If the oxidised and reduced forms of the analyte are present in solution under equilibrium conditions, $E_{1/2} \simeq E^0$.

Haptens Low-molecular compounds (in immunological tests: the analytes) that are not themselves capable of causing an immune reaction, but which do so when the hapten in question is bonded to a (usually endogenous) carrier protein. Hence the name: it is derived from the Greek word ἅπτειν *(haptein)* = "to grasp".

Heavy atom effect The more massive an atom, the more it can interfere with photophysical/chemical processes and, for example, promote an *intersystem crossing*; in the case of organic analytes, in particular the heavier halogens (Br, I) show this effect.

Hetero-atom In NMR, all atomic nuclei that are not carbon or hydrogen are considered hetero-nuclei.

high field shift If a stronger magnetic field is required in NMR to excite an atomic nucleus to resonance, the corresponding NMR signal appears in the spectrum at smaller δ-values ("further to the right").

HOMO (*Highest Occupied Molecular Orbital*) The orbital with the highest energy populated with one or two orbitals in the ground state of a multi-atom system.

Hydrate *See Water of crystallisation*

Hygroscopy Property of a substance to attract moisture from the environment (especially atmospheric humidity) and thus to change its chemical and physical properties; in gravimetric analysis, the increase in mass due to accumulated water must be taken into account (or ideally: avoided).

Hypergeometric distribution Discrete probability distribution in quality control; allows statements to be made about the probability of objects with a certain property being present within an aggregate.

IC *See Conversion*

Ideal solution A solution whose concentration is low enough to assume that all solution components do not interact with each other. In such cases, one assumes that $a(X) = c(X)$, i.e. activity and concentration may be equated. If the concentration of a solution is too high for all the particles present to be considered "interaction-free", then is not an ideal solution, but a *real solution*: In this case, the activity must be considered instead of the concentration.

In situ A substance is said to be "*in situ*" if it is generated within the reaction vessel and immediately brought to reaction; of particular importance in coulometric quantifications.

INADEQUATE (*Incredible Natural Abundance DoublE QUAntum Transfer Experiment*) 2D NMR technique, in which information is obtained from $^1J_{CC}$ couplings about which C atoms are directly bonded to one another (via a C-C single or multiple bond). Due to the low natural occurrence of the ^{13}C atoms required for this, this method is relatively insensitive, but very informative.

Incubation period Generally, the period of time that elapses between infection with a pathogen and the first appearance of symptoms of disease; in immunoassays (such as the RIA), the period of time during which labelled and unlabelled antigen compete for interaction with the antibody.

Indicator electrode In potentiometry: An electrode that responds to the analyte itself.

Inner conversion *See Conversion*

Integral In ^1H-NMR the area lying under an NMR signal; allows statements about the number of hydrogen atoms belonging to the respective signal. Please do not try to infer information from signal *heights*.

Interpolation If the available measured values y for an analyte content x allow the creation of a trend line (= calibration line), this allows the estimation of an expected measured value y for not-measured values for x lying between the two extremes of the measured data; at the same time, the resulting y value can be used to infer the x value belonging to a newly measured sample.

Intersystem Crossing (ISC) Radiationless transition of a system from one state to another, where the multiplicity state of the system changes; mostly symmetry-forbidden.

Ionic conductivity Electrical conductivity that occurs in crystalline solids due to the (limited) mobility of the cations and/or anions present in them; occurs primarily when the radii of cations and anions differ significantly.

Ionic strength (I, also: μ) The electric field strength resulting from the charge carriers (cations and anions) dissolved in a solvent; in the case of multivalent ions, the charge influences the ionic strength as an exponent.

Ionophore Technical term for ion transporter; an ionophore causes increased solubility of the particle complexed by it (or otherwise interacting with it) in a solvent in which said particle is not soluble per se or is only very slightly soluble.

ISC Abbrev. for Intersystem Crossing

ISE Abbrev. for Ion-Sensitive Electrodes

Isobaric interference In mass spectrometry, isobaric interference occurs when particles with approximately identical mass/charge ratios are produced so that their peaks coincide (if the resolution is insufficient).

Isothermal mode Performing a differential calorimetry experiment at a given (constant) temperature.

Isotope Atoms of the same kind differ in their mass number but not in their atomic number. Due to their different number of neutrons, different isotopes have different masses.

Isotope labelling The deliberate introduction of a specific isotope to label a functional group or a moiety of a molecule; of particular importance for the elucidation of reaction mechanisms and metabolic pathways. If the isotope used is radioactive, this simplifies its tracking; it is then referred to as a **tracer**.

IUPAC *International Union for Pure and Applied Chemistry;* generally recognised institution for all questions concerning the designation of chemical compounds (nomenclature) as well as with regard to the natural constants to be used in chemistry, etc. The recommendations of the IUPAC are not legally binding, but they are usually observed. (Occasionally, however, they are simply ignored.)

K_a Equilibrium constant of a homogeneous solution equilibrium, in which an ideal solution must not be assumed due to (quite) high concentrations so that the respective activities must be taken into account.

Karplus-Conroy relationship Indicates the relationship between the coupling constant of atoms bonded to neighbouring (C) atoms and their torsion angle; particularly important for $^3J_{HH}$ couplings. In general, a corresponding coupling has a minimum at a torsion angle of 90° and experiences a maximum at ecliptic (syn-periplanar) conformation (torsion angle = 0°) and in anti-position (anti-periplanar, torsion angle = 180°). For $^3J_{HH}$ couplings the difference can be more than 10 Hz. In the literature you will find corresponding Karplus-Conroy curves which show this relationship graphically. (Please do not be surprised: In some textbooks and technical dictionaries Harold Conroy's contribution is consistently omitted, although he was able to explain the findings presented by Martin Karplus mainly phenomenologically on the basis of molecular orbital interactions. Accordingly, the relationship between the torsion angle and the coupling constant is often only listed as *Karplus relationship*.)

K_c Equilibrium constant of a homogeneous solution equilibrium in which all concentrations involved are low enough to speak of an ideal solution.

K_p Equilibrium constant of a homogeneous gas phase equilibrium whose individual components i each have the partial pressure p_i.

Labelling, radioactive Targeted introduction of a radioactive isotope into a (physico)chemical process to obtain more detailed information; often used to elucidate metabolic pathways or reaction mechanisms.

Lactone Cyclic ester

Lambert-Beer law Basis of photometry of sufficiently dilute solutions; describes the relationship between the concentration of an analyte solution and the absorbance at a precisely defined wavelength.

Larmor frequency The frequency at which an atomic nucleus—if it has a permanent dipole moment—precesses around its axis in a magnetic field; the Larmor frequency of a nucleus is identical to the frequency of the radiation with which the nucleus in question can be excited to precession.

Law of mass action (LMA) The LMA quantitatively describes the state of a chemical equilibrium: it is the ratio of the mathematical product of the concentrations of the products of a reaction to the mathematical product of the concentrations of the reactants. Stoichiometric factors enter the mathematical expression as exponents. The LMA leads to the equilibrium constant K.

Least squares method A mathematical method used to calculate regression curves.

Lectins (Glyco) proteins that interact specifically with selected saccharides.

Libby half-life See *conventional ^{14}C age*

Limit theorem Central *see central limit theorem*

Line spacing The distance of the individual "sub-signals" of a multiplet to their respective neighbouring "sub-signals". Since these distances are independent of the measurement conditions (field strength, frequency of the excitation radiation), they are given in absolute values (in Hz).

LMA See *Law of Mass Action*

London dispersion forces Interactions between two non-polar but polarisable particles (atoms and/or molecules); usually subsumed under the generic term van der Waals forces.

Long-distance coupling Any coupling that occurs over more than three bonds; in high-resolution ^{1}H NMR, $^4J_{HH}$ and even $^5J_{HH}$ may contribute to structural elucidation.

Low-field shift If an atomic nucleus can be excited to resonance in NMR at comparatively low magnetic field strength (i.e. relatively easily), the associated NMR signal appears in the spectrum at larger δ-values ("further to the left").

LUMO (Lowest Unoccupied Molecular Orbital) The energetically lowest molecular orbital of a multi-atom system which in the ground state is not populated with one or two electrons.

MacLaurin's series Expansion of a function f(x) into a series/sum of power functions (constant, linear term, quadratic term and higher powers), with the development point $x_0 = 0$. For example, if you know the function value as well as slope, curvature and higher geometric properties at the point x = 0, you can use MacLaurin's series to construct a function that allows you to approximate the course of the function by power functions in a certain range around the point $x_0 = 0$—the so-called convergence range. The series is used to approximate complicated functions by simple functions; more on this in the further reading on mathematics. MacLaurin's series is a special case of the Taylor series, *see Taylor series*.

Macrolide The term macrolide is used to describe cyclic compounds with a large number of ring members, whereby the ring is held together by (at least) one ester group and can therefore also be regarded as a lactone (which is why the term macrolactone is not uncommon).

MALDI-TOF (Matrix-Assisted Laser Desorption/Ionisation—Time Of Flight) Mass spectrometric method for the investigation of analytes with a high molar mass (e.g. proteins, polymers). Together with a suitable matrix material, those are brought into the gaseous phase via laser-assisted ablation. Fragmentation is very rarely observed in MALDI-TOF; the time required for the analyte to reach the MS detector allows conclusions to be drawn about the respective analyte's molar mass.

Mass number (m) Nucleon number; sum of the number of all protons and neutrons of an atom; understandably, the mass number of each atom is always an integer.

Mass *See nominal mass, molar mass, relative atomic mass*

Mean error of the mean (σ_x) Standard deviation of the arithmetic mean.

Medium uncertainty *See Uncertainty*

Memory effect In mass spectrometry a common term for the reappearance of peaks that originate from substance residues of previously examined analytes and can usually be explained by condensation of minute substance residues on cooler components of the ion source.

Mercury drop electrode Working electrode (with associated reference electrode, usually a saturated calomel electrode) in which elemental mercury. Being added dropwise, comes into contact with the analyte solution. The analyte (a metal) is reduced to its elemental form and dissolves in the mercury, forming an amalgam. The mercury chemically modified this way is discarded and a new mercury droplet (from a reservoir) takes its place.

Metastable Compounds or mixtures of substances that do not (or: not noticeably) change chemically under the prevailing conditions (temperature, pressure, possible reaction partners, etc.), although they should react spontaneously for purely thermodynamic reasons (enthalpy of reaction), are referred to as being metastable. The reason for metastability is usually kinetic inhibition, which can be overcome by increasing the temperature or otherwise modifying the prevailing conditions. An example is a gas mixture of elemental hydrogen and elemental nitrogen: One would expect these two gases to react spontaneously to form ammonia, since ammonia is thermodynamically more energetically favourable than the starting materials (the reaction $3 H_2 + N_2 \rightarrow 2 NH_3$ is exothermic); however, the reaction does not take place because of the high activation energy. The same applies, for example, to the spontaneous elimination of water from carbohydrates, etc.

Methine group A carbon atom to which only one hydrogen atom is bonded (in addition to the three other bonds the carbon atom has formed).

Methyl group A carbon atom to which three hydrogen atoms are bonded in addition to another bonding partner; $^-CH_3$.

Methylene group A carbon atom to which two hydrogen atoms are bonded in addition to two other bonding partners; $^-CH_2^-$.

Micro-thermogravimetry (μ-TG) *See Thermogravimetry*

Mobility (u, also: μ) Mobility of charge carriers in an electric field; depends on the particle size and the charge density (and thus also the extent of the solvate shell). It is inversely proportional to the strength of the electric field applied.

Molecular ion In mass spectrometry: molecular cation M$^+$ obtained by knocking out an electron, the mass of which corresponds to that of the (neutral) initial analyte. Since multiple ionisations are also possible, it makes more sense to state the mass/charge ratio (m/z) instead of the mass of the corresponding ion, with z standing for the number of positive charges obtained in the course of the ionisation.

Mononuclidic/monotopic elements Elements of which only one naturally occurring stable isotope exists; however, other isotopes can be produced synthetically if necessary. The most important mononuclidic elements probably are fluorine (^{19}F), sodium (^{23}Na), phosphorus (^{31}P), iodine (^{127}I) and gold (^{197}Au).

Multi-line system Generally speaking, a multi-line system is one in which the NMR signal of the respective atomic nucleus (or several chemically equivalent atomic nuclei) is split into multiplets in a high-resolution spectrum.

Multiplet If an NMR signal is split into a multi-line system by coupling with other nuclei, a multiplet results, depending on the number of coupling nuclei. The relative intensities of the individual lines are characteristic on the one hand and follow Pascal's triangle on the other.

μ *See Mobility; see also Ion Strength*

Negative control A term commonly used in the life sciences for a blank sample that contains a substance (such as an enzyme or similar) that is very similar to the analyte but, unlike the analyte, does *not* lead to a characteristic (colour) reaction. Not to be confused with the blank sample.

Nernst equation Equation for determining the redox potential of a redox couple as a function of the concentrations present in each case.

Nernst's equation The Nernst's equation describes the dependence of the redox potential of a redox couple on the respective concentrations of the oxidised (better: more highly oxidised) and the reduced (better: less highly oxidised) form. Under standard conditions, it is:

$$E = E^0 + \frac{0.059V}{z} \cdot \lg\frac{[Ox]}{[Red]}$$

NHE *See Normal Hydrogen Electrode*

$^n J_{xy}$ Generally accepted symbol in NMR for any form of coupling. The number in superscript on the left tells us how many bonds separate the two coupling nuclei; the number in subscript on the right tells us the nature of the nuclei coupling with each other. So a $^1J_{CH}$ coupling would be the interaction between a hydrogen atom and the carbon to which it is directly bonded (C-H), a $^3J_{HH}$ coupling is the interaction of two H atoms bonded to two directly bonded (C) atoms. For larger distances, one usually speaks of long-distance couplings.

NMR *See nuclear magnetic resonance spectroscopy*

Noble Collective term for theoretically oxidisable substances which, however, do not release elemental hydrogen even in a strongly acidic aqueous medium; antonym: base metal.

NOESY (Nuclear Overhauser Enhancement and exchange SpectroskopY) A two-dimensional NMR technique to identify which carbon atoms are in close spatial proximity to each other, where it is not necessary for the nuclei concerned to be linked via a covalent bond. (CC couplings are not considered in this technique.) This technique is very useful for structure elucidation (especially with respect to the three-dimensional structure of (more complex) molecules like proteins), but also rather complex. (It is based on the nuclear Overhauser effect.) But since it is so effective, it should at least be mentioned.

Nominal mass If one describes (in mass spectrometry) the mass of a molecule or ion (obtained by ionisation or fragmentation) by the sum of the mass numbers of all atoms present in this molecule, one obtains its nominal mass (often equated with the value for m/z, especially in the case of simple ionisation). A methyl cation (CH_3^+), for example, has a nominal mass of 15 if only the most abundant isotopes are present (^{1}H,^{12}C).

Non-noble (often called base) Generic term for oxidisable substances with $E_0 < 0$; are oxidised accordingly in a strongly acidic medium (pH = 0) with the release of elemental hydrogen. Antonym: noble.

Non-rejection area *See acceptance area*

Normal distribution *See Gaussian distribution.*

Normal hydrogen electrode (NHE) Reference electrode for measuring the normal potentials (E^0) of other redox pairs. The redox potential of the associated reaction (2 H$^+$ + 2 e$^-$ ⇌ H$_2$) is thereby (arbitrarily) defined to be 0.00 V. The criteria p(H$_2$) = 1.013 bar, c(H$^+$) = 1.00 mol/L and T = 298.15 K apply.

Normalisation constant Number/constant required to make the total probability equal to 1.

Nuclear charge number (Z) Number of protons in the nucleus of an atom; corresponds to the atomic number.

Nuclear magnetic resonance spectroscopy (NMR) Analytical technique in which low-energy electromagnetic radiation in a sufficiently strong magnetic field causes suitable atomic nuclei to change their spin state; leads to nuclear magnetic resonance spectra.

Nuclear spin All atomic nuclei in which the number of protons and the number of neutrons is not even, have a nuclear spin with a nonzero value, i.e. they have a permanent dipole moment and rotate around their own axis in a magnetic field. The nuclear spin can have integer and half-integer values; depending on the nuclear spin, there are different possibilities for the corresponding nucleus to align itself in the magnetic field, and the resulting spin states differ in their energy content. For the two atomic nuclei most commonly used in NMR spectroscopy by far (^{1}H,^{13}C), spin = ½, i.e. these nuclei can only align themselves parallel (+½) or antiparallel (−½) to the magnetic field.

Nucleon number Number of all nucleons (nuclear building blocks) of an atom; corresponds to the mass number of the isotope concerned.

Null hypothesis (H_0) is the assumption to be tested in a hypothesis test.

nσ-environment The region of a normal distribution that contains a certain percentage of the distribution. For n = 1, the percentage is 68.3%, n = 2 comprises 95.5% and triple σ-environment contains 99.73% of the distribution.

Ohm's law The strength of the electrical current (I) flowing through an object to which an electrical voltage (U) has been applied depends on its electrical resistance (R); in other words, the ratio of voltage and current strength is always constant: U = R • I or R = U/I.

Ohm's potential Every (half) cell has an electrical resistance that follows Ohm's law. This resistance can be overcome by increasing the applied voltage. Thus, Ohm's potential contributes to the overvoltage of any electrolysis system.

Optode Short for: Optical electrode.

Ordinate The correct technical term for the y-axis of an x/y diagram.

Ordinate intercept The ordinate intercept is the constant value of a linear function: Here, x = 0.

Osmosis Movement of molecules (often: solvent molecules, mostly water) through a semi-permeable membrane. (The membrane of a living cell can also be understood as semipermeable.)

Outlier An outlier is a measured value that obviously does not match the other values found. The so-called outlier test is then used to clarify whether this measurement point is actually to be classified as an error—and is then also removed from the measurement series—or whether the value does belong to the series due to the scatter of the results.

Overvoltage The voltage that must be applied in addition to the voltage required by the redox potentials of the reactants in order to bring about electrolysis. The overvoltage required experimentally in each case depends not only on the activation energy of the respective redox reaction, but also on the current density (the higher this is, the more the overvoltage increases), on the type of electrodes used and, if applicable, on the gaseous (by-)products forming during the reaction.

Oxidation number Purely theoretical, but very helpful construct for setting up redox equations. In an oxidation, the oxidation number of the particle in question is increased; in a reduction, it is decreased. The following rules apply:
– Elementary substances always have the oxidation number ± 0.
– For monatomic cations or anions (e.g.: M^{m+}, X^{x-}) the oxidation number corresponds to the charge of the particle.
– In the case of polyatomic, covalently structured compounds (uncharged molecules or molecular ions), the oxidation number of the individual atoms can be determined by formally assigning the electrons involved in the bond completely to the more electronegative atom in each case, so that it (formally!) receives more electrons than it is entitled to from the periodic table, while the less electronegative atom formally (!) lacks these electrons.
– For atoms of the same electronegativity (mostly: atoms of the same element), formal homolytic bond cleavage occurs. Accordingly, in the case of methane (CH_4), because EN(C) > EN(H), the oxidation number for carbon is −4, while the (less electronegative) hydrogen atoms each have the oxidation number + 1; the central carbon atom of 2,2-dimethylpropane, on the other hand, has the oxidation number ± 0.
– Overall, the sum of all oxidation numbers must correspond to the charge of the particle under consideration: For methane this is ±0.

- This last rule is particularly clear in the case of molecular ions such as sulphate (SO_4^{2-}), where the (more electronegative) oxygen atoms are each assigned the oxidation number −2, and sulphur is assigned +6.
- In some textbooks, oxidation numbers are given in Roman numerals; however, IUPAC prefers Arabic numerals for ease of reading, and Roman numerals do not have a sign for ±0, anyway.

Oxidation Electron release; leads to an increase in the oxidation number.

Paramagnetism Behaviour of particles in the magnetic field that have at least one unpaired electron. Probably the most important paramagnetic substance is dioxygen (O_2).

Partial pressure The pressure that a gaseous component contributes to the total pressure of a gas mixture; the partial pressure p_i corresponds to the molar ratio of component i to the total amount of substance n present, multiplied by the total pressure: $p_i = x_i \cdot p_{total}$.

Peptide molecule Consisting of several amino acids linked by peptide bonds (-C(=O)-NH-). If a molecule, which is primarily held together by peptide bonds, also contains ester groups, it is referred to as a depsipeptide.

Pericyclic reaction Any concerted reaction in which bonds are changed by simultaneous shifting of electron density. Typical examples are the (retro) Diels-Alder and the En reactions. (You can learn more about this in advanced Organic Chemistry textbooks or courses.)

Pheromones Messenger substances; serve for (unconscious) communication between individuals of the same species; often (somewhat simplistically) referred to as (sexual) attractants.

Phosphorescence If, after photochemical excitation of a multi-electron system via electromagnetic radiation of wavelength λ_1, an *intersystem crossing* occurs, a part of the excitation energy is "stored" and only gradually released again in the form of photons of wavelength λ_2 (with $\lambda_2 > \lambda_1$): Phosphorescence occurs. In contrast to fluorescence, phosphorescence can last for several minutes, in special cases up to hours.

Photometry Analytical method with which the analyte content of a solution can be determined on the basis of the absorption of a precisely defined wavelength (or a narrowly defined wavelength range) from the visible light range.

Photon counter Variant of a scintillation counter that specifically responds to high-energy photons (gamma radiation).

Physisorption Adsorption of a substance (at the microscopic level: individual atoms or molecules) onto the surface of another substance, whereby *physical forces* are responsible for the cohesion, primarily London dispersion forces. Please do not confuse this with chemisorption.

Platinised platinum A sheet of elemenal platin electrochemically coated with elemental platinum generated *in situ*. This drastically increases the surface area, which in turn lowers the overvoltage.

Point estimation Designates an estimation function with which a certain property of the underlying probability measure is to be estimated. Often the quantity of interest is a parameter of the probability distribution, such as the expectation value.

Poisson distribution Discrete probability distribution used for Bernoulli experiments in which the probability of an event occurring is very small and at the same time the magnitude N is very large.

Polarography Variant of voltammetry in which a mercury drop electrode is used; only suitable for reduction processes, but can even be used for very base (non-noble) metals.

Polycondensation Production of macromolecules by reaction of the starting materials (monomers), whereby a small molecule is split off as a "by-product" each time a new bond is formed between the starting materials (usually water, but methanol or ammonia are also not uncommon).

Polymerisation Production of macromolecules by the reaction of the starting materials (monomers), whereby new bonds are formed *without* the formation of low-molecular by-products.

Population *See Ensemble*

Positive control A term commonly used in the life sciences for a sample definitely containing the analyte (often even with a precisely known concentration).

Positron Subatomic particle with the mass of an electron, but with the opposite charge: Usually symbolised by e^+, the notation β^+ is also found—especially in nuclear chemistry—because it represents the result of one of the three variants of β-decay.

Potentiometric titration Any titration whose course/progress is tracked by measuring the potential of the solution; the tracking of the change in pH by means of a glass electrode is strictly speaking a potentiometric titration.

Potentiometry Collective term for analytical methods based on the potential difference between a half-cell of known and a half-cell of unknown concentration.

ppb The abbreviation ppb, used like a unit, stands for *parts per billion*. 1 billion = 10^9.

ppm An abbreviation, used as a unit, stading for *parts per million*; analogous to an indication in percent (only a whole four powers of ten smaller); in NMR correspondingly the "unit" of the chemical shift δ (which of course is actually dimensionless, i.e. without a unit).

Precession If a rotating gyroscope is not positioned absolutely perpendicular, the gyroscopic motion results in a change of direction of the rotation axis; this motion is called precession. In nuclear magnetic resonance spectroscopy, *Larmor precession* is important.

Predissociation In fluorescence experiments: bond cleavage caused by the irradiated excitation radiation, thus being an indirect consequence of the excitation; does not lead to fluorescence and tends to increase with increasing energy content of the excitation radiation used (but of course also depends on the stability of the respective analyte).

Probability density (density function) Function describing a continuous probability distribution. The integration of the probability density over an interval indicates the probability that a random variable with this density function takes a value from this interval. The probability density can take on values greater than 1 and *should not be confused with the probability itself*.

Probability of error *See Significance figure*

Probability Grade of certainty for the occurrence of an event.

Proportion value Number of objects of a certain characteristic value in relation to the total.

Pseudo-multiplet If two (or more) coupling constants within a molecule are nearly identical (for whatever reason), one no longer obtains a clear "multiplet of multiplets" structure, but a multi-line system whose relative signal intensities no longer necessarily follow Pascal's triangle. From here one moves straight into the high art of spectrum evaluation—that would go beyond the scope of this introduction.

Pyrolysis Thermochemical decomposition of (mostly) organic compounds without the addition of oxygen (i.e. it is by no means a combustion process); any resulting gaseous compounds can subsequently be separated via gas chromatography and/or analysed by GC/MS.

Quantile (more precisely, **p-quantile**) is a value that splits a set of data into two such that at least one proportion p is less than or equal to the p-quantile, and at least one proportion 1-p is greater than or equal to the p-quantile.

Quantum yield/quantum efficiency (Q) A measure of how many fluorophores could theoretically be excited to fluorescence (or any other luminous phenomenon) and how many are actually excited. The ratio of the number of analytes excited to fluorescence to their total number is considered; $Q \leq 1$.

Quartet *See Multiplett*

Quenching Reduction of luminous phenomena by interaction of the system responsible for the formation of light with a reaction partner (atom, molecule, molecular ion), especially paramagnetic substances (such as dioxygen, O_2) cause strong, quantifiable quenching.

Quintet *See Multiplett*

Radioimmunoassay (RIA) Analytical method based on the interaction of antibodies with radioactive or radio-labelled antigens, whereby it is essential that unlabelled antigens interact in the same way and to the same extent with the corresponding antibodies, so that these and the labelled antigens compete for the binding sites.

Radionuclides Collective term for all radioactive (unstable) isotopes; which way they decay (and how quickly this occurs) depends entirely on the respective radionuclide (but is then also specific and invariable for that specific isotope).

Random error *See error*

RDA *See Retro-Diels-Alder-Reaction*

Real solution *See ideal solution*

Realisations Realisations are the values that a random variable X takes on.

Rearrangement Chemical reaction in which individual atoms or polyatomic groups/moieties (such as a methyl group or similar) change the bonding partner so that the entire molecular structure changes. A rearrangement particularly important in mass spectrometry is the En reaction or the variant derived from it, known as the McLafferty rearrangement, which also involves a β-cleavage.

Reduction Electron absorption; leads to lowered oxidation number.

Reference electrode (in potentiometry) A half-cell of exactly known composition and thus exactly known potential; the potential difference, which results together with the indicator electrode, then permits statements about the composition (concentration) of an unknown solution.

Reflected light microscope Microscope in which the specimen to be microscoped is not illuminated from below (i.e. transilluminated), but in which the light comes from the direction of the objective or even from the objective itself.

Relative molar mass Mass of an atom, molecule or ion, given in relation to (i.e. "relative to") a reference value. IUPAC has defined the mass of the carbon isotope ^{12}C as the reference value. Thus, if the molar mass of an atom is given in the unit g/mol, it implicitly always refers to the fact that, by definition, 1 mol of carbon atoms of the isotope carbon-12 has exactly the mass 12.000000 … g.

Relaxation In NMR spectroscopy, the usual term for the return of an atomic nucleus to its original spin state. In other interactions of an atom or multi-atom system the return of said system from an electronically, vibratorily and/or rotatorily excited system to a less excited state, which might even be the ground state.

Reporter enzyme An enzyme used in biochemical (or genetic) analytical methods that catalyses a reaction leading to quantifiable staining or a reaction with quantifiable luminescence (chemi luminescence, fluorescence).

Residual current The current that flows in a voltammetric measurement when the potential of the electrode is raised (far) above the current plateau in the resulting voltammogram; results from side reactions that occur due to impurities in the solution and due to other factors.

Retro-Diels-Alder reaction (RDA) Reversal of the Diels-Alder reaction; occurs in organic synthesis as well as in mass spectrometry.

Ring current The magnetic force current generates a secondary magnetic field in cyclically conjugated π-electron systems, which leads to the atomic nuclei bonded to this delocalised system being more easily excited to resonance and are therefore low-field shifted, i.e. have a larger δ value. (Strictly speaking, this is only true for delocalised systems that follow the Hückel rule (4N + 2), in which case *diamagnetic* ring current occurs. If cyclically conjugated π-electron systems do *not* follow this rule (i.e., if the number of π-electrons present can be described with the formula 4N), a *paramagnetic* ring current results, which has exactly the opposite effect, so that the electrons in question are high-field shifted, i.e. have a smaller δ value, which might even become negative.)

Saccharides The more technically correct term for "carbohydrates", i.e. "sugars" with the general molecular formula $C_nH_{2n-2x}O_{n-x}$.

Sample Random partial survey from a population for the investigation of a characteristic X.

Sample value Value that a characteristic X assumes in a sample.

Scattering radiation Radiation resulting from elastic collisions of analyte molecules with photons; known from Raman spectroscopy as Rayleigh scattering

SCE (Standard Calomel Electrode) *See Calomel Electrode*

Scintillation counter Measuring device for quantifying ionising radiation (i.e. α-, β-, γ- and neutron radiation); the impact of the radiation causes a measurable light pulse.

Second order/second type electrode Electrode with constant potential; often used as conductive electrode and/or as reference electrode.

Selective decoupling *See Decoupling*

Selectivity coefficient (K_{Sel}) A measure of how selectively an electrode responds to the analyte; selectivity coefficients always refer to the sensitivity of the electrode in question with respect to analyte A in direct comparison with a non-analyte X; accordingly, one always states: $K_{Sel(A, X)}$. For different non-analytes X, the respective selectivity coefficients may differ by several orders of magnitude.

Semicrystalline/partially crystalline *See crystalline*

Sensilla Hair-like sensory organs of arthropods (insects, crustaceans and arachnids, millipedes, etc.); depending on the associated receptor system, a sensillum react to mechanical, optical, thermal or also chemical stimuli. The latter are also triggered by pheromones.

Septet *See Multiplet*

Series expansion *See MacLaurin series, Taylor series*

Sextet *See Multiplet*

Shielding A reduced sensitivity of an atomic nucleus to the effect of the magnetic field caused by an increase (or at least a less pronounced reduction) in the electron density. Shielded atomic nuclei are less easily excited to resonance (i.e. by higher-energy excitation radiation or only in a somewhat stronger magnetic field); this leads to a *high-field shift* of the signal concerned. Opposite: de-shielding.

Theorem of Huygens-Steiner A calculation rule for determining the sum of squared deviations from the arithmetic mean. The shift theorem facilitates the calculation of the sample variance when the measured values occur continuously.

σ (sigma) Symbol for the shielding constant commonly used in NMR spectroscopy.

σ-cleavage Cleavage of a simple σ-bond without further (stabilising) influence of hetero-atoms or π-electrons. Occurs in mass spectrometry, although other cleavages (α-cleavage, etc.) are much more frequent (and accordingly lead to higher peaks in the spectrum).

Signal converter *See transducer*

Significance number (also: probability of error) Indicates how high the risk is of making a wrong decision. For most tests, a level of 0.05 or 0.01 is used.

Significant figures Significant figures of a number are the specified figures without *leading zeros*.

Silicone Organosiloxanes, consisting of silicon atoms—bridged by oxygen atoms—that additionally carry organic substituents (-R).

Single-rod electrode A glass electrode with the corresponding reference electrode inside.

Singlet state (S) In the singlet state of a multi-electron system, the total sum of all electron spins is 0. The reason for this can be that all electrons are spin-pair (↑↓), or that—in an excited state—two electrons populate orbitals of the same or different energy content, but still differ in their spin: (↑)(↓) and $_{(\uparrow)}^{(\downarrow)}$, respectively. If a singlet state is vibrationally/rotationally excited, it is denoted by an Asterisk (*): S*.

Singlet *See Multiplet*

Slope measure (m) In the straight line equation: parameter that determines the slope of the straight line.

Spin state In general: the current state of a particle with a spin >0. Electrons (*see above*) as well as the nuclei of the isotopes ^1H and ^{13}C have a spin of ½, so the two spin states +½ and −½ are possible. For nuclei with a spin >½ the number of possible spin states increases accordingly.

Spin state (for electrons) The spin state indicates the spin quantum number m_s with which the electron in question is described: If $m_s = +½$, the electron is represented graphically by (↑); if $m_{s=} - ½$, by (↓). A change in the spin state is called *spin reversal*.

Standard deviation (s) Measure of the dispersion of the values of a random variable around its expectation value.

Standard hydrogen electrode (SHE) Refined (more precise) version of the standard hydrogen reference electrode; with regard to pressure and temperature, the same values apply as for the NHE, but in addition the solution is not $c(H^+) = 1.00$ mol/L, but $a(H^+) = 1.00$.

Standard potential (E^0) The standard potential of any redox couple is defined as the electrical potential that results under standard conditions between the half-cell of the substance in question and a standard hydrogen electrode. In most cases, the value is given as *standard reduction potential*, i.e. the potential for the conversion of the oxidised form to a less highly oxidised form (for convenience usually referred to as "reduced"). Temperature and pressure are held constant at standard values, and it is assumed that the activity a of the reduced form (in solution) and of the oxidised form (even if it is present as a solid, e.g. as an elemental metal) is exactly a = 1. (The activity of a solid in contact with a solution is by definition a = 1).

Statistical certainty *See level of confidence*

Statistical errors *See error*

Stochastically subjected To chance, dependent on chance.

Stock solution Solution with a precisely defined concentration (or otherwise precisely defined content), which serves as a starting material for further solutions derived from it within the framework of a dilution series or similar; of course also applicable to "biosamples" such as antibodies.

Stokes shift In fluorescence processes, the wavelength of the emitted fluorescence photon λ_F is larger than the wavelength of the excitation radiation λ_E: $\lambda_F > \lambda_E$.

Stripping analysis Variant of voltammetry, mainly used in trace analysis. The analytes are first concentrated in elemental mercury by reduction and then anodically oxidised so that they return to solution. The current flow that can be measured in this process allows conclusions to be drawn about the original analyte concentration; several analytes (with different redox potentials) can also be detected in parallel.

Student t-distribution The probability distribution that must be used when you need to estimate the mean (due to a sample size that is too small). The basis for this is the sample variance. The t-distribution allows you to calculate the distribution of the difference *between the mean of the sample and the true mean of the population.*

Symmetry-forbidden Some electron transitions can be energetically far less favourable than others, so that they occur only (very) rarely; the reason for that can be found in the symmetrie properties of the orbitals involved. In such a case, one speaks of a symmetry-prohibited (or in short: forbidden) transition. Whether an electron transition is forbidden or allowed can be determined with the help of selection rules, but that would go much too far here.

Systematic errors *See error*

Tactile polarography Variant of polarography, in which the potential of the electrode is gradually increased; a staircase-like polarogram is obtained.

Taylor series Power series with which a function can be represented in the vicinity of a reference point x_0.

Temperature scanning Differential calorimetric measurement

Test distribution The distribution that is tested for using statistical methods.

Tetramethylsilane (TMS) $Si(CH_3)_4$; reference substance for the shift values in both ^1H and ^{13}C-NMR. In both cases, TMS is assigned $\delta = 0$ (ppm).

TG/IR Combination of (micro-) thermogravimetry and infrared spectroscopy, often abbreviated TG/FTIR, because the measuring instruments usually work with a Fourier transformation.

TG/MS Combination of (micro-) thermogravimetry and mass spectrometry.

Thermogravimetry (TG; also: thermogravimetric analysis TGA) Gravimetric analytical technique to determine the mass loss of a sample during a process with increasing temperature; such mass losses may be due to evaporating water of crystallisation of hydrates or to thermally induced decomposition processes such as loss of carbon dioxide from carbonates, etc. The variant of thermogravimetry known as **micro-thermogravimetry (μ-TG)** allows the investigation of even tiny sample quantities (<1 μg).

TISAB (Total Ionic Strength Adjustment Buffer) Buffer system is frequently used in potentiometric measurements to compensate for fluctuations in the ionic strength of the respective analyte solutions and to keep the pH value largely constant.

TMS *See Tetramethylsilane*

Torsion angle The angle at which adjacent substituents are twisted against each other in the course of free rotatability about a single bond. In $^nJ_{HH}$ couplings, the torsion angle affects the resulting coupling constant; this effect can be estimated quantitatively by the Karplus-Conroy relationship.

Tracer Alternative term for (radioactive) markers for substances or molecule components, e.g. in order to clarify reaction mechanisms or trace (biochemical) metabolic pathways; tracers are also increasingly used in medicine in the context of nuclear medical diagnostics.

Transducer Any device that converts a measurable physical quantity into an electrical signal that then provides the "actual" measured value.

Transmission (T) The ratio of the intensities of light passing through an analyte solution on the one hand and a blank sample on the other; expressed in %.

TRFIA (time-resolved fluoroimmunoassay) Time-resolved fluorimetric immunoassay; combination of the two analytical techniques that give it its name; now a real standard in the life sciences.

Triplet *See Multiplett*

Triplet state (T) A triplet state exists when two electrons populate two orbitals of the same or different energy content, but do *not* differ in their spin: (↑)(↑) and (↓)(↓), respectively. If the ground state of the system under consideration is a singlet state, the associated triplet state is in any case electronically excited, and a vibratory/rotational excitation may still be added. In both cases, the fact that it is an excited state should be denoted by an asterisk, e.g.: T*.

True value (µ) The value that would have to be determined in the case of a completely error-free measurement, but which can in principle never be achieved.

Two-point distribution Simple discrete probability distribution defined on a two-element set {a,b}. The best known special case is the Bernoulli distribution, which is defined on {0,1}.

u (unified atomic mass unit) 1 u = 1.660539×10^{-27} kg; corresponds by definition to 1.000000 g/mol; in life sciences the u is often called **Dalton** (with the unit sign **Da**). In (mostly English-language) older texts, the unit **amu** (for **atomic mass unit**) can also be found, but actually there is as small numerical difference between the values for u and amu.

u Formula characters for mobility

Uncertainty, mean The mean uncertainty delimits a range of values within which the true value of the measurand lies with a probability to be specified (ranges for about 68% and about 95% are common). The estimated value or single measured value used as the measurement result should already be corrected for known systematic deviations. The measurement uncertainty is positive and is given without a sign. Measurement uncertainties are themselves also only estimated values. The measurement uncertainty can also be called uncertainty for short.

Upconversion Complex process in fluorescence: If two molecules that have reached an excited state in the course of the various conversions of a fluorescence process collide, the "excess energy" of one molecule occasionally is transferred to the other. While the former falls back to the ground state (or at least a less excited state), the energy content of the latter is increased even further. This now even higher "excess energy" can then be emitted in the form of a fluorescence photon whose wavelength is shorter than that of the excitation radiation. However, the quantum yield of this process is rather low.

Vacant Related to orbitals: not occupied by electrons; in solid state chemistry: mainly used to describe "vacant lattice sites".

Value range The set of numbers that the function values of a function can assume.

Variance (s^2) Important measure of dispersion of the probability distribution of a real random variable. It describes the expected squared deviation of the random variable from its expected value. The square root of the variance is called the standard deviation of the random variable.

vib./rot. Excitation When a polyatomic system is excited to a higher energy level of vibration or rotation, a vibratory/rotatory excited system is present.

Vibration relaxation Process in which a vibrationally/rotationally excited system assumes a more energetically favourable state; this can be the ground state or also a less strongly vibratory/rotationally excited state.

Vicinal Vicinal substituents (or vicinal functional groups) are those that are bonded to two immediately adjacent carbon atoms. Ethylene glycol (ethane-1,2-diol, HO-CH_2-CH_2-OH), for example, is a vicinal diol; not to be confused with *geminal*.

VIS range The wavelength range of the electromagnetic spectrum that can be detected by the human eye and interpreted by the brain as (coloured) light.

Voltammetry Generic term for various analysis techniques in which information is obtained from the relationship between voltage and current in electrochemical processes; effectively a "special form" of electrolysis.

Voltammetry, cyclic (cyclovoltammetry) Variant of voltammetry in which the voltage is not applied continuously but cyclically rising and then falling back to 0. The resulting voltammograms have cathodic and anodic peaks that are near the half-step potential. Cyclic voltammetry is e.g. used to investigate complex multi-stage redox processes.

Water of crystallisation Water bound in the crystalline solid; integral part of the crystal structure itself, often (but not always!) coordinatively bonded. For this reason, there is usually an integer number of water molecules per formula unit of the substance in question, i.e. monohydrates (A_xB_{y-} H_2O), dihydrates (A_xB_{y-} 2 H_2O) etc.; occasionally, hemihydrates (A_xB_{y-} ½ H_2O) and sesquihydrates (A_xB_{y-} 1.5 H_2O) are also found.

Weibull distribution A continuous probability distribution over the set of positive real numbers. Depending on the choice of parameters, it resembles a normal distribution or an exponential distribution (or other asymmetric distributions). The Weibull distribution describes the lifetime and failure frequency of electronic components or (brittle) materials such as ceramics.

Working electrode The electrode of an electrolysis system at which the redox reaction that is decisive for the analyte takes place; in electrogravimetry, therefore, the electrode (cathode) at which the analyte is deposited in elemental form.

Index

A

Absolute 14, 28, 43, 61, 82, 84, 101, 161, 205, 223, 225, 243, 294, 327, 328
Absorption 28, 30–32, 166, 182, 206, 210–212, 214, 318, 331
Acceptance area 299
Acid error 104
Activity 85–87, 89–91, 95, 96, 101, 102, 104, 105, 107, 108, 157, 160, 164, 165
Activity coefficient 85–87, 89, 90, 95, 101, 107, 129
Adjustment of fit parameters 241–243
Age determination, radioactive 159–161, 184, 259
Alkali error 104, 106, 127, 131
Allyl cleavage 19, 20
α-cleavage (alpha cleavage) 17–19
Alpha decay 154, 184
Amperometry 118, 119, 128, 131, 139, 192, 194, 197–201
Analysis
– gravimetric 145
– radiochemical 155–157
Analyte, electroactive 94–102, 127
Analyte plug 210–212, 216
Anisotropy effects 36, 44–46, 77
Antibody 161–164, 166–168, 198
Antigen 161–164, 167, 182, 184, 185, 198, 199, 212, 214
Arbitrary distribution 198, 295, 298, 310
Aromatics 24, 44, 45
ATP content, determination 118, 192, 197
Autoimmune disease 162
Azo coupling 211, 212
Azo dyes 211, 212, 214

B

Background fluorescence 180
Balancing line 241, 315
Bell curve 226, 270
Benzyl cleavage 20, 67
Bernoulli experiment 246–251, 254, 259, 287–289
Beta decay 154, 155, 161, 184, 187
Binding cleavage 17, 20, 67
Binomial distribution 246, 249–254, 259, 264, 285, 287, 288, 296, 297, 325
Biosensor 111, 118, 127, 128, 192, 197–201, 214
Blank sample 166, 168, 310, 313
Blood glucose meter 118, 119, 206
Broadband decoupling 58

C

Calibration series 310, 311, 318, 326
Calomel electrode, saturated 96, 122, 123, 127
Calorimetry 150–151, 183, 186
Cambridge half-life 160, 187
Carbon electrode, enzyme coated 197
CC-INADEQUATE 62
Cell, galvanic 84, 85, 90, 91, 94, 95, 97, 129
Cell symbols 91, 103, 129
Charge, electrical 82, 114
CH-COLOC 62
Chemical equivalence 33, 46–48

Chemical ionisation (CI) 12–15, 66
Chemical shift 34, 39–45, 58, 63, 64, 69, 74, 75
Chemisorption 141, 142
Chi-square distribution 278–283, 330
Clark electrode 118, 192, 194, 214
Classical 31, 69, 97, 115, 181, 194–197, 246
^{13}C NMR spectrum 28, 37, 39–41, 46, 50, 54, 57–64, 67–69, 75, 77
CO_2 gas electrode 111, 200
Coefficient of determination 316, 317
Coin toss 249
Compensation calculation 310–312, 314, 315, 326, 330
Compensation curve 317
Compensation parabola 243
Completeness relation 264, 265
Complex 41, 42, 52, 54, 55, 61, 63, 65, 77, 82, 109, 110, 114, 118, 126, 128, 144, 148, 157, 160–162, 164, 168, 174, 175, 180, 184, 198–200, 204, 205, 207, 212, 214, 281
Components 86–89, 94, 100, 103, 107, 108, 111, 118, 179, 181, 182, 199, 204, 214, 220, 313
Composite core 268
Composite electrodes 106, 111, 118, 127, 128, 131
Concentration 84–89, 91, 94, 95, 97, 100–105, 107, 108, 111, 114, 116, 118, 119, 122–129, 131, 139, 142, 143, 156, 160, 166, 167, 176, 177, 185, 198, 200, 207, 211, 213–216, 234, 307, 310, 312–316, 318, 331
– chain 85, 195
– dependence 84, 85, 102
– potential 139, 142, 143, 183, 185
Conductometric 196
Confidence interval 222, 227, 232–234, 273, 285, 291–298, 307, 324, 326, 330
Confidence level 232, 291, 293–298, 326
Configuration isomerism 48
Continuous 123, 144, 157, 210, 213, 223, 246, 264, 265, 267, 288
Continuous uniform distribution 264, 325
Continuous wave (CW) 32, 33
Conversion 85, 115, 119, 131, 143, 150, 160, 170–172, 175, 183, 195, 198, 210, 212–214
Correlation spectroscopy (COSY) 61, 62
Coulometry 114–116, 119, 127, 128, 131, 151
Counter electrode 138, 140
Couplings 37, 43, 46, 48–62, 64, 69–74, 76, 77, 163, 173
– constants 43, 51–55, 59–62, 64, 69–72, 77
Cramer's rule 241, 242
Crystal membrane electrodes 106
Crystal water 143–145, 185
Current 44, 45, 69, 82, 84, 97, 107, 114, 116, 118, 119, 122–126, 128, 131, 132, 139–144, 156, 159, 170, 185, 194, 196–198, 204, 205, 214, 215, 227
CW NMR spectroscopy 32, 33, 42
Cyclical 44, 128
Cyclic voltammetry 124, 126

D

Dalton (Da) 9
Deactivation processes 172, 184, 205
Decay 8, 145, 154–156, 159, 161, 180, 184, 187, 259, 262, 264
– radiation 155, 184
Decomposition voltage 139, 183
Decoupling, selective 59, 62
Density function 264, 269, 270, 273, 278, 281, 288, 289, 325, 326

Detection reaction 212
Diamagnetic 44, 45
Diaphragm 89, 91, 94, 103, 104
Diels-Alder reaction 21
Differential calorimetry 150, 151, 183
Differential scanning calorimetry (DSC) 150, 151, 184, 186
Differential thermal analysis (DTA) 148, 150, 151, 183, 184, 186
Diffusion potential 97, 99, 101, 104, 125–127, 130, 142
Direct dissociation (DD) 171, 173
Direct potentiometry 97, 127
Discrete 246, 247, 264, 269, 288, 325
Discrete uniform distribution 246, 247, 325
Dissociation, direct 171, 173
Distribution function 246, 247, 250, 253–256, 258, 260–262, 264, 265, 267–270, 273–275, 278–281, 285, 293, 325
Distribution, hypergeometric 246, 254–256, 258, 259, 325
Dixon Q-test 320–322, 326
DSC, see Differential scanning calorimetry
DTA, see Differential thermal analysis
Duplicate 166
Dye, voltage sensitive 182
Dynamic 157, 184

E

Electroactive analytes 94–102, 127
Electroanalytical 82, 85, 139, 182
Electrochemical 83, 94, 96, 102, 116, 122, 126, 128, 139, 194–201, 204, 213, 214
Electrode 83–85, 91, 94–97, 101–111, 114, 118, 119, 122–128, 131, 138, 140–142, 183, 192, 194–196, 200, 201, 213
Electrogravimetry 138, 142, 183
Electrolysis 114, 116, 118, 122, 128, 138–142, 183
Electrolyte 84, 85, 95, 97–99, 101, 102, 114, 127, 130, 138, 215
Electromagnetic radiation 6, 28, 30, 32, 35, 40, 42, 77, 155, 213
Electron density 33–36, 39, 42–44, 57, 63, 67, 68, 77
Electron impact ionisation (EI) 12–14, 17, 18, 66
Electron spray ionisation (ESI) 14
Emission 30, 31, 155, 171, 174, 177, 179, 180, 182, 187, 207
Enzyme-coated 197
Enzyme immobilisation 200
Enzyme linked immunosorbent assay (ELISA) 163–168
Equal distribution 262
Equality of parameters 302
Equilibrium constant, thermodynamic 86, 88, 89
Equivalence, chemical 33, 46–48
Error 13, 104–106, 127, 131, 161, 176, 220, 223, 225–227, 230, 236, 237, 242, 243, 270, 291, 295–305, 313, 324, 327–329, 331
– analysis 220
– of the first kind 300, 301, 304, 305
– medium 127
Estimation methods 281, 284–287, 326
Estimator 33, 285–291, 293, 298
Evaluation 8, 28, 57, 59, 61, 62, 64, 77, 123, 160, 180, 216, 266, 268, 310–312, 314
– statistical 222, 314
Expected value 104, 139, 226, 228, 229, 248, 256, 257, 260–262, 269, 273, 275, 279, 280, 284–286, 291, 293, 294, 297, 298, 302, 325, 329
Experimental 83, 84, 95, 116, 138, 143, 145, 150, 177, 198, 210, 213, 220, 227, 238
Experimental errors 220, 323
Exponential distribution 264, 267–269, 325
External 29, 102, 103, 159, 162, 171, 215, 312–314
External conversion (EC) 171, 175

External standard 312–318, 326
Extinction 173, 176, 177

F

Faraday constant 82, 84, 91, 114
Faraday laws 114, 127
Fast atom bombardment (FAB) 15
Field desorption (FD) 14
Field ionisation (FI) 14
Figure, significant 231, 237, 249, 327
1st order electrode 95, 127
Flow injection analysis (FIA) 192, 210–216
Fluorescein 171, 172, 177, 179
Fluorescence 163, 164, 170–177, 179–188, 198, 204–208, 210, 212, 214
– intensity 177, 204, 205
– method 184
– microscopy 181–183
– quantum yield 171, 172, 175, 177, 184, 187
– spectrometry 176, 177, 179–181, 184, 185
– spectrum 177, 178, 180
– system, matrix-immobilised 204
Fluorescent dye 181, 206, 207
Fluorescent marker 173, 182
Fluoride electrode 106–108, 195
Fluorophore 177, 179–181, 185, 207
^{19}F NMR spectrum 63–64
Fourier transform (FT) 33, 145
4σ-environment 320, 322, 327
14C age, conventional 160
Fragmentation 8, 10, 12–25, 66, 67
FT NMR spectroscopy 33, 42

G

Galvani voltage 84, 91, 129
Galvanostatic 114
Gamma function 278
γ-radiation 155, 163, 184
Gaussian algorithm 241
Gaussian distribution 223, 226, 227, 232, 269, 274, 292, 294, 322, 325
Gaussian error propagation 236–243
General 6, 8, 16, 21, 29, 42, 44, 46, 82–91, 95, 100, 106, 109, 115, 118, 122, 143, 154, 155, 159, 163, 171, 172, 174, 177, 192, 211, 222, 241, 269, 288, 313, 326, 327
Gibbs energy 90
Glass electrode 97, 102–106, 127, 130, 131, 200, 201, 215
Glass temperature 149, 183
Glass transition 148–150, 186
Glucose content, determination 192, 197, 201, 206
Glucose sensor 206
Grubbs test 320, 322–323, 327, 331
Gyromagnetic ratio 30–32, 40

H

Half-cells 84, 85, 89–91, 94, 95, 97, 127–129, 183
Half-life 155–157, 159–161, 184, 187
Hapten 163, 164
Heavy atom effect 173, 179
High-field shift 35, 41, 44–46, 57, 63
Histogram 222

Index

¹H NMR 28, 33, 38–57, 59, 60, 62, 63, 68, 69, 72–74, 76, 158
Hydrates, complex 144
Hydrogen on platinum electrode 94, 102, 138, 140, 141, 196
Hydrogen overvoltage 140–142, 183, 185
Hypergeometric distribution 246, 254–256, 258, 259, 325
Hypothesis 294, 299–305, 326, 330

I

Implementation 213
Incubation time 163, 167
Indicator electrode 94, 97, 101
Influence of the magnetic field strength 32, 33, 35, 39, 41–44
Influence on multiplet 36, 43
Injection analysis 210, 213, 214
Inorganic 8, 13, 82, 96, 103, 106, 115, 122, 149, 154, 173, 192, 194–197, 200, 204–208, 211, 213, 214
Interference, isobaric 11
Internal 39, 43, 72, 84, 102, 103, 107, 129, 170, 171, 182, 195, 208, 215, 312, 313
Internal conversion (IC) 170, 171, 179
Internal standard 39, 43, 72, 208, 312–318, 326
Intersystem crossing (ISC) 170
Ionic conductivity 107
Ionic strength 85, 86, 99–101, 107, 108, 130
Ionisation
chemical 12–15, 66
methods 12–16
Ion-selective 94, 100, 102, 105, 109, 127, 192
Ion selective electrode (ISE) 94, 100, 102, 105, 109, 127, 192
Isotopes 8–11, 16, 18, 19, 38, 46, 63–67, 77, 154, 155, 157–159, 161, 163, 184
– labelling 158

K

Karl Fischer titration 115, 116, 127
Karplus-Conroy relationship 55
Kern-Overhauser effect 59

L

Lambda probe 195, 196, 214
Lambert-Beer law 176, 184, 312
Lanthanoid phosphorescence 180
Larmor frequency 31, 32, 39, 40
Leading electrode 85
Libby half-life 160
Lifetime, excited states 173, 175
Limit theorem, central 298
Linear regression 237–241, 328
Link 106, 167, 284
Liquid membrane electrode 108, 109
Logarithmic 298
Logarithmic normal distribution 274, 275, 325

M

MacLaurin series expansion 260
Magnetic field strength 32, 33, 35, 39, 41–44
Magnetic resonance imaging (MRI) 63
MALDI-TOF 15
Marking, radioactive 157
Mass 8–20, 22, 25, 54, 76, 77, 86–88, 114, 131, 139, 143–145, 150, 154, 155, 158, 183, 185, 186, 192, 195, 225, 236, 237
Mass/charge ratio 77
Mass effect Law 86–88
Mass number 9, 11, 18, 24, 64–66, 154, 155
Mass spectrometer 8, 11–12, 14, 15, 18
Mass spectrometry (MS) 6, 8, 9, 11–18, 20–22, 24, 25, 76–77, 145, 154, 158, 160, 183
Mass spectrum 9, 10, 17–23, 25
Maximum likelihood procedure 288–290, 326
McLafferty rearrangement 21, 22
Mean 222, 225, 226, 230, 231, 266, 268, 273, 285, 293
Mean, arithmetic 222–226, 285, 324, 329
Mean value 222–227, 230, 231, 238, 240, 248, 255, 265–267, 274, 278, 279, 284, 285, 289, 295, 298, 302–305, 307, 310, 311, 315, 317, 322, 324, 327–331
Measured value distribution 246–262, 264–270, 273–275, 278, 284, 324
Mercury dropping electrode 122, 123
Metal oxide semiconductor gas (MOX) sensor 196, 197
Method of least squares 232, 238, 294, 316
Methods 6, 8, 12, 14, 32, 48, 61, 82, 97, 114, 122, 124, 126–128, 138, 139, 143–145, 150, 156, 157, 161, 163, 170–177, 179–188, 192, 199–201, 210, 213, 241, 268, 270, 281, 284–287, 298, 299, 303, 304, 307, 310, 312, 326
Micro thermogravimetry 144, 183
Molecular ion 13–17, 19, 24, 66, 67, 91, 128, 170
Molecular spectroscopy 6, 29
Multifunction sensor 201
Multiple bonds 44, 45, 57
Multiplets 36–39, 43, 48, 50–57, 59–62, 64, 70–77
Multiplicity 36, 37, 49, 50
Multiplicity state 50
Multi-stage experiment 174, 249

N

Negative controls 166, 168
Nernst's equation 85, 94, 97, 123–126, 139, 142, 185, 215
Neutron activation method 155–157
Neutron capture 155, 159, 160
Neutron sources 156
Nominal 9–11, 14, 16, 64, 65
Normal distribution 222, 226, 232, 234, 254, 264, 269–275, 278, 281, 284, 289, 291–293, 295–298, 301, 325, 326
Normal hydrogen electrode (NHE) 83, 95
Nuclear magnetic resonance (NMR) spectroscopy 6, 28–64, 66–77, 158
Nucleon number 9
Nucleus 28–32, 34, 35, 37, 39, 40, 49, 57, 59, 63, 64, 77, 154, 155, 184
Null hypothesis 299–303, 326, 329

O

Ohmic potential 142, 143, 183
Optical 199, 204, 205, 207, 208, 214
Optodes 204, 205, 207, 208, 214
Oregon Green 488 206, 207
Organosiloxane polymer membrane 194
Outlier tests 320–323, 326–332
Overvoltage 139–142, 183, 185
Oxidation 82, 90, 91, 96, 115, 116, 119, 122, 123, 125–129, 138, 140, 148, 194, 206

Oxidation number 82, 96, 115, 122, 128, 168
Oxygen content, determination 118, 128, 192, 194–197, 205, 207, 214
Oxygen detector 195
Oxygen sensor 194, 204
Oxygen-specific 194

P

Paramagnetic 44, 45, 124, 173, 174, 204
Parameter estimation 278–281, 284–302, 304, 305, 307, 326
Parameter tests 278, 299–307, 326
Pascal's triangle 36, 48, 50, 54, 71, 77
pH measurement 102–105, 127
Phosphorescence 170, 173, 179, 180, 185, 187, 188
Photon counter 155, 163
Physical basics 28
Physisorption 141, 142
Planning 301
Platinum electrodes 94
Platinum, platinum-plated 141
^{31}P NMR spectrum 64
Point estimate 285, 291
Poisson distribution 246, 259–264, 289, 325
Polarography 122
Polymers 148–150, 183, 186, 200
Positive controls 166, 168
Positron 154, 161
Potential 61, 83–85, 89–91, 94–99, 101–106, 109, 114, 118, 122–130, 139, 142, 143, 182, 183, 185, 194–198, 213, 248, 322, 327, 331
Potential difference 84, 89–91, 94, 97–99, 101–103, 105, 106, 108, 109, 127, 129, 130, 139, 142, 183, 195, 196, 198
Potentiometric/potentiometry 94–111, 122, 127, 130, 192, 194, 195, 198, 199, 201, 214
Potentiometric titration 97, 101–102, 122, 127, 130
Potentiostatic 114
Precession 31, 40, 63, 77
Predissociation (PD) 171, 173
Probability density 264, 265, 267, 275, 278, 279, 281, 284
Probability function 226, 246, 247, 250, 254, 256–260, 262, 264, 279, 288, 289, 326
Probability of default 66, 171, 173, 179, 226, 232, 246–250, 253–255, 258, 259, 261, 262, 264, 265, 268, 270, 273, 275, 285, 288, 291, 296, 297, 305, 307, 320, 322, 323, 325, 329, 331
Process, thermal 148–151, 183
Pseudomultiplets 56

Q

Quadrupole mass spectrometer 12
Quantum yield 171, 172, 174, 175, 177, 184, 187, 188, 205
Quartet 36, 38, 43, 49–54, 59–62, 68, 69, 71, 72, 249
Quenching 173, 175, 204–208, 214

R

Radiation, electromagnetic 6, 28, 30, 32, 35, 40, 42, 77, 155, 213
Radioactive 9, 154–168, 184, 186, 187, 259, 262
Radiocarbon dating 159
Radio frequency range 28
Radioimmunoassay (RIA) 161–168, 184, 198
Radionuclide 154–156, 163, 184
Ratio, gyromagnetic 30–32, 40
Redistribution 254
Redox reaction 82–84, 94, 96, 126, 131, 140

Reduction 82, 89–91, 96, 122, 123, 125, 129, 138, 142, 185, 220
Reference electrode 94–97, 99, 102–104, 106, 108, 111, 114, 119, 122–125, 127, 130, 192, 194, 213
Reflected light microscopes 181
Regression line 240, 312, 316, 317
Regression, linear 237–241, 328
Relative molar 8
Relative sensitivity 205
Relaxation 28, 32, 170, 179
Reporter enzymes 164, 165
Residual current 123
Resolution 11, 33, 36, 38, 42, 43, 50, 65, 183
Resonance 20, 28–50, 52–55, 57–64, 66–73, 75–77, 128
Retro-Diels-Alder reaction 21, 25
Ring current 44, 45, 47, 69
Role of glass 103

S

Sample 39, 101, 107, 108, 115, 116, 126, 131, 143–145, 148, 150, 151, 156, 157, 159–161, 163, 167, 168, 173, 176, 177, 179–181, 183, 184, 186, 187, 192, 195, 197, 201, 205, 206, 211, 213, 214, 216, 222–229, 231, 254, 255, 258, 259, 281, 284, 285, 287–291, 293–299, 301, 302, 304, 307, 310, 312–315, 317, 318, 323, 326, 329–332
Scattered radiation 180
Scintillation counter 155, 160, 163, 184, 198
2nd order electrode 95, 102, 127
Second type 301, 302
Selectivity coefficient 106, 108, 127, 131, 201
Sensors 175, 192, 194–201, 204, 205, 207, 208, 213–215
– optical 192, 204, 205, 207, 208, 214
Separation, electrolytic 114
Sequential 210, 213, 214
Series development 166, 168, 211, 237, 260, 303, 305–307, 310, 326
Shielding
– constant 39, 40
– effect 34, 46
σ-cleavage (sigma cleavage) 20, 66
Signal converter 204
Significance number 299
Silver/silver chloride electrode 95, 96, 102, 106, 118, 119, 194, 213
Single-rod electrodes 102
Singlet 36, 37, 43, 47, 58, 59, 62, 170, 174, 175, 205
Solid-state electrode 106, 111, 131, 195, 214
Solvent 14, 28, 38, 62, 86, 98, 99, 110, 115, 138–140, 170, 171, 174, 214, 215, 313
Sources of error 104, 302, 313
Spectrofluorophotometer 177, 178
Spectrometry 6, 76–77, 154, 158, 176–181, 184, 185, 187, 188
Spectroscopy 6, 28–50, 52–55, 57–64, 66–73, 75–77, 126, 158, 170, 180, 223, 249
Spin-spin coupling 37, 49
Spin states 29–32, 36, 37, 49, 50, 59, 64, 170
– in magnetic field 29, 31
Spreadsheet program 228, 232, 271–272, 294
Standard 89, 129, 185, 305, 314, 324
Standard addition 310, 313, 326
Standard deviation 222–224, 226, 227, 229–231, 234, 246, 248, 249, 251, 256, 262, 270, 273, 285, 293–298, 302, 303, 305, 307, 311, 317, 320, 322, 323, 327, 329–331
Standard hydrogen electrode (SHE) 95
Standard potential 83, 84, 89, 95, 124, 125, 128, 129
State, excited – lifetime 173–174
Statistical 57, 187, 220, 222–232, 234, 269, 273, 278, 291, 296, 299, 314, 320, 323

Index

Statistics 8, 11, 156, 223, 224, 231, 232, 238, 269, 278, 281, 286, 294, 310, 322
Steady 183, 269
Stokes shift 170, 177, 180, 187
Stripping analysis 126, 128
Student t-distribution 281–284, 294, 295, 298, 304, 305, 326, 330
Systematic 177, 220, 222, 226, 230, 313, 323

T

Tactile polarography 123, 125, 126, 128
Taguchi sensor 197
Tetramethylsilane (TMS) 39, 40, 57, 77
Thermogravimetry (TG) 138, 143–145, 183
Thermospray process 15, 16, 66
3D NMR 63
Time-resolved 179–181, 185, 187, 188
Titanium dioxide ceramic 196
Titration 97, 101, 114–116, 127, 130, 223, 234, 307
Torsion angle 55
Total Ionic Strength Adjustment Buffer (TISAB) 107, 108
Total spin 50
Trace analysis 101, 126, 128, 157
Tracer, radioactive 157–158, 163
Transducer 192, 198, 199, 204, 214
Transmission 176
Tree diagram 249
Tris-(1,10-phenanthroline) ruthenium (II) cation 204
Two-dimensional 54, 59, 61–63
2D NMR 61, 62
Two-point distribution 246–249, 325

U

Uncertainty, medium 230
Units 8, 9, 20, 42, 43, 82, 86–88, 98, 114, 130, 154, 156, 194, 207, 215, 249, 295, 296
Upconversion fluorescence 174
Uranium isotope, radioactive 154, 161
Urea quantification 200, 201

V

Validation of methods 310, 312–318, 326
Valinomycin 109–111
Value, true 223–226
Variance 227–229, 231, 246, 248, 249, 251, 253, 255, 257, 261, 262, 264–266, 268, 270, 275, 278, 279, 284–288, 290, 291, 293–296, 298, 307, 316, 325, 326, 330
Vibration relaxation 170
Voltammetric/voltammetry 122–132, 192, 194, 197–201, 213, 214
Voltammogram 122, 124–126, 128

W

Weibull distribution 275
Working electrode 114, 118, 119, 123, 126, 138, 139, 213

Z

Zirconia membrane 195, 196

GPSR Compliance

The European Union's (EU) General Product Safety Regulation (GPSR) is a set of rules that requires consumer products to be safe and our obligations to ensure this.

If you have any concerns about our products, you can contact us on ProductSafety@springernature.com

In case Publisher is established outside the EU, the EU authorized representative is:

Springer Nature Customer Service Center GmbH
Europaplatz 3
69115 Heidelberg, Germany

Batch number: 08329930

Printed by Printforce, the Netherlands